Implicit and Explicit Semantics Integration
in Proof-Based Developments of Discrete Systems

Yamine Ait-Ameur · Shin Nakajima ·
Dominique Méry

Editors

Implicit and Explicit Semantics Integration in Proof-Based Developments of Discrete Systems

Communications of NII Shonan Meetings

Springer

Editors
Yamine Ait-Ameur 🆔
Department of Computing and Applied
Mathematics
IRIT/INPT-ENSEEIHT
Toulouse, France

Shin Nakajima
National Institute of Informatics
Tokyo, Japan

Dominique Méry 🆔
Telecom Nancy
University of Lorraine
Vandœuvre-lès-Nancy, France

ISBN 978-981-15-5056-0   ISBN 978-981-15-5054-6   (eBook)
https://doi.org/10.1007/978-981-15-5054-6

This Springer imprint is published by the registered company Springer Nature Singapore Pte Ltd.
The registered company address is: 152 Beach Road, #21-01/04 Gateway East, Singapore 189721, Singapore

Preface

Formal methods have been proposed for improving system design and system verification. One can cite state-based techniques, proof-based techniques, model checking, type systems, static analyses and abstract interpretation, algebraic methods, CSP (constraint solving problem) techniques, ... These techniques are well established on a solid formal basis and their domain of efficiency, their strengths and weaknesses are well acknowledged by the formal methods community. The semantics of the used formal method is *implicit* and encoded in the modelling language of this formal method.

Although some *ad hoc* formalisation of design contexts and domain knowledge within formal methods is possible, none of these techniques offers built-in mechanisms to make *explicit* some features like hypotheses, assumptions, expectations and properties related to the context and to the domain knowledge associated to the system under design. These features are usually hidden in the requirements handling explicit semantics. Moreover, these methods do not offer mechanisms to relate domain knowledge features to system design models concepts nor reasoning mechanisms to exploit these knowledges in order to enhance the quality of the developed systems and of the development processes.

For instance, an integer variable (typed by an integer in the the design model) may denote a temperature expressed in Celsius degrees, whilst another one may denote a pressure measured in bars at the extreme limit of the left wing of an aircraft in the landing phase. In general, these knowledges are omitted or abstracted by the produced formalisations or their formalisation is hard encoded in the designed formal model. Those knowledges carried by the concepts manipulated in these formal models are still *in the mind* of the model designer, it is not explicitly formalised and therefore, it is not shared.

Several approaches aim at formalising mathematical theories that are applicable in the formal system developments. These theories are helpful for building complex formalisations, expressing and reusing proof of properties. Usually, these theories are defined within contexts, that are imported and/or instantiated. They are a powerful way to represent and to model explicitly the external knowledge (e.g. domain knowledge) related to the system under development. They are expressed using types, logics, algebras, ... based approaches.

Providing features in formal methods to make knowledge models explicit, by external formal domain models like ontologies, is still a major challenge. One of the main benefits of offering such mechanisms preserves the systematic aspect of the method. Indeed, the formal models are defined in the used formal method language and explicit reference and/or annotation mechanisms are provided for associating explicit semantics to formal models concepts.

Once this integration is realised, the formalisation and verification of several properties related to the heterogeneous models' integration become possible. Specific properties like interoperability, adaptability, dissimilarity, re-configurability and identification of degraded modes, etc. can be addressed. Refinement/ instantiation and composition/decomposition will play a major role for specifying and verifying these properties.

Nowadays, no formal method or formal technique provides explicit means for handling such an integration.

In November 2016, Yamine Ait-Ameur (IRIT Toulouse, France), Shin Nakajima (NII, Tokyo Japan) and Dominique Méry (LORIA, Nancy, France) organised the Shonan meeting number 090 (https://shonan.nii.ac.jp/seminars/090/) in Shonan Village in the subrubs of Tokyo.

Entitled *Implicit and explicit semantics integration in proof based developments of discrete systems,* this seminar gathered, during a whole week, 30 researchers issued from both academia and industry from America, Asia, Australia and Europe to discuss solutions to address the problem of making explicit domain knowledge in formal developments.

This book results from discussions and meetings held during this seminar. It contains contributions from the participants expressing and summarising their point of view and solutions on this problem. The following contributions have been included.

Contributions

- Domain Modelling

    - *Modelling an e-voting domain for the formal development of a Software Product Line: when the implicit should be made explicit* by *J Paul Gibson and Jean-Luc Raffy* reports on an approach based on the specification of Event-B contexts to model an e-voting ontology, and its integration with an e-voting features model tree which formally specifies a software product line (SPL). The importance of making the implicit explicit in two different ways—domain experts need to explicitly model implicit knowledge, and Event-B modellers need to explicitly communicate the semantics of the formal model constructs to the domain experts—has been identified.

- *Domain-Specific Developments Using Rodin Theories* by *Thai Son Hoang, Laurent Voisin and Michael Butler* exploit domain theories expressed in Event-B in order to capture domain-specific Abstract Data Types (ADTs) and build dynamic systems using the developed structures. The approach is applied on an industrial example of developing a train control system.
- *Integrating Domain Modelling within a Formal Requirements Engineering Method* by *Steve Tueno, Régine Laleau, Amel Mammar and Marc Frappier* describes a metamodel for a domain modelling language built from OWL and PLIB. The language is part of the SysML/KAOS requirements engineering method which also includes a goal modelling language. Moreover, the formal semantics of SysML/KAOS models is specified, verified and validated using the Event-B method.

- Knowledge-Based Modelling

  - *Operations Over Lightweight Ontologies and Their Implementation* by *Marco A. Casanova and Rômulo C. Magalhães* defines an algebra of operators to build new ontologies from other ones ensuring sound propagation of their constraints.
  - *Formal Ontological Analysis for Medical Protocols* by *Neeraj Kumar Singh, Yamine Ait-Ameur and Dominique Méry*. This chapter shows the use of explicit domain-specific knowledge in a system model and how it is related to a system model using an annotation mechanism. A proof-based formal approach is set up to evaluate a medical protocol. An assessment of the proposed approach is given through a case study, relative to a real-life reference protocol (electrocardiogram (ECG) interpretation), which covers a wide variety of protocol characteristics related to different heart conditions.
  - *Deriving Implicit Security Requirements in Safety-Explicit Formal Development of Control Systems* by *Inna Vistbakka and Elena Troubitsyna*. This work presents a formal approach that allows the designers to uncover the implicit security requirements that are implied by the explicit system level safety goals. It relies on modelling and refinement in Event-B to systematically uncover mutual interdependencies between safety and security and derives the constraints that should be imposed on the system to guarantee its safety in the presence of accidental and malicious faults
  - *Towards an Integration of Probabilistic and Knowledge-Based Data Analysis Using Probabilistic Knowledge Patterns* by *Klaus-Dieter Schewe and Qing Wang*. In this chapter, an extension to probabilistic knowledge patterns is defined, where the rules become clauses in probabilistic logic. Using maximum entropy semantics for the probabilistic logic, the fixed-point construction can be extended resulting in a probabilistic model, i.e. distributions for the randomised relations.

- Proof-Based Modelling

    - *An Explicit Semantics for Event-B Refinements* by *Pierre Castéran* discusses a shallow embedding of the Event-B semantics in the CoQ proof assistant. This formalisation offers reasoning capabilities at the meta-level on machines and their behaviours, considered as first-class citizens.
    - *Contextual Dependency in State-Based Modelling* by *Souad Kherroubi and Dominique Méry*. This chapter recalls and details preliminary results on contextualisation and dependency state-based modelling using the Event-B modelling language. The contextualisation of Event-B models is based on knowledge provided from domains classified into constraints, hypotheses and dependencies, according to truthfulness in proofs.
    - *Configuration of Complex Systems—Maintaining Consistency at Runtime* by *Azadeh Jahanbanifar, Ferhat Khendek and Maria Toeroe* addresses consistency maintenance during dynamic system reconfiguration, i.e. satisfying the defined constraints, and adjusting the system configuration at runtime. The overall model-based framework for tackling these important issues is described in this chapter.

- Assurance Cases

    - *Towards Making Safety Case Arguments Explicit, Precise, and Well Founded* by *Valentin Cassano, Thomas S.E. Maibaum and Silviya Grigorova* discusses the use of safty cases to assess the safety of a system. Via a 'safety argument', a safety case aims to explicate, and to provide some structure for, the kind of reasoning involved in demonstrating that a system is safe. To date, there are several notations for writing down safety arguments. These notations suffer from not having a well founded semantics, making them deficient w.r.t. the requirements of a serious approach to engineering. With this goal in mind, the authors take some steps towards constructing a logical calculus for safety arguments by exploring some of the features of this calculus. The authors consider that their work establishes a framework for discussing safety arguments in a more rigorous manner.
    - *The Indefeasibility Criterion for Assurance Cases* by *John Rushby*. In this paper, the author adopts a criterion from epistemology and argues that assurance should be "indefeasible", meaning that we must be so sure that all doubts and objections have been attended to that there is no (or, more realistically, we cannot imagine any) new information that would cause us to change our evaluation. Then, application of this criterion is explored to the interpretation and evaluation of assurance cases and derive a strict but practical characterisation for a sound assurance case.

- Refinement-Based Modelling

    - *An Event-B Development Process for the Distributed BIP Framework* by *Badr Siala, Jean-Paul Bodeveix, Mamoun Filali and Mohamed Tahar Bhiri* defines as approach that makes explicit the transition from formal

requirements to a distributed executable model. It is based on a refinement
chain specified as Event-B models.

- *Explicit Exploration of Refinement Design in Proof-Based Approach:*
  Refinement Engineering in Event-B by *Fuyuki Ishikawa, Tsutomu Kobayashi*
  and Shinichi Honiden discusses the problem of refinement design and presents an approach for explicitly exploring and manipulating possible
  refinement designs. Specifically, experiences on refinement planning and
  refactoring to support engineering activities on refinement are reported.
- *Constructing Rigorous Sketches for Refinement-Based Formal Development:*
  An Application to Android by *Shin Nakajima*. In this chapter, the author
  addresses the problem of discovering refinement plans. The approach
  introduces an iterative process to use Alloy to produce an under-constrained,
  but unambiguous Alloy descriptions, which acts as a rigorous sketch of the
  target for us to make a refinement plan that is deployed as an Event-B
  refinement. The proposed modelling method was assembled as educational
  materials for Event-B.

Toulouse, France   Yamine Ait-Ameur
Tokyo, Japan   Shin Nakajima
Vandœuvre-lès-Nancy, France   Dominique Méry

Contents

Domain Modelling

Modelling an E-Voting Domain for the Formal Development of a Software Product Line: When the Implicit Should Be Made Explicit

J. Paul Gibson and Jean-Luc Raffy

Abstract There has been much recent interest in the development of electronic voting (e-voting) systems, but there remain many outstanding research challenges for software and system engineers. Software product line (SPL) techniques offer many advantages for the practical development of reliable and trustworthy e-voting systems, but the composition of system features poses significant problems that can be addressed satisfactorily only through the use of formal methods. When such systems are used in government elections, then they are obliged to follow legal standards and/or recommendations written in natural language. For the formal development of e-voting systems, it is necessary to build a domain model which is consistent with the legal requirements. We have already demonstrated that Event-B models can be used to verify critical requirements for e-voting system components. However, the refinement-based approach needs to be applied to the engineering of a complete e-voting system. We report on our approach, using Event-B contexts to model an e-voting ontology, and its integration with an e-voting features model tree which formally specifies the SPL. During this work, we identified the importance of making the implicit explicit in two different ways—domain experts need to explicitly model implicit knowledge, and Event-B modellers need to explicitly communicate the semantics of the formal model constructs to the domain experts. If either of these tasks is not adequately carried out, then this compromises validation of the requirements model (instance of the SPL).

J. P. Gibson (✉)
SAMOVAR UMR 5157, Télécom Sud Paris, 9 rue Charles Fourier, 91011 Evry cedex, France
e-mail: paul.gibson@telecom-sudparis.eu

J.-L. Raffy
Télécom Sud Paris, Evry, France
e-mail: jean-luc.raffy@telecom-sudparis.eu

© Springer Nature Singapore Pte Ltd. 2021
Y. Ait-Ameur et al. (eds.), *Implicit and Explicit Semantics Integration in Proof-Based Developments of Discrete Systems*,
https://doi.org/10.1007/978-981-15-5054-6_1

1 Introduction

Electronic voting systems are those which depend on some electronic technology—including both software and hardware—for them to function as required [13]. Initial adoption of electronic voting systems focused on *direct recording electronic* (DRE) machines, where the voting is under supervision in a controlled environment, commonly known as a voting station. These systems drew much criticism, and many problems were reported around the world [35]. Much progress has been made and there is general agreement that it is possible to develop DRE that provides a reliable, trustworthy and secure e-voting system [13]. However, there are continued reports of problems with such systems [26], which are due to poor engineering practices and lack of understanding of the complex interaction between e-voting system requirements [32]. The state-of-the-art in such machines now promotes the use of a form of voter verified printed audit trail (VVPAT) [47], and a risk-limiting audit or manual recount [30].

E-voting systems make up a family of products which share common functionality, but where each system has its own unique requirements [18]. As a consequence, they are well-suited to the development of a software product line [14]. Such techniques offer many advantages for the practical development of reliable and trustworthy e-voting systems, but the composition of system features poses significant problems that can be addressed satisfactorily only through the use of formal methods. A formal domain model for e-voting systems is a critical part of any such formal development [15] and this correspond to the standard notion of ontology [11].

Previous work has demonstrated that Event-B models can be used to verify critical requirements for e-voting system components [5, 6]. Recently, it has been shown how the refinement-based approach can be applied to the engineering of a complete e-voting system [12]. The next step in our research is the integration of the SPL approach for formal specification of system requirements, and the refinement of such a specification to a concrete implementation. We report on our ongoing work, using Event-B contexts to model an e-voting ontology, and its integration with an e-voting features model tree which formally specifies the SPL.

Our research has identified the key issue of making explicit that which may be implicit—domain experts need to explicitly model implicit knowledge and Event-B modellers need to explicitly communicate the semantics of the formal model constructs to the domain experts. If either of these tasks is not adequately carried ou,t then this compromises validation of the requirements model (instance of the SPL).

In general, "explicit" means clearly expressed or readily observable while "implicit" means implied or expressed indirectly. However, there is some inconsistency regarding the precise meaning of these adjectives [1]:

- **Logic and belief models** [29]—"a sentence is explicitly believed when it is actively held to be true by an agent and implicitly believed when it follows from what is believed."

- **Semantic web** [42]—"semantics can be implicit, existing only in the minds of the humans […]. They can also be explicit and informal, or they can be formal."
- **Requirements engineering community** [45]—use the terms to distinguish between declarative (descriptive) and operational (prescriptive) requirements, where they acknowledge the need for "a formal method for generating explicit, declarative, type-level requirements from operational, instance-level scenarios in which such requirements are implicit".

We propose a more formal (explicit) treatment of the adjectives implicit and explicit when engineering electronic systems.

The remainder of the paper is structured as follows. Section 2 provides an overview of e-voting machines: the complexity of the requirements, the different types of implementations, and the legal aspects. Section 3 is concerned with building a formal domain model as a type of ontology and examines the question of when the explicit should be implicit (if ever). Section 4 reviews some real-world examples of e-voting systems where problems have arisen because of the implicit–explicit duality. Section 5 reports on our ongoing research and development of a SPL for e-voting: using Event-B to specify a domain ontology, a SPL feature tree and a generic system architecture, in order to support a feature-driven refinement process towards a correct-by-construction implementation. The paper concludes in Sect. 6.

2 E-Voting Machines

2.1 Complex, Interacting Requirements

Since the earliest analysis of e-voting systems [20], it has been argued that there are many complex interactions between the different requirements that these systems may be required to meet. Much of the current research in this area is concerned with a better understanding of these interactions, designing and implementing systems that meet certain combinations of requirements and evaluating the use of such systems during elections:

1. Authentication [10]—how to guarantee that the person who wishes to record a vote is the person they claim to be?
2. Anonymity/Privacy/Secrecy [7]—I should be able to vote without anyone knowing how I have voted.
3. Verifiability/Auditability [50]—I should be able to check that the voting process was executed correctly.
4. Accuracy [9]—whether the votes were tabulated/counted following the election rules.
5. Usability [3]—if the voting interface is easy to use for the voters and election administrators.

6. Understandability/Trustability [39]—can voters understand how the system works, and can they trust that their understanding is correct?
7. Fault Tolerance/Security from attack [37]—is the system secure against attacks and tolerant of faults in its component parts (electronic or otherwise).
8. Availability [28]—can access to the e-voting system be attacked, leading to a denial of service?
9. Maintainability [18]—as requirements change, can the system evolve in order to maintain correct behaviour?
10. Cost/Lifetime [35]—will the machines cost more/less than the traditional systems (over the lifetime of the system), including development and maintenance costs?
11. Openness/transparency [38]—is difficult to guarantee when parts of the system are outsourced.

There are numerous documented complex interactions between these different requirements. For example, it is difficult to provide a usable interface when applying complex cryptographic protocols for security. Verifiability and anonymity appear to be inconsistent—how can a voter demonstrate that their vote was counted incorrectly if they cannot demonstrate how they have voted? Authentication and anonymity are difficult to guarantee when a voter has to identify themselves to the same machine that will be used to record their vote. Understandability is compromised when complex cryptographic protocols are used to provide verifiability. Fault tolerance and auditability both increase the cost of system development and maintenance. Making the system code open source may conflict with other security requirements.

2.2 Remote Electronic Voting

Remote electronic voting (REV) permits the voter to record a vote without having to be physically present in a supervised environment [27]. The voter must use unsupervised mechanisms for recording and transmitting their vote [19] In the modern world, this will most likely be an electronic computer/device that is connected to the Internet [24]. Coercion is the biggest risk [8], and authentication is a major challenge [34]. Denial of service attacks has already been observed [41]. Computer viruses and malware provide powerful attack mechanisms, that have already been developed [22].

2.3 End-to-End Verifiable Systems

With End-to-end verifiable systems (E2E-V) [25], voters have an opportunity to verify that their vote is cast as they intended and correctly recorded (individual verifiability) Anyone can verify that all recorded votes were properly included in the tally (universal verifiability). Such systems provide a high degree of evidence that the outcome is correct, assuming that the voters correctly performed the verifications. E2E-V systems typically use sophisticated cryptographic techniques for providing privacy (although this is not a requirement). Such protocols should guarantee that voters do not need to blindly trust any component of the system; all components can be scrutinised so that their computation can be verified if their trustworthiness is in doubt. However, even requiring the use of E2E-V systems does not guarantee that the system will meet all the requirements.

2.4   Laws Standards and Recommendations

We must ask whether a given e-voting system is lawful, as a system that does not comply with international law should not be used in democratic elections. The fundamental principles of elections are firmly stated in *article 25 of the International Covenant on Civil and Political Rights and article 21 of the Universal Declaration of Human Rights*. Voting systems are normally required to comply with laws at other levels of governance, for example, constitutional, national, state, regional, etc. These laws often make reference to international and national standards that must be followed. An e-voting system has a myriad of inter-related legal requirements to meet; these multiple layers do not provide solid foundations upon which to build a system—none of the layers are fixed and the texts are open to different interpretations. In some cases, there is no consistent interpretation of system requirements. When problems arise with a particular e-voting system, it is for judges to decide if these were due to some aspect which could be said to be illegal. The final problem to consider is that each voting system has to meet specific needs that are not directly addressed by the laws and standards. The requirements of the system must somehow integrate these specific needs with multiple layers of laws and standards.

There are four main actors in the specification and use of e-voting system requirements:

1. The standards bodies establish the requirements that all e-voting systems (within a certain geopolitical space) must meet.
2. The procurement offices establish the requirements that a specific machine must meet in order for it to be purchased for use in a specific election.
3. The manufacturers develop machines that meet the generic requirements specified by the standards bodies and the specific requirements stipulated by procurement offices.
4. The Independent Testing Authorities test the delivered machines to ensure that they meet the requirements.

Unfortunately, there is evidence that the communication and co-ordination between these actors are poorly managed [33].

3   Need for a Formal Domain Model: Ontology

3.1   Terminology: Dictionaries and Glossaries

The European Council of Europe e-voting recommendations [46] recognises that
consistent use of terminology is key, and states *"In this recommendation the following terms are used with the following meanings: ..."* The terms that it chooses
to include are *authentication, ballot, candidate, casting of the vote, e-election or
e- referendum, electronic ballot box, e-voting, remote e-voting, sealing, vote, voter,
voting channel, voting options and voters registrar*. The list is very incomplete, but
it is a first important step towards developing an e-voting domain ontology. It should
be noted that, even in this short set of definitions, fundamental terms are used incon-
sistently. A formal ontological model would facilitate automated validation of model
consistency.

A number of other countries, outside of Europe, have also developed glossaries for
their particular voting systems and requirements, for example, Canada,[1] Australia[2]
and USA.[3] The level of detail in such glossaries varies from a short list of terms to
hundreds of pages. There is a clear need for a standard ontological domain model.

3.2   When the Explicit Should Be Implicit: Ontologies and
   Domain-Specific Languages

In all forms of communication, implicit shared understanding improves signal rate,
and is often necessary in achieving an acceptable communication mechanism. Shared
implicit understanding is good when it is coherent. There should always be an explicit
representation of the implicit knowledge as a base reference, if needed. This is the
role of an ontology [21].

When a community of developers shares much common knowledge, then the next
step is the development of a domain-specific language (DSL) [44]. Many such DSLs
are structured in terms of domain features [43], and this provides a strong link to the
SPL modelling and development approach. Building a DSL is a complex task [36],
and there has been much recent research on using formal approaches [4]. Integration
of DSLs and ontologies requires the use of formal methods [48].

With respect to the implicit/explicit dichotomoy, such DSLs bring the best of both
worlds. Shared domain knowledge is implicit when the DSL is used to describe and
synthesise systems within the domain, but is also explicit when used to reason about
and analyse such systems. The implicit aids human–human interaction and commu-

---

[1] www.elections.ca.

[2] www.aec.gov.au/footer/Glossary.htm.

[3] www.eac.gov/voting-equipment/voluntary-voting-system-guidelines/.

nication. The explicit aids automation and tool development. Both are necessary for model validation.

4 E-Voting: Examples of When the Implicit Should Be Explicit

4.1 DUALVOTE—E-Pen

The authors have been involved in the development of a novel e-voting system (DUALVOTE) that provides an innovative interface for e-voting using an electronic pen [31]. The advantages of the system arise out of the way in which it generates a paper vote (for audit) and an electronic vote simultaneously. However, there were some problems that arose when testing the system (during real elections) because of the unpredictable behaviour of the voters [17].

A major issue was that the developers had made implicit assumptions about voter behaviour which had never been explicitly stated to the voters. The correct functionality of the system was dependent on these assumptions being true. Unfortunately, this was not the case:

- Some voters recorded their vote using their own pen rather than the e-pen that was provided. We wrongly assumed that all voters would record their votes using the e-pen.
- Some voters recorded their votes on a surface other than the electronic surface provided. The instructions explicitly stated that they should write on the surface provided, but it was wrongly assumed that all voters would follow the instructions.
- It was explicitly stated that any identifying mark on the ballot paper which could uniquely identify the voter would render the vote invalid. However, there was no explicit statement of how a machine could automate the identification of such invalid votes. The system failed to function correctly because the election administrators had implicit (domain-specific) knowledge of how voters could mark their vote which had never been explicitly stated. As an additional complication, this knowledge varied from election-to-election and from constituency-to-constituency. There were even disagreements between election officials in the same voting station as to what would render a vote invalid.
- The electronic voting interface was a limited resource and tying up the resource could lead to a denial of service type attack. We implicitly, and wrongly, assumed that all voters would spend a reasonable amount of time to vote. This was based on us wrongly assuming that voters would not deliberately attack the voting system. A simple attack, in this case, would be to stay at the voting interface for a long (unreasonable) amount of time. The implicit need to timeout a voter should have been explicitly stated and defined before the voting process started.

Our DUALVOTE experience highlighted the need to make the implicit explicit when developing voting systems.

4.2   Implicit Programming Language Semantics

While analysing an e-voting system, we identified issues that arise when system components are developed independently, and they make inconsistent assumptions about the global system and its environment. A good example of this arises when the system software is written using different languages. The meaning of software behaviour is now dependent on the semantics of the programming language concepts. Unfortunately, concepts that share the same syntax (in different languages) do not always share the same semantics.

This can happen with concepts as simple as arrays [16]. In this study, two components of the system were developed using different modelling/programming languages/techniques. Each of the developers had a correct implicit understanding of the semantics of arrays in the language they were using (and an explicit statement of these semantics was available). However, the semantics of arrays was different in each of the two languages. As votes moved from being recorded in one language to being counted in another language, their representations changed. The inconsistency was caught late in development (during testing), but it would be better if such issues had been avoided at the beginning of development.

4.3   Negative Counts—Can Anything Be Too Obvious?

It is obvious that a vote count for a specific option (or candidate) should not be negative. Returning a negative count is clearly an error. Such negative counts have been reported in real elections. We must ask how such a stupid error could have happened. Should non-negativity of counts be explicit in the requirements specification? In order to answer this question, we need to examine a major difference between modelling languages and programming languages with respect to integer representation.

In Event-B, there are three in-built types that can be used for counting: integers, naturals (including 0) and naturals (excluding 0). If we model a count as a natural (NAT), then the model guarantees that the count has a non-negative value. There should be no need to explicitly state this as it is implied by the semantics of NAT. However, if the client (the person specifying the system requirements) does not know/understand the formal semantics of Event-B, then how can they validate the model as being correct? After validation, the model has to be implemented. Most programming languages do not contain the notion of a natural number as a primitive type. As a consequence, it is not surprising that a count could be implemented as an integer (which does permit negative values). In such a case, a program invariant (stating that the count is always non-negative) would have solved the problem. This property of NATs is implicit in the Event-B model; thus the the non-negative count invariant is not explicitly stated and may be overlooked in an integer-based implementation. Of course, using a refinement-based development method will prohibit an incorrect implementation.

A similar issue occurs with many other model language semantics. A good example is that of sets. The modeller knows that a collection of entities modelled as a set implicitly guarantees that the collection contains no repeated elements; but the client may not know this. In e-voting, it is often very important that collections have such set-like behaviour. How should we validate such models?

4.4 Vote Coercion in a Typical Voting Station

Formal methods are generally used in the development of electronic systems. However, they can also be used in modelling and analysing traditional voting systems (with no electronic components). Insight from the formal development of e-voting systems has helped us to identify previously undocumented issues with the traditional systems. Without a formal model—explicitly modelling assumptions—the correct functioning of a traditional system can be guaranteed only if the implicit assumptions are valid. Unfortunately, many of these implicit assumptions have not been explicitly documented and are true only because the actors in the system behave in a certain way. A simple example from voting a polling stations in France illustrates this case.

In remote voting, there is a major threat of man-in-the-middle attacks on the communication network between the voting booth and the ballot box. Such attacks have been modelled using formal methods and we have a good understanding of how systems can be defended against such attacks. By explicitly modelling such attacks, we can verify that our systems are protected against them. It has been claimed that no such threats exist when using traditional paper voting at a controlled voting station, as there is no underlying network to be attacked. Such reasoning is based upon a false assumption that is implicit in the correct functioning of a traditional voting station; namely, that voters do not interact with other parties between recording their vote anonymously in the booth and the submission of the recorded vote in the ballot box. In France, and in other countries, the passage between the booth and the box is not strictly controlled. The implicit need for voters to transfer their vote directly from the booth to the box is, in general, not explicitly enforced. This could lead to problems of coercion and vote buying. The problem is mitigated by the fact that the polling station is a public area where voters can be observed; but there is no guarantee that invalid behaviour of voters will be witnessed.

5   A SPL for E-Voting

In this section, we provide a brief overview of our current work in developing a
formal SPL for e-voting.

5.1   A Feature Tree Model for E-Voting

In Fig. 1, we illustrate part of the feature tree for an early version of our SPL prototype.

From the diagram (which shows only part of the full feature tree model) we
see that VoteCounting is a mandatory abstract feature of our e-voting SPL.
ValidityChecking is a mandatory (sub) feature of VoteCounting which can
be done either automatically or manually (which are modelled as concrete features).
The details of individual requirements associated to each feature are not important.
We note that the number of features (in the complete tree) was chosen to provide a
SPL which was simple enough to develop, but complex enough to provide a proof-of-
concept. In our current version, our SPL can be configured into hundreds of different
concrete instances.

5.2   Formalisation in Event-B

The tree is specified formally in an Event-B generic context, which is generated
automatically from a feature tree graphical editor. The configuration of the specific
instance of the SPL can also be done interactively using a graphical editor. Again,
we generate automatically the Event-B context corresponding to the instance. At this

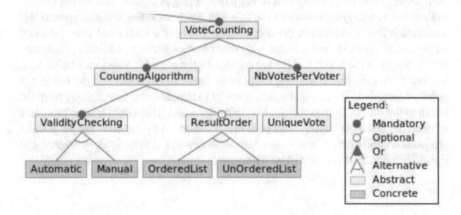

Fig. 1 The (partial) feature tree for an e-voting SPL

CONTEXT PresidentialRound2
SETS
      BULLETINS
      REGISTEREDVOTERS
CONSTANTS
      BLANKVOTE
      NUMBEROFOPTIONS
      CANDIDATES
      INVALIDVOTE
      POSSIBLEOPTIONCHOICE
      MAXNUMBEROFPREFERENCES
      MINNUMBEROFPREFERENCES
AXIOMS
    axm1: $BLANKVOTE \in BULLETINS$
    axm2: $NUMBEROFOPTIONS \in \mathbb{N}$
    axm3: $CANDIDATES \subseteq REGISTEREDVOTERS$
    axm4: $INVALIDVOTE \in BULLETINS$
    axm5: $POSSIBLEOPTIONCHOICE = 1$
    axm6: $NUMBEROFOPTIONS = 2$
    axm7: $MAXNUMBEROFPREFERENCES = 1$
    axm8: $MINNUMBEROFPREFERENCES = 1$
    axm9: ⟨theorem⟩ $MAXNUMBEROFPREFERENCES \leq NUMBEROFOPTIONS$
    axm10: ⟨theorem⟩
      $MINNUMBEROFPREFERENCES \leq MAXNUMBEROFPREFERENCES \wedge$
      $MINNUMBEROFPREFERENCES > 0$
END

Fig. 2 The instantiated context for Presidential Election Round 2

stage, the formal methods guarantee that the tree instantiation is a valid instance of the SPL, respecting the constraints specified in the feature tree model. We now have to associate behaviour with each of the features in the tree. Currently, we follow a 3-tier approach:

1. Each feature has an associated context where static relationships between sets and constants are specified. These sets and constants are taken from an e-voting domain ontology context. (Fig. 2 illustrates the context that is generated for an instance corresponding to the second round of a presidential election in France.)
2. Each feature has an associated state, which is a union of all system variables that it is concerned with. Features specify an invariant over the relevant state. The state variables are also part of the domain ontology.
3. Each feature can be associated to one or more events that correspond to changes to its relevant state variables. The events are also modelled in the domain ontology.

As such, the domain ontology groups together all the concepts (static and dynamic) that are shared between the SPL features. More details of the formalisation of the feature tree can be found in [2].

Combining these three tiers into a single abstract Event-B machine is the main challenge that we are currently addressing.

5.3 A Pipeline Design Pattern

The initial abstract machine is modelled as a simple pipeline, of 4 phases—set up, voting, counting and audit (see Fig. 3). We have identified a useful formal design pattern, where a state variable in one phase becomes a constant in the next phase:

- During election set up, the list of candidates and list of electors are variable (as they must register before being added to the lists). However, once the set up is completed, these lists must be fixed.
- During the voting step, the list of electors who have voted is variable (names get added as votes are recorded). However, once the polling station closes, this list is fixed before the count begins.
- Counting, in phase 3, will be complete when the list of ballots counted is complete, and then we can start the final audit phase.

We note that the precise semantics of the pipeline operator have recently been published [12]. We also note that the same paper addresses a correct-by-construction refinement approach using the same architecture. However, the initial abstract

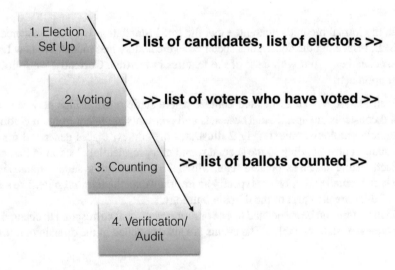

Fig. 3 The pipeline architecture

machine in this work is not initially generated from the SPL feature tree. The development of the feature tree model and the validation of the pipeline approach were done in parallel, and, as such, are not yet fully integrated.

5.4 Feature-Driven Refinement Towards Implementation

Developing features as a sequence of refinements guarantees correctness if such a sequence can be found. The order in which features/refinements are added is very important. We propose that such refinement should be driven by the feature configuration. Developing features in parallel poses many additional issues with respect to feature interactions [14]. We can usefully classify feature pairs in terms of the way in which they interact following the approach for composing fair objects [23]. As we identify (enemy) features that have contradictory requirements, we update the feature tree with a constraint to preclude their composition. As we identify (politician) features that need special co-ordination in order for them to work correctly, we refine the system in such a way that guarantees such a co-ordination. Features that require no special co-ordination in order to work correctly together are known as friends and their refinements can be safely developed in parallel (unlike with politicians). Development of a feature-driven refinement method is a major challenge in the formal development of software product lines [49], and is the major challenge in our future research.

6 Conclusions

We have presented our ongoing research and development of a formal SPL for e-voting. We have placed this work within the context of domain ontologies and demonstrated the need to better model and understand the implicit–explicit semantic dichotomy. Even if we trust the systems that we develop formally using our approach, we still do not know how we can get the public to trust them. We argue that if the SPL is trustworthy, then it guarantees the trustworthiness of every machine that is built using it. Using our approach does not guarantee the absence of unwanted feature interactions. However, it will aid in detecting such interactions and make explicit to the client the incompatibility between certain features. In the future, we would like to develop a prescriptive assurance case mechanism (such as seen in [40]) based around the SPL model. As we configure and develop more and more e-voting systems using our approach, we may consider the SPL e-voting feature tree as a type of Domain-Specific Language which evolves as we improve our understanding of feature interactions. As future work, we are interested in whether the formal SPL domain language ontological approach generalise to other problem domains.

References

1. Y. Ait-Ameur, J.P. Gibson, D. Méry, On implicit and explicit semantics: integration issues in proof-based development of systems, in *Leveraging Applications of Formal Methods, Verification and Validation. Specialized Techniques and Applications*. Lecture Notes in Computer Science, vol. 8803, ed. by T. Margaria, B. Steffen (Springer, Berlin, Heidelberg, 2014), pp. 604–618
2. A. Ait Wakrime, J.P. Gibson, J.-L. Raffy, Formalising the requirements of an e-voting software product line using event-B, in *27th International Conference on Enabling Technologies: Infrastructure for Collaborative Enterprises*, Paris, France, June 2018 (IEEE, 2018), pp. 78–84
3. B.B. Bederson, B. Lee, R.M. Sherman, P.S. Herrnson, R.G. Niemi, Electronic voting system usability issues, in *Proceedings of the SIGCHI Conference on Human Factors in Computing Systems* (ACM, 2003), pp. 145–152
4. J.-P. Bodeveix, M. Filali, J. Lawall, G. Muller, Formal methods meet domain specific languages, in *Integrated Formal Methods* (Springer, 2005), pp. 187–206
5. D. Cansell, J.P. Gibson, D. Méry, Formal verification of tamper-evident storage for e-voting, in *SEFM* (IEEE Computer Society, 2007), pp. 329–338
6. D. Cansell, J.P. Gibson, D. Méry, Refinement: a constructive approach to formal software design for a secure e-voting interface. Electr. Notes Theor. Comput. Sci. **183**, 39–55 (2007)
7. C.-L. Chen, Y.-Y. Chen, J.-K. Jan, C.-C. Chen, A secure anonymous e-voting system based on discrete logarithm problem. Appl. Math. Inf. Sci. **8**(5), 2571 (2014)
8. J. Clark, U. Hengartner, Selections: internet voting with over-the-shoulder coercion-resistance, in *Financial Cryptography*, vol. 7035 (Springer, 2011), pp. 47–61
9. D. Cochran, J.R. Kiniry, Formal model-based validation for tally systems, in *International Conference on E-Voting and Identity* (Springer, 2013), pp. 41–60
10. S. Falkner, P. Kieseberg, D.E. Simos, C. Traxler, E. Weippl, E-voting authentication with qr-codes, in *International Conference on Human Aspects of Information Security, Privacy, and Trust* (Springer, 2014), pp. 149–159
11. D. Fensel, *Ontologies* (Springer, 2001)
12. J.P. Gibson, S. Kherroubi, D. Méry, Applying a dependency mechanism for voting protocol models using event-B, in *Formal Techniques for Distributed Objects, Components, and Systems - 37th IFIP WG 6.1 International Conference, FORTE 2017, Held as Part of the 12th International Federated Conference on Distributed Computing Techniques, DisCoTec 2017, Neuchâtel, Switzerland, June 19-22, 2017, Proceedings*, ed. by A. Bouajjani, A. Silva. Lecture Notes in Computer Science, vol. 10321 (Springer, 2017), pp. 124–138
13. J.P. Gibson, R. Krimmer, V. Teague, J. Pomares, A review of e-voting: the past, present and future. Ann. Telecommun. **71**(7), 279–286 (2016)
14. J.P. Gibson, E. Lallet, J.-L. Raffy, Feature interactions in a software product line for e-voting, in *Feature Interactions in Software and Communication Systems X*, ed. by Nakamura and Reiff-Marganiec. Lisbon, Portugal, June 2009 (IOS Press, 2009), pp. 91–106
15. J.P. Gibson, E. Lallet, J.-L. Raffy, Engineering a distributed e-voting system architecture: meeting critical requirements, in *Architecting Critical Systems, First International Symposium, ISARCS 2010, Prague, Czech Republic, June 23-25, 2010, Proceedings*, ed. by H. Giese. Lecture Notes in Computer Science, vol. 6150 (Springer, 2010), pp. 89–108
16. J.P. Gibson, E. Lallet, J.-L. Raffy, Formal object oriented development of a voting system test oracle. Innov. Syst. Softw. Eng. (Spec. issue UML-FM11) **7**(4), 237–245 (2011)
17. J.P. Gibson, D. MacNamara, K. Oakley, Just like paper and the 3-colour protocol: a voting interface requirements engineering case study, in *Proceedings of 2011 International Workshop on Requirements Engineering for Electronic RE-Vote 2011*, Trento, Italy, August 2011 (IEEE, 2011), pp. 66–75
18. J.P. Gibson, M. McGaley, Verification and maintenance of e-voting systems and standards, in *8th European Conference on e-Government*, ed. by D. Remenyi. Lausanne, Switzerland, July 2008 (Academic Publishing International, 2008), pp. 283–289

19. G.S. Grewal, M.D. Ryan, L. Chen, M.R. Clarkson, Du-vote: remote electronic voting with untrusted computers, in *Computer Security Foundations Symposium (CSF), 2015 IEEE 28th* (IEEE, 2015), pp. 155–169
20. D. Gritzalis, Principles and requirements for a secure e-voting system. Comput. Secur. **21**(6), 539–556 (2002)
21. T.R. Gruber, Toward principles for the design of ontologies used for knowledge sharing? Int. J. Hum. Comput. Stud. **43**(5-6), 907–928 (1995)
22. J.A. Halderman, Practical attacks on real-world e-voting. *Real-World Electronic Voting: Design, Analysis and Deployment* (2016), pp. 145–171
23. G. Hamilton, J.P. Gibson, D. Méry, Composing fair objects, in *International Conference on Software Engineering Applied to Networking and Parallel/Distributed Computing (SNPD '00)*, ed. by Fouchal and Lee. Reims, France, May (2000), pp. 225–233
24. D. Jefferson, A.D. Rubin, B. Simons, D. Wagner, Analyzing internet voting security. Commun. ACM **47**(10), 59–64 (2004)
25. R. Joaquim, P. Ferreira, C. Ribeiro, Eviv: an end-to-end verifiable internet voting system. Comput. Secur. **32**, 170–191 (2013)
26. T. Kohno, A. Stubblefield, A.D. Rubin, D.S. Wallach, Analysis of an electronic voting system, in *IEEE Symposium on Security and Privacy (S&P04)* (IEEE, 2004), pp. 27–40
27. R. Krimmer, M. Volkamer, Bits or paper? comparing remote electronic voting to postal voting, in *EGOV (Workshops and Posters)* (2005), pp. 225–232
28. T.W. Lauer, The risk of e-voting. Electron. J. E-Gov. **2**(3), 177–186 (2004)
29. H.J. Levesque, A logic of implicit and explicit belief, in *AAAI*, ed. by R.J. Brachman (AAAI Press, 1984), pp. 198–202
30. M. Lindeman, P.B. Stark, A gentle introduction to risk-limiting audits. IEEE Secur. Priv. (5), 42–49 (2012)
31. D. MacNamara, J.P. Gibson, K. Oakley, A preliminary study on a dualvote and prltvoter hybrid system, in *International Conference for E-Democracy and Open Government 2012*, Danube University Krems, Austria, May (2012), pp. 77–89. Edition Donau-Universitt Krems
32. M. McGaley, J.P. Gibson, E-Voting: A Safety Critical System. Technical Report NUIM-CS-TR-2003-02, NUI Maynooth, Computer Science Department (2003)
33. M. McGaley, J.P. Gibson, A critical analysis of the council of Europe recommendations on e-voting, in *EVT'06: Proceedings of the USENIX/Accurate Electronic Voting Technology Workshop 2006 on Electronic Voting Technology Workshop*, Berkeley, CA, USA (2006), pp. 1–13. USENIX Association
34. B. Meng, A secure non-interactive deniable authentication protocol with strong deniability based on discrete logarithm problem and its application on internet voting protocol. Inf. Technol. J. **8**(3), 302–309 (2009)
35. R. Mercuri, A better ballot box? IEEE Spectr. **39**(10), 46–50 (2002)
36. M. Mernik, J. Heering, A.M. Sloane, When and how to develop domain-specific languages. ACM Comput. Surv. (CSUR) **37**(4), 316–344 (2005)
37. P.G. Neumann, Security criteria for electronic voting, in *16th National Computer Security Conference*, vol. 29 (1993)
38. A.-M. Oostveen, Outsourcing democracy: losing control of e-voting in the Netherlands. Policy Internet **2**(4), 201–220 (2010)
39. J. Pomares, I. Levin, R.M. Alvarez, G.L. Mirau, T. Ovejero, From piloting to roll-out: voting experience and trust in the first full e-election in Argentina, in *2014 6th International Conference on Electronic Voting: Verifying the Vote (EVOTE)* (IEEE, 2014), pp. 1–10
40. T. Rhodes, F. Boland, E. Fong, M. Kass, Software assurance using structured assurance case models. J. Res. Nat. Inst. Stand. Technol. **115**(3), 209 (2010)
41. D. Springall, T. Finkenauer, Z. Durumeric, J. Kitcat, H. Hursti, M. MacAlpine, J.A. Halderman, Security analysis of the Estonian internet voting system, in *Proceedings of the 2014 ACM SIGSAC Conference on Computer and Communications Security* (ACM, 2014), pp. 703–715
42. M. Uschold, Where are the semantics in the semantic web? AI Mag. **24**, 25–36 (2003)

43. A. Van Deursen, P. Klint, Domain-specific language design requires feature descriptions. CIT. J. Comput. Inf. Technol. **10**(1), 1–17 (2002)
44. A. Van Deursen, P. Klint, J. Visser, Domain-specific languages: an annotated bibliography. ACM Sigplan Not. **35**(6), 26–36 (2000)
45. A. van Lamsweerde, L. Willemet, Inferring declarative requirements specifications from operational scenarios. IEEE Trans. Softw. Eng. **24**, 1089–1114 (1998)
46. Venice Commission, Code of good practice in electoral matters. CDL-AD **23**, 2002 (2002)
47. A. Villafiorita, K. Weldemariam, R. Tiella, Development, formal verification, and evaluation of an e-voting system with VVPAT. Trans. Info. For. Sec. **4**(4), 651–661 (2009)
48. T. Walter, J. Ebert, Combining DSLS and ontologies using metamodel integration, in *DSL* (Springer, 2009), pp. 148–169
49. J. White, J.A. Galindo, T. Saxena, B. Dougherty, D. Benavides, D.C. Schmidt, Evolving feature model configurations in software product lines. J. Syst. Softw. **87**, 119–136 (2014)
50. X. Zou, H. Li, F. Li, W. Peng, Y. Sui, Transparent, auditable, and stepwise verifiable online e-voting enabling an open and fair election. Cryptography **1**(2), 13 (2017)

Domain-Specific Developments Using Rodin Theories

Thai Son Hoang, Laurent Voisin, and Michael Butler

1 Introduction

The Theory plug-in [4] for the Rodin Platform [3] enables modellers to extend the mathematical modelling notation for Event-B [2], with accompanying support for reasoning about the extended language. We consider in this presentation using Rodin theories to capture domain-specific Abstract Data Types (ADTs) and build dynamic systems using the developed structures. In particular, we proposed the notion of *theory instantiation* to incorporate more concrete representation of the ADTs. At the same time, the dynamic systems are refined further with respect to the changes of the underlying ADTs. We illustrate our approach with an industrial example of developing a train control system. We anticipate theory instantiation to be a promising direction for reusing theories via abstraction.

Structure

The rest of the chapter is structured as follows. Section 2 presents some background information on the Event-B modelling method and the Theory extension. We propose our approach for theory instantiation for system development in Sect. 3. The proposed mechanism is illustrated using a simplified train control system in Sect. 4. We conclude and discuss the advantage of our approach in Sect. 5.

T. S. Hoang (✉) · M. Butler
University of Southampton, Southampton, UK
e-mail: T.S.Hoang@ecs.soton.ac.uk

M. Butler
e-mail: mjb@ecs.soton.ac.uk

L. Voisin
Systerel, Aix-en-Provence, France
e-mail: laurent.voisin@systerel.fr

© Springer Nature Singapore Pte Ltd. 2021
Y. Ait-Ameur et al. (eds.), *Implicit and Explicit Semantics Integration in Proof-Based Developments of Discrete Systems*,
https://doi.org/10.1007/978-981-15-5054-6_2

19

2 Background

2.1 Event-B

Event-B [2] is a formal method for system development. The main features of Event-B include the use of *refinement* to introduce system details gradually into the formal model. An Event-B model contains two parts: *contexts* and *machines*. Contexts contain *carrier sets*, *constants*, and *axioms* that constrain the carrier sets and constants. Machines contain *variables* v, *invariants* I(v) that constrain the variables, and *events*. An event comprises a guard denoting its enabling-condition and an action describing how the variables are modified when the event is executed. In general, an event e has the following form, where t are the event parameters, G(t, v) is the guard of the event, and v := E(t, v) is the action of the event. Note that actions in Event-B are, in the most general case, nondeterministic [6]. We only use deterministic actions in this chapter.

$$e \mathrel{\hat{=}} \text{any } t \text{ where } G(t, v) \text{ then } v := E(t, v) \text{ end}$$

A machine in Event-B corresponds to a transition system where *variables* represent the states and *events* specify the transitions. Contexts can be *extended* by adding new carrier sets, constants, axioms, and theorems. Machine M can be *refined* by machine N (we call M the abstract machine and N the concrete machine). The state of M and N are related by a glueing invariant J(v, w) where v, w are variables of M and N, respectively. Intuitively, any "behaviour" exhibited by N can be simulated by M, with respect to the glueing invariant J. Refinement in Event-B is reasoned event-wise. Consider an abstract event e and the corresponding concrete event f. Somewhat simplifying, we say that e is refined by f if f's guard is stronger than that of e and f's action can be simulated by e's action, taking into account the glueing invariant J. More information about Event-B can be found in [6]. Event-B is supported by the Rodin platform (Rodin) [3], an extensible toolkit which includes facilities for modelling, verifying the consistency of models using theorem proving and model checking techniques, and validating models with simulation-based approaches.

2.2 Theory Plug-in

The Theory plug-in [4] enables developers to define new (polymorphic) datatypes and operators upon those datatypes. These additional modelling concepts might be defined directly (including inductive definitions) or axiomatically. Theories provide the encapsulation of datatypes and enrich the modelling language for the developers.

An (inductive) datatype can be directly defined using several constructors. Each constructor can have zero or more destructors. A datatype without any definition is axiomatically defined. Note that axiomatic datatypes are not recursive. We focus on axiomatic data types in this paper. By convention, an axiomatic datatype satisfies the *non-emptiness* and *maximality* properties, i.e., for an axiomatic type S we have S $\neq \varnothing$ and $\forall e \cdot e \in$ S. (Types are maximal by definition since there are no sub-types in Event-B.) As an example, an axiomatic type for stacks (without any additional axioms) is as follows.

```
1  theory Stack(T)
2  types STACK_TYPE
3  operators
4      STACK: ℙ(STACK_TYPE)
5  end
```

In the declaration of the theory, T denotes the type parameter of the Stack theory. Here STACK_TYPE denotes the "type" of stacks, whereas STACK denotes the set of stack instances. The reason for having a separate STACK_TYPE and the set of stack instances STACK will be discussed further in Sect. 3 about instantiation.

Operators can be defined *directly*, *inductively* (on inductive data types) or *axiomatically*. An operator defined without any definition will be defined axiomatically. Operator notation is prefix by default. Operators with two arguments can be infix. Further properties can be declared for operators including *associativity* and *commutativity*. In the following, we show the declaration for some stack operators: emptyStack, top, pop, and push.

```
1  operators
2      emptyStack: STACK
3      top(st : STACK): T
4      pop(st : STACK): STACK
5          for st ≠ emptyStack
6      push(st : STACK, e : T): STACK
7      ≺ (e : T, st : STACK) infix
8  axioms
9      @push_not_empty: ∀st, e · st ∈ STACK ∧ e ∈ T ⇒ push(st, e) ≠ emptyStack
10     @pop_push: ∀st, e · st ∈ STACK ∧ e ∈ T ⇒ pop(push(st, e)) = st
11     @top_push: ∀st, e · st ∈ STACK ∧ e ∈ T ⇒ top(push(st, e)) = e
12     @top_inStack: ∀st · st ∈ STACK ∧ st ≠ emptyStack ⇒ top(st) ≺ st
13     theorem @push_inStack: ∀st, e · st ∈ STACK ∧ e ∈ T ⇒ e ≺ push(st, e)
```

An additional infix operator ≺ defines a predicate (without any returning type) specifying whether an element e is in the stack st or not. The axioms are the assumptions about these operators that can be used to define proof rules. Note that @push_inStack is a theorem which is derivable from the axioms defined previously. We omit the presentation of proof rules in this paper.

Operators can be "partial", that is not defined for all values of parameters. *Well-definedness* is used to specify the condition (on the parameters) such that the operator is defined. In the Stack theory, operator pop is only defined when the parameter st is non-empty. The well-definedness condition (using keyword for) for pop is st ≠ emptyStack.

Finally, theories can be constructed in a hierarchical manner: a theory can *extend* one or more other theories by adding more data types, operators, and axioms.

3 Theory Instantiation

A possible instantiation of the stack ADT is one where a stack is represented as an array, where arrays are defined by an *array* ADT. Specifically, a stack is represented by a pair $(f \mapsto n)$, where n is the stack's size and f is an array of size n representing its contents.

```
1  theory Array
2  operators
3     ARRAY = {f ↦ n | n ∈ ℕ ∧ f ∈ 0 .. n− 1 → T }
4  end
```

Operations of the array ADT are defined using direct definition as follows:

- **append**: takes an array **a** and an element **e** and returns a new array where **e** is appended to the end of **a**.
- **front**: takes an array **a** and returns a new array where **a**'s last element is removed.
- **last**: takes an array **a** and returns **a**'s last element.
- **inArray**: a predicate indicating if an element **e** is in the array **a** or not.

```
1  append(f ↦ n: ARRAY, e: T)
2     = (f ⩤ {n ↦ e}) ↦ (n + 1)
3  front(f ↦ n: ARRAY) for n ≠ 0
4     = ({n − 1} ⩤ f) ↦ (n − 1)
5  last(f ↦ n: ARRAY) for n ≠ 0
6     = f(n − 1)
7  inArray(e: T, f ↦ n: ARRAY) infix
8     = e ∈ ran(f)
```

To instantiate the Stack ADT using the Array ADT, we use the following instantiation mappings. In the rest of the paper, we use the following notation $A \longleftarrow C$ to denote a mapping where an abstract element (e.g., a set or a constant) A is instantiated by a concrete element C.

```
1  STACK_TYPE ⟵ ℙ(ℤ × T) × ℤ
2  STACK ⟵ ARRAY
3  emptyStack ⟵ ∅ ↦ 0
4  push ⟵ append
5  pop ⟵ front
6  top ⟵ last
7  ≺ ⟵ inArray
```

Here, for consistency, **STACK_TYPE** has to be instantiated using a "type expression". This type expression is derived from the instantiation of **STACK** by **ARRAY**. Other instantiations are straightforward linking operators of the Stack ADT with those of the Array ADT. To prove that this instantiation is consistent, we have to prove that all the axioms associated with the Stack ADT are derivable as (instantiated) theorems of the Array ADT. For instance, the instantiated theorem for **pop_push** is as follows.

```
1  @pop_push: ∀st, e · st ∈ ARRAY ∧ e ∈ T ⇒ front(append(st, e)) = st
```

The above theorem can be proved by expanding the definition for front and append and gives rise to the following sub-goals (assuming that $st = f \mapsto n$):

$$\{n+1 - 1\} \vartriangleleft (f \Leftarrow \{n \mapsto e\}) = f$$
$$n+1 - 1 = n$$

Given that $f \mapsto n \in \mathsf{ARRAY}$, we have $n \in \mathbb{N}$ and $f \in 0.. \, n-1 \to \mathsf{T}$. As a result the above sub-goals are trivially discharged.

This paragraph is added about instantiation of WD condition. Note that if the operators are partial, i.e., having well-definedness conditions, we need to prove that the well-definedness conditions of the instantiating and the instantiated operators are *equivalent* (modulo of the instantiation mapping). For example, regarding the pop operator, we need to prove that for all $st \in \mathsf{ARRAY}$, we have

$$_1 \quad f \mapsto n \neq \varnothing \mapsto 0 \Leftrightarrow n \neq 0$$

The proof is trivial and is omitted here.

Given that the Array ADT correctly instantiates the Stack ADT, we can *reuse* any theorem and proof rule of the Stack ADT (with the correct instantiation). Theory instantiation allows the users to start with some abstract ADTs and to gradually introduce more concrete data types.

4 Case Study

4.1 System Descriptions

We illustrate our approach as a case study of a simplified train control system. This is the example used in [5]. The system is intended to keep all trains in a railway network at a safe distance apart to prevent collisions. The network consists of tracks divided into sections, and of points connecting these tracks. An interlocking system switches the points to connect different tracks together, and results in a dynamically changing track layout. Instead of light signals, the train control system uses radio communication to send the trains the permission to move or stop.

An overview of the interacting system components is given in Fig. 1. The trains themselves determine their positions and send them to the train control system by radio. Based on information about which parts of the network are occupied, the controller calculates for each train the area in which it can safely move without collisions. This area is called the MA and represents the permission for a train to move as long as it does not leave this area. The calculated MAs are then directly sent to the train where an onboard unit interprets it to calculate the location where the permission to drive ends, called the Limit of Authority (LoA). To prevent driving over the LoA, the onboard unit regularly determines a speed limit and applies the train's emergency brakes if necessary.

The most important properties of the train system are collision-freeness and non-derailment.

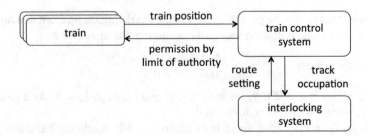

Fig. 1 System architecture

REQ1   There must be no collision between any two (different) trains in the system.
REQ2   Every train in the system must stay on the tracks of the network.

Collision-freeness between trains, *i.e.,* REQ 1, is guaranteed by the overall system and relies on two conditions: (C1) The trains are always within their assigned movement authorities, and (C2) the controller ensures that the MAs issued to the trains do not overlap. In fact, (C1) is implementable only if the MAs issued by the controller are never reduced at the front of the trains. Non-derailment, *i.e.,* REQ 2, is ensured by condition (C1) mentioned before, and condition (C3) stating that the controller only grants MAs over the active network.

4.2   Development Strategy

In our development of the train control system, we develop the ADTs used for modelling the trains and MAs, as follow.

- We start with an ADT that represents *regions* within the network.
- We subsequently instantiate the region ADT with an ADT representing *sequences* of sections.
- Finally, we instantiate the sequence ADT with the ADT corresponding to *arbitrary-based arrays*.

Another entity of the system that we model using ADTs is the active network that is controlled by the interlocking system.

- We start with an ADT representing the *network.*
- We subsequently instantiate the network ADT with an ADT representing *graphs*, where nodes are sections, and edges correspond to connections between sections.

These ADTs allow us to encapsulate the concepts useful for modelling, such as a region within a network. Moreover, we used the same ADT to model different aspects of the system. For example, both trains and MAs are represented abstractly by the *Region* ADT. Finally, important high-level properties of the system can be specified and reasoned about easily using the ADTs.

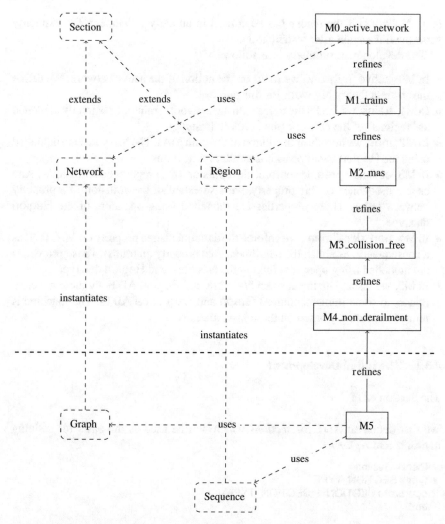

Fig. 2 The instantiation-chain of developments

In the following, we describe the ADTs with their usage in our formal model in the order that we defined them.

4.3 Development Using ADTs

Our model using ADTs involves different developments forming an *instantiation-chain*. The hierarchy of the development can be seen in Fig. 2. The model is available at [7]. For clarity, we omit the link between a machine (e.g., M2_mas) and a theory (e.g., Network) if the theory has been used in an early refinement. We explicitly show the links again after instantiation.

The development summary is as follows.

- In M0_active_network, we focus on the notion of the active network. We define the abstract theory Network for this purpose.
- In M1_trains, we model the trains with the network. Trains are abstractly modelled as "regions" of the network (the Region datatype).
- In M2_mas, we introduce the notion of the train MAs. The MAs are also modelled using the Region datatype as it is similar to the trains.
- In M3_collision_free, we enforce the collision-free property between MAs. And as a consequence of this property, we also establish the collision-free property between trains. These properties are modelled using operators of the Region datatype.
- In M4_non_derailment, we enforce the derailment-free property on MAs (and as a consequence, establish the derailment-free property on trains). These properties are modelled using operators linking the Network and Region datatype.
- In M5, we instantiate the abstract Network and Region ADTs by more concrete (closer to some implementation) Graph and Sequence ADTs. The machine is refined accordingly based on these instantiations.

4.3.1   The Initial Development

The Section ADT

We first define an ADT for *sections* which will be used as the basis for defining *networks* and *regions*.

```
1  theory Section
2  types SECTION_TYPE
3  operators SECTION: ℙ(SECTION_TYPE)
4  end
```

The Network ADT

The Network ADT imports Section and declares the type NETWORK_TYPE and a set of instances NETWORK.

```
1  theory Network
2  imports Section
3  types NETWORK_TYPE
4  operators NETWORK: ℙ(NETWORK_TYPE)
5  axioms
6     @NETWORK_non_empty: NETWORK ≠ ∅
```

Two predicate operators enlarge_WD and contract_WD are defined to capture the *well-definedness* of the operators (to be defined later) to enlarging and contracting a network. A well-definedness predicate of an operator are conditions specifying the "valid" values of the operator input. In our example, both well-definedness operators take a network n, a set of connections between sections g, and a set of sections s to be added or removed from the network n. Below, we show the declaration of enlarge_WD and its axioms. The declaration of contract_WD is similar.

```
1  operators
2    enlarge_WD(n : NETWORK_TYPE,
3      g: SECTION_TYPE ↔ SECTION_TYPE, s: ℙ(SECTION_TYPE))
4  axioms
5    @enlarge_WD_n: ∀n,g,s · enlarge_WD(n,g,s) ⇒ n ∈ NETWORK
6    @enlarge_WD_g: ∀n,g,s · enlarge_WD(n,g,s) ⇒ g ∈ SECTION ⤕ SECTION
7    @enlarge_WD_s: ∀n,g,s · enlarge_WD(n,g,s) ⇒ s ∈ ℙ(SECTION)
```

The axioms ensure that enlarging a network requires a network instance n, an injective connection between section instances g, and a set of sections s. Note that the axioms only state some *necessary conditions* for extending a network. The exact definition for the enlarge_WD is not given at this level.

The declaration for enlarge and contract use the above enlarge_WD and contract_WD operators as their well-definedness condition and are as follows.

```
1  operators
2    enlarge(n : NETWORK_TYPE,
3      g: SECTION_TYPE ↔ SECTION_TYPE, s: ℙ(SECTION_TYPE)): NETWORK
4      for enlarge_WD(n,g,s)
5
6    contract(n : NETWORK_TYPE,
7      g: SECTION_TYPE ↔ SECTION_TYPE, s: ℙ(SECTION_TYPE)): NETWORK
8      for contract_WD(n,g,s)
```

At the moment, there are no further axioms constraining enlarge and contract. In the next section, we will specify the constraints on enlarge and contract in relation with *regions*.

The Region ADT

Each train on a network occupies some *region* of the network. Furthermore, MAs also correspond to regions of the network. Relationship between regions include "sub-region" ⊑ (to state that trains must be always within their MAs) and "disjointness" ⊕ (to state that there is no collision between trains). As a result, we start the modelling of the Region ADT as follows.

```
1  theory Region
2  types
3    REGION_TYPE
4  operators
5    REGION: ℙ(REGION_TYPE)
6    ⊑ (r1: REGION, r2: REGION) infix
7    ⊕ (r1: REGION, r2: REGION) infix
```

Properties of ⊑ and ⊕ include that ⊑ is transitive (⊑_transitive) and reflexive (@⊑_reflexive), ⊕ is symmetric (@disjoint_symmetric), and a relationship between ⊑ and ⊕ (@⊑_⊕). In the axioms below, r1, r2, r3 are regions, i.e., members of REGION.

1   **axioms**
2   @⊑ _transitive: ∀r1,r2,r3· r1 ⊑ r2 ∧ r2 ⊑ r3 ⇒ r1 ⊑ r3
3   @⊑ _reflexive: ∀r· r ⊑ r
4   @⊕ _symmetric: ∀r1,r2· r1 ⊕ r2 ⇒ r2 ⊕ r1
5   @⊑ _⊕ : ∀r1,r2,r3· r1 ⊑ r2 ∧ r2 ⊕ r3 ⇒ r1 ⊕ r3

Two operators are declared for extending and reducing regions. Similar to **enlarge** and **contract** for the Network ADT, we also declare abstractly some well-definedness conditions for **extend** and **reduce** accordingly.

1   **operators**
2   extend_WD(r: REGION_TYPE, s: SECTION_TYPE)
3   reduce_WD(r: REGION_TYPE)
4   extend(r: REGION_TYPE, s: SECTION_TYPE): REGION **for** extend_WD(r,s)
5   reduce(r: REGION_TYPE): REGION **for** reduce_WD(r)

Operator **extend** takes a region **r** and a section **s**, and returns a region where **r** is extended to include **s**. Operator **reduce** takes a region **r** and returns a region where the last section of **r** is removed. Properties of **extend** and **reduce** related to ⊑ are as follows. While **extend** is monotonic and strengthening with respect to ⊑, **reduce** is monotonic and weakening.

1   **axioms**
2   @extend_⊑ _strengthening: ∀r,s· extend_WD(r,s) ⇒ r ⊑ extend(r,s)
3   @reduce_⊑ _weakening: ∀r· reduce_WD(r) ⇒ reduce(r) ⊑ r

Relationship between the Network ADT and Region ADT is captured by the relationship "a-part-of" ≪.

1   **operators**
2   ≪ (r : REGION, n: NETWORK) infix
3   **axioms**
4   @enlarge_≪ _monotonic:
5   ∀r,g,s,n· r ∈ REGION ∧ enlarge_WD(n,g,s) ∧ r ≪ n ⇒ r ≪ enlarge(n,g,s)
6
7   @⊑ _≪ _weakening:
8   ∀r1,r2,n· r1 ∈ REGION ∧ r2 ∈ REGION ∧ n ∈ NETWORK ∧ r1 ⊑ r2 ∧ r2 ≪ n
9   ⇒ r1 ≪ n

The axioms state that enlarging network will preserve ≪ relationship while ⊑ is "≪-weakening".

System Model Using the Network and Region ADTs

Given the Network and Region ADTs, we can develop an abstract system model of the train control system. This model contains the following refinement steps.

- **M0:** Model the active network using variable **active_network** as an instance of the Network ADT. Events **NRK_enlarges** and **NRK_contracts** are declared for updating the active network. For example, the **NRK_enlarges** event is as follows. Note that the variable **active_network** is updated by using the corresponding operator from the Network ADT.

```
1   NRK_enlarges
2   any
3     grp
4     scts
5   where
6     @grd1: enlarge_WD(active_network,grp,scts)
7   then
8     @act1: active_network := enlarge(active_network,grp,scts)
9   end
```

- **M1:** Model the active trains on the network using the Region ADT. Variables
  active_trains and train_reg model the set of trains and their layout on the network
  (which are regions).

```
1   invariants
2     @typeof−active_trains: active_trains ∈ ℙ(TRAIN_ID)
3     @typeof−train_reg: train_reg ∈ active_trains → REGION
```

New events related to trains include TRN_enters and TRN_leaves for trains to
enter and leave the network, and TRN_extends and TRN_reduces to model the
movement of trains on the network. An example of TRN_extends event modelling
the movement of train trn forward by extending to a section sct is as follows.

```
1   TRN_extends
2   any trn sct where
3     @grd1: trn ∈ active_trains
4     @grd2: extend_WD(train_reg(trn), sct)
5     theorem @thm1: extend(train_reg(trn), sct) ∈ REGION
6   then
7     @act1: train_reg(trn) := extend(train_reg(trn), sct)
8   end
```

- **M2:** Model the MAs using the Region ADT. A new variable train_ma is introduced to capture the active trains' MAs. An important property relating the trains
  and their MAs is that the trains must always be within their MA (see invariant
  @movement_authority).

```
1   invariants
2     @typeof−train_ma: train_ma ∈ active_trains → REGION
3     @movement_authority: ∀trn · trn ∈ active_trains ⇒ train_reg(trn) ⊑ train_ma(trn)
```

Event TRN_extends has an additional guard to ensure that the train trn will not
overrun its MA, i.e.,

```
1     @grd3: extend(train_reg(trn), sct) ⊑ train_ma(trn)
```

A new event updates_train_ma is introduced to change the MA of a train trn to
a new movement authority ma.

```
1   updates_train_ma
2   any
3     trn
4     ma
5   where
6     @grd1: trn ∈ active_trains
```

```
7    @grd2: ma ∈ REGION
8    @grd3: train_ma(trn) ⊑ ma
9  then
10   @act1: train_ma(trn) := ma
11 end
```

Guard @grd3 ensures that the train's MA can only be extended. The fact that updates_train_ma maintains invariant @movement_authority relies on the transitivity of ⊑, i.e., axiom @⊑_transitive of the Region ADT.

- **M3**: State and prove the collision-free property REQ 1 using the ⊕ operator. Here, we first add an invariant stating that the trains' MAs are non-overlapping. The collision-free property is a consequence (i.e., a theorem) of this invariant.

```
1  @ma_collision_free:
2    ∀trn1, trn2 · trn1 ∈ active_trains ∧ trn2 ∈ active_trains ∧ trn1 ≠ trn2
3      ⇒ (train_ma(trn1) ⊕ train_ma(trn2))
4  theorem @collision_free:
5    ∀trn1, trn2 · trn1 ∈ active_trains ∧ trn2 ∈ active_trains ∧ trn1 ≠ trn2
6      ⇒ (train_reg(trn1) ⊕ train_reg(trn2))
```

The proof of the theorem relies on the invariant (@movement_authority) stating that a train is always within its movement authority and the relationship between ⊕ and ⊑, i.e., axiom @⊑_⊕.

- **M4**: State and prove the derailment-free property REQ 2 using the ≪ operator. In this refinement, we first add an invariant stating that the trains' MAs are always part of the active network. The derailment-free property is the consequence of this invariant.

```
1  @ma−derailment_free:
2    ∀trn · trn ∈ active_trains ⇒ train_ma(trn) ≪ active_network
3  theorem @derailment_free:
4    ∀trn · trn ∈ active_trains ⇒ train_reg(trn) ≪ active_network
```

The proof of the theorem relies on the fact that a train is always within its movement authority (invariant @movement_authority), and ⊑ operator is "≪-weakening" (axiom @ ⊑_≪_weakening).

4.3.2 The First Instantiation Development

As mentioned in Fig. 2, we are going to instantiate the Network and Region ADTs with Graph and Sequence ADTs. The system model will be developed according to the newly introduced ADTs. Compared to the "abstract" DTs, i.e., Network and Region, the "concrete" ADTs, i.e., Graph and Sequence are closer to implementation, with more details about their internal representation.

The Graph ADT

We first specify the Graph ADT which will be used as an instantiation for the Network ADT. The Graph ADT contains the type GRAPH_TYPE and operators GRAPH, edges, and nodes. Operator GRAPH denotes the set of valid graph instances while edges and nodes return the edges and nodes of a graph accordingly. Axioms @edges_domain and @edges_range ensure the consistency between edges and nodes: the edges of a graph g must connect only nodes of that graph g.

```
1  types
2    GRAPH_TYPE
3  operators
4    GRAPH: ℙ(GRAPH_TYPE)
5    edges(g : GRAPH): SECTION ⇸ SECTION
6    nodes(g : GRAPH): ℙ(SECTION)
7  axioms
8    @GRAPH_non_empty: GRAPH ≠ ∅
9    @edges_domain: ∀g · g ∈ GRAPH ⇒ dom(edges(g)) ⊆ nodes(g)
10   @edges_range: ∀g · g ∈ GRAPH ⇒ ran(edges(g)) ⊆ nodes(g)
```

Two additional operators add and remove (together with their well-definedness operators add_WD and remove_WD) are defined to adding to a graph or removing from a graph.

```
1  operators
2    add(g : GRAPH, e : SECTION ⇸ SECTION, n : ℙ(SECTION)): GRAPH
3      for add_WD(g, e, n)
4    remove(g : GRAPH, e : SECTION ⇸ SECTION, n : ℙ(SECTION)): GRAPH
5      for add_WD(g, e, n)
6  axioms
7    @add_edges: ∀g,e,n · add_WD(g,e,n) ⇒ edges(add(g,e,n)) = edges(g) ∪ e
8    @add_nodes: ∀g,e,n · add_WD(g,e,n) ⇒ nodes(add(g,e,n)) = nodes(g) ∪ n
9    @remove_edges: ∀g,e,n · add_WD(g,e,n) ⇒ edges(remove(g,e,n)) = edges(g) \ e
10   @remove_nodes: ∀g,e,n · add_WD(g,e,n) ⇒ nodes(remove(g,e,n)) = nodes(g) \ n
```

The axioms specify the effect of adding (resp. removing) a set of connections e and a set of nodes n to a graph g: e and n are added to (resp. removed from) the edges and the nodes graph accordingly.

The Sequence ADT

The Sequence ADT specifies a datatype for representing sequences of sections.

```
1  types
2    SEQUENCE_TYPE
3  operators
4    SEQUENCE: ℙ(SEQUENCE_TYPE)
5    extend_head_WD(seq: SEQUENCE, s: SECTION)
6    front_WD(seq: SEQUENCE)
7    extend_head(seq: SEQUENCE, s: SECTION)
8      for extend_head_WD(seq, s)
9    front(seq: SEQUENCE)
10     for front_WD(seq)
```

For a sequence of sections, we are interested in the *link* between sections and other aspects such as *head*, and *rear*. The following operators are introduced for that purpose.

```
1  operators
2    link(seq: SEQUENCE): SECTION ⇸ SECTION
```

3   head(seq: SEQUENCE): SECTION
4   rear(seq: SEQUENCE): SECTION
5   middle(seq: SEQUENCE): $\mathbb{P}$(SECTION)
6   members(seq: SEQUENCE): $\mathbb{P}$(SECTION)
7   = {head(s)} $\cup$ middle(s) $\cup$ {rear(s)}
8   **axioms**
9   @empty_head_rear:
10    $\forall$seq · seq $\in$ SEQUENCE $\wedge$ link(seq) = $\varnothing$ $\Rightarrow$ head(seq) = rear(seq)
11
12 @link_domain:
13    $\forall$seq · seq $\in$ SEQUENCE $\wedge$ link(seq) $\neq$ $\varnothing$
14    $\Rightarrow$ dom(link(seq)) = {head(seq)} $\cup$ middle(seq)
15
16 @link_range:
17    $\forall$seq · seq $\in$ SEQUENCE $\wedge$ link(seq) $\neq$ $\varnothing$
18    $\Rightarrow$ ran(link(seq)) = middle(seq) $\cup$ {rear(seq)}
19
20 @head_notin_middle: $\forall$seq · seq $\in$ SEQUENCE $\Rightarrow$ head(seq) $\notin$ middle(seq)
21 @rear_notin_middle: $\forall$seq · seq $\in$ SEQUENCE $\Rightarrow$ rear(seq) $\notin$ middle(seq)

Note that **members** is an operator directly defined based on other operators. The axioms state the consistency between different aspects of a sequence. The meaning of the axioms are as follows.

- @empty_head_rear: If the connection between sections is empty, the sequence is within a section, hence its head and rear must be identical.
- @link_domain: If the connection is non-empty, then the *domain* of the connection is the head and the middle of the sequence.
- @link_range: If the connection is non-empty, then the *range* of the connection is the middle and the rear of the sequence.
- @head_notin_middle: The head of a sequence must not be in the middle of a sequence.
- @rear_notin_middle: The rear of a sequence must not be in the middle of a sequence.

These axioms constrain the sequence ADT to what is suitable to model the trains and their movement authorities. They capture the *well-formness* properties of sequence that formed by trains and/or movement authorities.

Other relevant operators related to the **Sequence** ADT are to extend the head of the sequence (**extend_head**) and remove the rear of the sequence (**front**). These operators are used to model the movement of the trains.

1   **operators**
2   extend_head_WD(seq: SEQUENCE, s: SECTION)
3   front_WD(seq: SEQUENCE)
4   extend_head(seq: SEQUENCE, s: SECTION)
5   **for** extend_head_WD(seq, s)
6   front(seq: SEQUENCE)
7   **for** front_WD(seq)
8   **axioms**
9   @extend_head_link:

10   ∀seq, s · extend_head_WD(seq, s)
11   ⇒ link(extend_head(seq, s)) = link(seq) ∪ {s ↦ head(seq)}
12
13   @extend_head_head:
14   ∀seq, s · extend_head_WD(seq, s) ⇒ head(extend_head(seq, s)) = s
15
16   @extend_head_rear:
17   ∀seq, s · extend_head_WD(seq, s) ⇒ rear(extend_head(seq, s)) = rear(seq)
18
19   @extend_head_middle:
20   ∀seq, s · extend_head_WD(seq, s)
21   ⇒ middle(extend_head(seq, s)) = (middle(seq) ∪ {head(seq)}) \ {rear(seq)}
22
23   **theorem** @extend_head_members:
24   ∀seq, s · extend_head_WD(seq, s)
25   ⇒ members(extend_head(seq, s)) = member(seq) ∪ {s}

The effect of **extend_head** and **front** with respect to other operators, i.e., **link**, **head**, **rear**, and **middle** are captured by the associated axioms. For example, axiom **extend_head_link** states that by extending the head of a sequence **seq** to section **s**, the link between sections corresponding to the sequence is extended by a new connection from **s** to the current head of the sequence, i.e., **head(seq)**. Note that axiom **extend_head_middle** ensures that the middle of a sequence is updated correctly even in the situation where the original sequence **seq** is within a single section (i.e., **link(seq)** $= \varnothing$). In this case, according to axiom **@empty_head_rear**, we have **head(seq)** = **rear(seq)**. If the sequence is extended to include section **s**, the middle of the new sequence will still be empty, this is ensured by removing the rear of the sequence. Similar axioms related to **front** are omitted.

Instantiation of Network and Region ADTs by Graph and Sequence ADTs

We now present the instantiation of the Network and Region ADTs by Graph and Sequence ADTs. The instantiation mappings are as follows.

1   NETWORK_TYPE ⟵ GRAPH_TYPE
2   NETWORK ⟵ GRAPH
3   enlarge, enlarge_WD ⟵ add, add_WD
4   contract, contract_WD ⟵ remove, removed_WD
5   REGION_TYPE ⟵ SEQUENCE_TYPE
6   REGION ⟵ SEQUENCE
7   extend, extend_WD ⟵ extend_head, extend_head_WD
8   reduce, reduce_WD ⟵ front_head, front_head_WD

Note that the (abstract) WD conditions are instantiated by the concrete WD conditions, and hence are trivially equivalent.

Predicate operators for **REGION** ADT, i.e., sub-region (⊑) and disjoint (⊕) are now defined (in terms of **SEQUENCE** ADT) as follows.

1   **operators**
2   ⊑ (r1: SEQUENCE, r2: SEQUENCE) infix
3   = members(r1) ⊆ members(r2) ∧ link(r1) ⊆ link(r2)
4   ⊕ (r1: REGION, r2: REGION) infix
5   = members(r1) ∩ members(r2) = ∅

We have to prove that the definition given for ⊑ and ⊕ operators satisfies the axioms as specified in the REGION ADT, e.g., @ ⊑_transitive, @ ⊑_reflexive, @disjoint_symmetric, and @ ⊑_⊕. They can be proved by expanding the definition accordingly. For example, the proof for @ ⊑_⊕ is as follows.

```
 1    ∀r1,r2,r3· r1 ⊑ r2 ∧ r2 ⊕ r3 ⇒ r1 ⊕ r3
 2
 3   ⇐ "Definition of ⊑ and ⊕ "
 4    ∀r1,r2,r3· members(r1) ⊆ members(r2) ∧ link(r1) ⊆ link(r2) ∧
 5        members(r2) ∩members(r3) = ∅
 6   ⇒
 7       members(r1) ∩members(r3) = ∅
 8
 9   ⇐ "Property of ⊆ and ∩"
10    ⊤
```

Further constraints in the REGION ADT relating ⊑ with operators extend and reduce (i.e., extend_⊑_strengthening and reduce_⊑_ weakening) can be proved (after instantiation) in a similar fashion. For example, the proof of the instantiated axiom extend_⊑_strengthening is as follows.

```
 1    ∀r,s· extend_head_WD(r,s) ⇒ r ⊑ extend_head(r,s)
 2
 3   ⇐ "Definition of ⊑ "
 4    ∀r,s· extend_head_WD(r,s) ⇒
 5    members(r) ⊆ members(extend_head(r,s)) ∧
 6    link(r) ⊆ link(extend(r,s))
 7
 8   ⇐ "Axioms @extend_head_link and @extend_head_members"
 9    ∀r,s· extend_head_WD(r,s) ⇒
10    members(r) ⊆ member(seq) ∪ {s} ∧
11    link(r) ⊆ link(seq) ∪ {s ↦ head(seq)}
12
13   ⇐ "Property of ⊆ and ∪ "
14    ⊤
```

Finally, the predicate operator ≪ (a-part-of) relating the Network and Region ADTs is now defined in terms of the Graph and Sequence ADTs as follows.

```
 1  operators
 2   ≪ (r : SEQUENCE, n: GRAPH) infix
 3   = link(r) ⊆ edges(n) ∧ members(r) ⊆ nodes(n)
```

Here, a sequence r is a part of a graph n if r's links are edges of n and the r's members are nodes of n. The axioms related to the *a-part-of* relation, i.e., @enlarge_≪_monotonic and @ ⊑_≪_weakening are proved straightforwardly.

Refinement of the System Model

Given the more concrete datatypes Graph and Sequence, we refine our system model (after the instantiation) accordingly. We give an example of how the system model can be refined as follows. Consider event NRK_contracts which reduces the current active network as follows.

```
1   NRK_contracts
2   any grp scts where
3     @grd1: remove_WD(active_network,grp,scts)
4     @grd2: ∀trn · trn ∈ active_trains
5       ⇒ train_ma(trn) ≪ remove(active_network, grp, scts)
6   then
7     @act1: active_network := remove(active_network,grp,scts)
8   end
```

Focus on the guard @grd2 specifying that the reduced current network must still contain all the active trains' movement authorities. We can replace this guard by the following new guards.

```
1     @grd3: ∀trn · trn ∈ active_trains ⇒ link(train_ma(trn)) ∩grp = ∅
2     @grd4: ∀trn · trn ∈ active_trains ⇒ members(train_ma(trn)) ∩scts = ∅
```

Guard @grd3 states that the connections grp being removed from the network does not overlap with any connection of the trains' MAs. Guard @grd4 states that the sections scts being removed from the network do not contain sections that belong to any trains' MAs. The correctness of this refinement (i.e., guard strengthening of @grd2) relies on the definition of ≪ and the properties of remove, i.e., remove_edges and remove_nodes. Expanding the earlier definition of aPartOf, we have to prove that for any trn ∈ active_trains, we have

```
1   link(train_ma(trn)) ⊆ edges(remove(active_network, grp, scts))
2   members(train_ma(trn)) ⊆ nodes(remove(active_network, grp, scts))
```

From the properties of remove, i.e., in the form of axioms @remove_edges and @remove_nodes, we have

```
1   edges(remove(active_network, grp, scts)) = edges(active_network) \ grp
2   nodes(remove(active_network, grp, scts)) = nodes(active_network) \ scts
```

From invariant @ma − derailment_free, we have

$$\text{train\_ma(trn)} \ll \text{active\_network,}$$

which (by definition of ≪) is equivalent to

```
1   link(train_ma(trn)) ⊆ edges(active_network)
2   members(train_ma(trn)) ⊆ nodes(active_network)
```

Together with @grd3 and @grd4, we can prove that @grd2 indeed holds. Other system events are refined according to the definition provided by the instantiation.

Overall, the refinement and its proof are tightly coupled with the ADT instantiation, relying on (1) the properties of the ADTs, e.g., of remove, and (2) instantiation mapping, e.g., the definition of ≪.

5   Conclusion

5.1   Summary

In this chapter, we present the theory instantiation mechanism for developing datatypes using the Theory extension of Event-B. Together with refinement, theory instantiation allows developers to model systems at a high level of abstraction (both in terms of datatypes and system details) and to gradually introduce system details into the formal models. We illustrate the mechanism using the example of a train control system.

5.2   Advantage of This Approach

When modelling large systems, one has to capture in the model some concepts of the application domain. In other terms, one needs to make implicit (expressed in the modelling language) some explicit concepts of the system.

Traditionally, this is done by defining a direct encoding of the explicit concept in the modelling language. However, this approach can lead to quite clumsy modelling artefacts that are difficult to manipulate and make the model obscure.

In the approach advocated here, the explicit concepts are not directly translated to the modelling language, but they rather captured a piece at a time. At each abstraction level of the model, we only capture the parts of the explicit concepts that we need to express our model and prove it correct. Then, when we refine the model, we also refine the concepts that are captured, using theory instantiation.

Theory instantiation offers a parallel refinement process which is driven by the modelling needs. The approach is thus quite similar to data refinement in Classical B [1], where one refines at the same time the program code and the data structure on which it operates.

This is very visible on the case-study example. Usually, people [2, 8] model such systems by first defining the notion of the graph to describe the network topology, then all the models are expressed using this notion of graph, which makes it very complicated. Here, we start with some abstract and incomplete notions that capture some aspects of a graph and the graph is only fully introduced in the last refinements. This makes the model much more readable in the first refinements where we use simple abstract operators that are sufficient to express our model at that level of abstraction. In particular, the proofs are simpler with very abstract datatypes. Proofs are inherited through refinement and instantiation. Developers just have to prove the theory instantiation, and the refinement of data on the state-variables are "guided" by the instantiation.

References

1. J.-R. Abrial, *The B-book - Assigning Programs to Meanings* (Cambridge University Press, Cambridge, 2005)
2. J.-R. Abrial, *Modeling in Event-B: System and Software Engineering* (Cambridge University Press, Cambridge, 2010)
3. J.-R. Abrial, M. Butler, S. Hallerstede, T.S. Hoang, F. Mehta, L. Voisin, Rodin: an open toolset for modelling and reasoning in Event-B. Softw. Tools Technol. Transf. **12**(6), 447–466 (2010)
4. M.J. Butler, I. Maamria, Practical theory extension in event-b, in *Theories of Programming and Formal Methods - Essays Dedicated to Jifeng He on the Occasion of His 70th Birthday*, eds. by Z. Liu, J. Woodcock, H. Zhu. Lecture Notes in Computer Science, vol. 8051 (Springer, Berlin, 2013), pp. 67–81
5. A. Fürst, T.S. Hoang, D. Basin, N. Sato, K. Miyazaki, Large-scale system development using abstract data types and refinement. Sci. Comput. Program. **131**, 59–75 (2016)
6. T.S. Hoang, An introduction to the Event-B modelling method, in *Industrial Deployment of System Engineering Methods* (Springer, Berlin, 2013), pp. 211–236
7. T.S. Hoang, L. Voisin, M. Butler, Developments using theory instantiation (2017). http://users. ecs.soton.ac.uk/tsh2n14/developments/Shonan2017/
8. C. Metayer, M. Clabaut, DIR 41 case study, in *Abstract State Machines, B and Z, First International Conference, ABZ 2008, London, UK, September 16-18, 2008. Proceedings*, eds. by E. Börger, M.J. Butler, J.P. Bowen, P. Boca. Lecture Notes in Computer Science, vol. 5238 (Springer, Berlin, 2008), p. 357

Integrating Domain Modeling Within a Formal Requirements Engineering Method

Steve Tueno, Régine Laleau, Amel Mammar, and Marc Frappier

Abstract One way to build safe critical systems is to formally model the requirements formulated by stakeholders and to ensure their consistency with respect to domain properties. This paper describes a metamodel for a domain modeling language built from *OWL* and *PLIB*. The language is part of the *SysML/KAOS* requirements engineering method which also includes a goal modeling language. The formal semantics of SysML/KAOS models is specified, verified, and validated using the *Event-B* method. Goal models provide machines and events of the Event-B specification while domain models provide its structural part (sets and constants with their properties and variables with their invariant). Our proposal is illustrated with a case study dealing with the specification of a localization component for an autonomous vehicle.

1 Introduction

Computer science is a relatively young science, but it does not prevent it from tackling huge challenges such as implementation of critical and complex software or cyber-physical systems. Such systems require careful analysis and design to ensure they do

S. Tueno
Université Paris-Est Créteil, 94010 Créteil, France
e-mail: steve.tuenofotso@univ-paris-est.fr

Université de Sherbrooke, Sherbrooke, QC J1K 2R1, Canada

R. Laleau (✉)
Université Paris-Est Créteil, 94010 Créteil, France
e-mail: laleau@u-pec.fr

A. Mammar
Télécom SudParis, 91000 Evry, France
e-mail: amel.mammar@telecom-sudparis.eu

M. Frappier
Université de Sherbrooke, Sherbrooke, QC J1K 2R1, Canada
e-mail: Marc.Frappier@usherbrooke.ca

© Springer Nature Singapore Pte Ltd. 2021
Y. Ait-Ameur et al. (eds.), *Implicit and Explicit Semantics Integration in Proof-Based Developments of Discrete Systems*,
https://doi.org/10.1007/978-981-15-5054-6_3

not cause disasters. Literature is full of disasters caused by failures at one of these stages [16]. The purpose of the *ANR FORMOSE* project [4] is to design a formally grounded, model-based requirements engineering method, for critical and complex systems, supported by an open-source environment. Modeling a system according to the defined requirements engineering method requires the representation of its requirements as well as of entities and properties of its application domain. This representation implicitly implies a semantics that must be defined explicitly through a formal method in order to be verified and validated and thus to prevent potential failures. The *SysML/KAOS* goal modeling language [15] focuses on modeling of functional and non-functional requirements through goal hierarchies. Furthermore, Matoussi et al. [18] report on the explicit representation of the semantics of SysML/KAOS goal models with *Event-B* [1].

This paper complements the aforementioned studies with the definition of a domain modeling language. We first synthesize the body of knowledge related to the concrete representation of the semantics of SysML/KAOS goal models. Then, we analyze existing domain modeling approaches and describe the defined SysML/KAOS domain modeling language. The illustration is performed on *TACOS* [3], a case study dealing with the specification of a localization software component that uses *GPS*, *Wi-Fi*, and sensor technologies for the real-time localization of the *Cycab* vehicle [24], an autonomous ground transportation system.

The remainder of this paper is structured as follows: Sect. 2 briefly describes Event-B and SysML/KAOS. Section 3 summarizes existing work [17, 18] on the explicit representation of the semantics of SysML/KAOS models. Section 4 presents the relevant state of the art on domain modeling in requirements engineering and defines our expectations regarding the SysML/KAOS domain modeling language. Finally, Sect. 5 describes and illustrates the domain modeling language while Sect. 6 reports our conclusions and discusses future work.

2 Background

This section provides a brief overview of the Event-B formal method and of the SysML/KAOS requirements engineering method.

2.1 Event-B

Event-B [1] is a formal method created by *J. R. Abrial* for *system modeling*. It is used to incrementally build a specification of a system that preserves a set of properties expressed through invariants. Event-B is mostly used to model closed systems: the modeling of the system is accompanied by that of its environment and of all interactions likely to occur between them. An Event-B model includes static parts called *contexts* and dynamic parts called *machines*. Contexts contain declarations of abstract and enumerated sets, constants, axioms, and theorems. Machines contain variables, invariants, and events. Moreover, a machine can access the definitions of a context. Each event has a *guard* and an *action*. The guard is a condition that must be satisfied for the event to be triggered and the action describes updates of state variables. The system specification can be constructed using stepwise refinement, by refining machines. Proof obligations are defined to prove invariant preservation by events (invariant has to be true at any system state), event feasibility, convergence, and machine refinement [1]. We use *B System* [8], a variant of Event-B proposed by *ClearSy*, an industrial partner in the *FORMOSE* project, in its integrated development environment *Atelier B* [5]. *B System* and *Event-B* share the same semantics but are syntactically different [27].

2.2 SysML/KAOS

SysML/KAOS [15] is a requirements engineering method which combines the traceability provided by *SysML* [12] with goal expressiveness provided by KAOS [16]. It allows the representation of requirements to be satisfied by a system and of expectations with regards to the environment through a hierarchy of goals. The goal hierarchy is built through a succession of refinements using different operators: *AND*, *OR*, and *MILESTONE*. An *AND refinement* decomposes a goal into subgoals, and all of them must be achieved to realize the parent goal. An *OR refinement* decomposes a goal into subgoals such that the achievement of only one of them is sufficient for the accomplishment of the parent goal. A *MILESTONE refinement* is a variant of the AND refinement which allows the definition of an achievement order between goals.

KAOS captures domain entities and properties within a model called the **object model** which is a *UML* class diagram. Its expressiveness is however considered insufficient by *FORMOSE* industrial partners [4], regarding the complexity and the criticality of the systems of interest.

Within SysML/KAOS, a functional goal describes the *expected behavior* of the system once a certain condition holds [17]: *[if **CurrentCondition** then] sooner-or-later **TargetCondition***. A functional goal can also be defined without specifying a *CurrentCondition*. In this case, the expected behavior can be observed from any system state.

Figure 1 represents a SysML/KAOS goal diagram for the *Cycab* localization component. Its main purpose is vehicle localization.

To achieve the root goal, which is the localization of the vehicle (**LocalizeVehicle**), raw localizations must be captured from vehicle sub components (**CaptureRaw Localizations**) which can be *GPS* (**CaptureGPSlocalization**) or *Wi-Fi* (**Capture WIFILocalization**), be validated using a vehicle sensor (**ValidateRawlocalizations**) which has to be either a speed sensor (**ValidateUsingSpeedSensor**) or an accelerometer (**ValidateUsingAccelerometer**), and used to compute the vehicle's accurate localization (**ComputeAccuratedlocalization**).

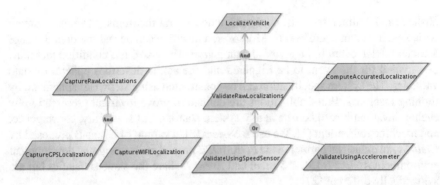

Fig. 1 Excerpt from the localization component goal diagram

3 Expression of the Semantics of SysML/KAOS Models in Event-B

3.1 Semantics of Goal Models

The formalization of SysML/KAOS goal models is detailed in [18]. Each refinement level of a goal diagram gives an Event-B machine. Each goal gives an event. The semantics of refinements links between goals is explicited using proof obligations that complement classic proof obligations for **invariant preservation** and for **event actions feasibility** defined in [1]. The other classic Event-B proof obligations are not relevant for our purpose [18]. Regarding the added proof obligations, they depend on the refinement pattern used. For an abstract goal G and two concrete goals G_1 and G_2,[1]

- For an *AND* refinement, the proof obligations are

  - $G_1\_Guard \Rightarrow G\_Guard$ $- G_2\_Guard \Rightarrow G\_Guard$
  - $(G_1\_Post \wedge G_2\_Post) \Rightarrow G\_Post$

- For an *OR* refinement, they are

  - $G_1\_Guard \Rightarrow G\_Guard$ $- G_2\_Guard \Rightarrow G\_Guard$
  - $G_1\_Post \Rightarrow G\_Post$ $- G_2\_Post \Rightarrow G\_Post$
  - $G_1\_Post \Rightarrow \neg G_2\_Guard$ $- G_2\_Post \Rightarrow \neg G_1\_Guard$

- For a *MILESTONE* refinement, they are

  - $G_1\_Guard \Rightarrow G\_Guard$ $- G_2\_Post \Rightarrow G\_Post$
  - $\Box(G1\_Post \Rightarrow \Diamond G2\_Guard)$ (each system state, corresponding to the post condition of G_1, must be followed, at least once in the future, by a system state enabling G_2)

---

[1]For an event G, G_Guard represents the guards of G and G_Post represents the post condition of its actions.

Fig. 2 Formalization of the
root level of the goal
diagram of Fig. 1

SYSTEM
    localizationComponent
SETS
CONSTANTS
PROPERTIES
VARIABLES
INVARIANT
INITIALISATION
EVENTS
    LocalizeVehicle=
    BEGIN
        // localization of the vehicle
    END
END

Figures 2 and 3 represent the *B System* components obtained, respectively, from the root level of the goal diagram of Fig. 1 and from its first refinement level. The structural part of the *B System* specification (constants constrained by properties and variables constrained by an invariant) and the body of events must be manually provided. The objective of our study is to automatically derive the structural part from a rigorous modeling of the domain of the system.

Proof obligations related to the *AND* refinement link between the root and the first refinement levels are

$$CaptureRawlocalizations\_Guard \Rightarrow LocalizeVehicle\_Guard \quad (1)$$

$$ValidateRawlocalizations\_Guard \Rightarrow LocalizeVehicle\_Guard \quad (2)$$

$$ComputeAccuratedlocalization\_Guard \Rightarrow LocalizeVehicle\_Guard \quad (3)$$

$$CaptureRawlocalizations\_PostCondition \wedge ValidateRawlocalizations\_PostCondition \wedge$$
$$ComputeAccuratedlocalization\_PostCondition \Rightarrow LocalizeVehicle\_PostCondition$$
$$(4)$$

3.2 Toward an Event-B Expression of the Semantics of Domain Models

A domain model is a conceptual model capturing the topics related to a specific problem domain [7]. The main difference between requirements and domain models is that domain models are independent of stakeholders. They must conform to the operational context of the system. In [6], *a domain description primarily specifies semantic entities of the domain intrinsics, semantic entities of support technolo-*

Fig. 3 Formalization of the
first refinement level of the
goal diagram of Fig. 1

REFINEMENT
  localizationComponentRef1
REFINES
  localizationComponent
SETS
CONSTANTS
PROPERTIES
VARIABLES
INVARIANT
INITIALISATION
EVENTS
  CaptureRawlocalizations=
  BEGIN
      // capture raw localizations
  END;
  ValidateRawlocalizations=
  BEGIN
      // validate raw localizations
  END;
  ComputeAccuratelocalization =
  BEGIN
      // compute vehicle accurate localization
  END
END

gies already "in" the domain, semantic entities of management and organization domain entities, syntactic and semantic of domain rules and regulations, syntactic and semantic of domain scripts, and semantic aspects of human domain behavior. In [23], Pierra defines a domain model as a set of categories represented as classes, their properties, and their logical relationships. Modeling the domain of a system consists in giving a representation of the set of concepts that the system will be called upon to manipulate and the set of properties and constraints associated with them.

A first attempt at modeling domains within SysML/KAOS is achieved in [17]. Domain modeling involves *UML* class diagrams, *UML* object diagrams, and ontologies. The case study presented reveals the use of ontologies for domain knowledge representation; the model obtained is the *domain model*. Furthermore, UML object and class diagrams are used to represent the system structure and constraints in a model known as the *structural model* which must conform to the domain model. A set of rules is proposed to translate some domain model elements to Event-B. However, the proposal involves *UML* diagrams which are semi-formal graphical representations [19, 20]. Moreover, it uses several languages which is an extra source of complexity.

4 State of the Art on Domain Modeling in Requirements Engineering

4.1 Existing Domain Modeling Approaches

In *KAOS* [16], the domain of a system is specified with an *object model* using *UML* class diagrams. An object within this model can be (1) an *entity* if it exists independently of the others and does not influence the state of any other object, (2) an *association* if it links other objects on which it depends, (3) an *agent* if it actively influences the system state by acting on other objects, or (4) an *event* if its existence is instantaneous, appearing to impulse an update of the system state. This approach, which is essentially graphic and semi-formal, as argued in [19], is difficult to exploit in case of critical systems [20].

In [10], Devedzic proposes to model the domain knowledge through either formulae of first-order logic or ontologies. He considers ontologies as a more structured and extensible representation of domain knowledge.

In [14], domain models are built around concepts and relationships: each definition of a domain model consists of an assertion linking two instances of Concept through an instance of Relationship. A categorization is proposed for concepts and relationships: a concept can be a function, an object, a constraint, an actor, a platform, a quality, or an ambiguity, while a relationship can be a performative or a symmetry, reflexivity, or transitivity relation. However, the proposed metamodel misses some relevant domain entities such as datasets, predicates to express domain constraints, and relation cardinalities. Moreover, it does not propose modularisation mechanisms between domain models.

In [20], ontologies are used not only to represent domain knowledge, but also to model and analyze requirements. The proposed methodology is called *Knowledge-Based Requirements Engineering (KBRE)* and is mainly used for the detection and processing of inconsistencies, conflicts, and redundancies among requirements. In spite of the fact that *KBRE* proposes to model domain knowledge with ontologies, the proposal focuses on the representation of requirements. A similar approach called *GOORE* is proposed in [26].

In [9], Dermeval et al. proposes a systematic literature review related to usages of ontologies in requirements engineering. They end up describing ontologies as a standard form of formal representation of concepts within a domain, as well as of relationships between those concepts.

These approaches suggest that ontologies are relevant for modeling the domains of systems.

4.2 A Study of Ontology Modeling Languages

An ontology can be defined as a formal model representing concepts that can be grouped into categories through generalization/specialization relations, their instances, constraints, and properties as well as relations existing between them. Ontology modeling languages can be grouped into two categories: *Closed World Assumption (CWA)* for those considering that any fact that cannot be deduced from what is declared within the ontology is false and *Open World Assumption (OWA)* for those considering that any fact can be true unless its falsity can be deduced from what is declared within the ontology. As [2], we consider that accurate modeling of the knowledge of engineering domains, to which we are interested, must be done under the *CWA* assumption. Indeed, this assumption improves the formal validation of the consistency of system's specifications with respect to domain properties. Moreover, systems of interest to us are so critical that no assertion should be assumed to be true until the consensus is reached on its veracity. Similarly, we also advocate *strong typing* [2] because our domain models must be translatable to Event-B specifications.

Several ontology modeling languages exist. The main ones are **OWL (Ontology Web Language)** [25], **PLIB (Part LIBrary)** [22], and **F-Logic (Frame Logic)** [13]. A summary of similarities and differences between these languages is described in Table 1:

- *PLIB*, *OWL*, and *F-Logic* implement modularisation mechanisms. *PLIB* supports partial import: a class of an ontology *A* can extend a class of an ontology *B* and explicitly specify the properties it wishes to inherit. Moreover, if nothing is specified, no property will be imported. On the other hand, *OWL* and *F-Logic* use

Table 1 Comparative table of the three main ontology modeling languages

Characteristics	OWL	PLIB	F-Logic
Modularity	Total	Partial	Total
CWA versus OWA	OWA	**CWA**	**CWA**
Inheritance	Multiple	**Simple**	Multiple
Typing	Weak	**Strong (any element belongs to one and only one type)**	weak
Expressivity	**Strong**	Weak	Weak
Contextualization of a property (parameterized attributes)	–	+	+
Different views for an element	–	+	–
Graphic representation	+	–	–
Domain Knowledge (static versus dynamic)	Static	Static	Static

the total import: when an ontology *A* refers to an ontology *B*, all elements of *B* are accessible within *A*.

- *PLIB* and *F-Logic* use the *CWA* assumption for constraint verification, *OWL* uses the *OWA* assumption.
- *OWL* and *F-Logic* implement multiple inheritance and instantiation while *PLIB* implements simple inheritance and instantiation. On the other hand, with the *is_case_of* relation, a *PLIB* class can be *a case of* several other classes, each class bringing some specific properties.
- *PLIB* and *F-Logic* allow the definition of parameterized attributes using context parameters, which is not possible with *OWL*.
- *PLIB* allows several representations or view points for a concept while neither *OWL* nor *F-Logic* do.
- The knowledge modeled using *OWL*, *PLIB*, and *F-Logic* is always considered static because there is no distinguishing mechanism. It is for instance impossible to specify that the localization of a vehicle can change dynamically while its brand cannot.

As stated in [29], *all the studied languages emphasize more on modeling static domain knowledge*. None of these languages allows to specify that a knowledge described must remain unchanged or that it is likely to be updated. Moreover, none of the languages fully meet our requirements. For instance, *OWL* assumes the *OWA* assumption, *PLIB* is weakly expressive, etc. The most aligned are *OWL* and *PLIB*.

5 Our Approach for Domain Modeling

We choose to represent domain knowledge using ontologies since they are semantically richer and therefore allow a more explicit representation of domain characteristics. Thus, in this Section, we propose a metamodel, based on that of *OWL* and *PLIB* while filling their shortcomings, to represent the domain of a system whose requirements are captured using the SysML/KAOS method. The domain modeling language makes the *Unique Name Assumption (UNA)* [2]: the name of an element is sufficient to uniquely identify it among all others. Furthermore, the metamodel is designed to allow the specification of knowledge that are likely to evolve over time.

5.1 Presentation

Figures 4, 5, 6, and 7 present the main part of the metamodel associated with the SysML/KAOS domain modeling language. The yellow elements are those that have an equivalence in *OWL*, while the red ones are the ones that have been inserted or customized. In addition, some constraints and associations, such as the *parentCon-*

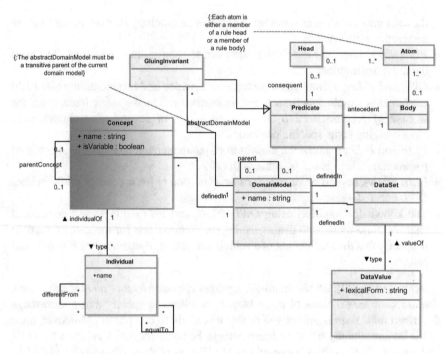

Fig. 4 First part of the metamodel associated with the domain modeling language

cept association, come from the *PLIB* metamodel. Due to space consideration, we will not highlight all the elements and constraints of the metamodel.

5.1.1 Concepts and Individuals, Datasets, and Data Values

Domain models are built around instances of **Concept** which represent sets of individuals sharing common characteristics (Fig. 4). A *concept* can be *variable* (*isVariable=true*) when the set of its individuals is likely to be updated through addition or deletion of individuals. Otherwise, it is *constant* (*isVariable=false*). A concept can be associated with another one, known as its parent concept, through the *parentConcept* association, from which it inherits properties. As a result, any individual of the child concept is also an individual of the parent concept. It should be noted that when a variable concept CO is a subconcept of another variable concept PCO, the set of elements that CO can contain, over its whole existence, is included in the set of elements that PCO can contain. However, this version of the domain modeling language allows that, at some point, because of the variability of CO and PCO, an element present in CO is not present in PCO.

Datasets (instances of **DataSet**) are used to group data values (instances of **DataValue**) having the same type (Fig. 5). Default datasets are INTEGER, NATURAL

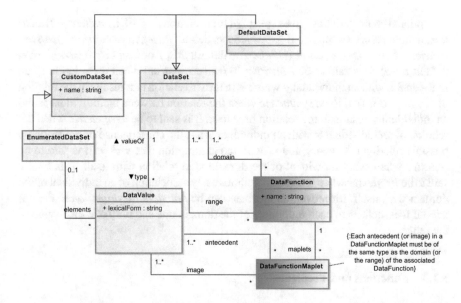

Fig. 5 Fourth part of the metamodel associated with the domain modeling language

for positive integers, FLOAT, STRING, or BOOL for booleans. The easiest way to build a dataset is to list its elements. This can be done by defining instances of EnumeratedDataSet.

5.1.2 Relations and Attributes

Relations (instances of Relation) are used to capture links between concepts (Fig. 6) while attributes (instances of Attribute) capture links between concepts and datasets (Fig. 7). A relation (Fig. 6) or an attribute (Fig. 7) can be *variable* if its set of maplets can be updated through addition or deletion. Otherwise, it is *constant*. Relations are characterized by their *cardinalities*: DomainCardinality and RangeCardinality (Fig. 6). Each instance of DomainCardinality (respectively RangeCardinality) makes it possible to define, for a relation re, the minimum and maximum limits of the number of individuals, having the domain (respectively range) of re as *type*, that can be put in relation with one individual, having the range (respectively domain) of re as *type*. The following constraints are associated with these limits: $(minCardinality \geq 0) \wedge (maxCardinality = \infty \vee maxCardinality \geq minCardinality)$, knowing that if $maxCardinality = \infty$, then there is no maximum limit. Relation maplets (instances of RelationMaplet) define associations between individuals through relations. In an identical manner, attribute maplets (instances of AttributeMaplet) define associations between individuals and data values through attributes.

Optional characteristics can be specified for a relation (Fig. 6): *transitive* (*isTransitive*, default *false*), *symmetrical* (*isSymmetric*, default *false*), *asymmetrical* (*isASymmetric*, default *false*), *reflexive* (*isReflexive*, default *false*), or *irreflexive* (*isIrreflexive*, default *false*). It is said to be *transitive* (*isTransitive=true*) when the relation of an individual x with an individual y which is in turn in relation to z results in the relation of x and z. It is said to be *symmetric* when the relation between an individual x and an individual y results in the relation of y to x. It is said to be *asymmetric* when the relation of an individual x with an individual y has the consequence of preventing a possible relation between y and x, with the assumption that $x \neq y$. It is said to be *reflexive* when every individual of the domain is in relation with itself. It is finally said to be *irreflexive* when it does not authorize any association of an individual of the domain with itself. Moreover, an attribute can be *functional* (*isFunctional*, default *true*) if it associates to each individual of the domain one and only one data value of the range.

5.1.3 Functions and Predicates

Data functions (Instances of **DataFunction**) (Fig. 5) define operations which allow to determine data values at the output of a set of processes on some input data values. At each tuple of data values of the domain, the `data function` assigns a tuple of data values of the range, and this assignement cannot be changed dynamically. *Example*: a data function named `multiply` can be defined to produce, given two integers (individuals of `INTEGER`) x and y, the integer representing $x * y$. On the other side, predicates (instances of **Predicate**) (Fig. 4) represent constraints between different elements of the domain model as *horn clauses*: each predicate has a body which represents its *antecedent* and a head which represents its *consequent*, body and head designating conjunctions of atoms. A *typing atom* defines the type of a term: **ConceptAtom** for individuals and **DataSetAtom** for data values (Fig. 12). An *association atom* defines an association between terms: **RelationAtom** for associations through instances of **Relation**, **AttributeAtom** for associations through instances of **Attribute**, and **DataFunctionAtom** for associations through instances of **DataFunction** (Fig. 12). For each case, types of the related terms must correspond to domains/ranges of the considered link. A *comparison atom* defines comparison relationships between terms: **EqualityAtom** for equality and **InequalityAtom** for difference (Fig. 12). Built in atoms are specialized atoms, characterized by identifiers captured through the **AtomType** enumeration, and used to represent special constraints between terms (Fig. 12) such as arithmetic constraints between several integers (e.g., $a + b < c$). Predicates can also be used to represent constraints required for *parameterized/dependent* relations or attributes. For example, knowing that each material resistance depends on medium temperature, resistance and temperature are dependent attributes.

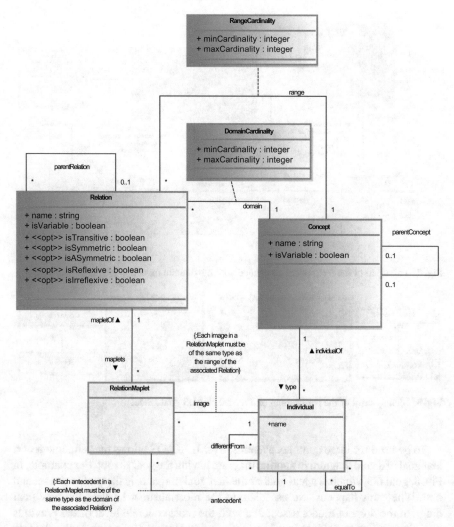

Fig. 6 Second part of the metamodel associated with the domain modeling language

5.1.4   Domain Model and Goal Model

Each domain model is associated with a refinement level of the SysML/KAOS functional goal model and can have, as its parent, another domain model (Fig. 4). This allows the child domain model to access and extend some elements defined within the parent domain model. It should be noted that the parent domain model must be associated with the refinement level directly above the one to which the child domain model is associated.

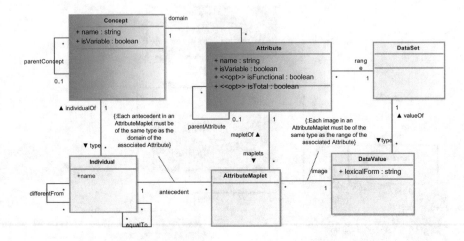

Fig. 7 Third part of the metamodel associated with the domain modeling language

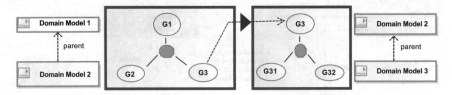

Fig. 8 Management of the partitioning of a SysML/KAOS goal model

To be used for large complex systems, SysML/KAOS allows the refinement of a leaf goal of a goal diagram in another diagram having the goal as root. For example, in Fig. 8, goal *G3*, which is a leaf goal of the first goal diagram, is the root of the second one. When this happens, we associate to the most abstract level of the new goal diagram the domain model associated with the most concrete level of the previous goal diagram as represented in Fig. 8: Domain Model 2, which is the domain model associated with the most concrete level of the first diagram, is also the domain model associated with the root of the second one.

5.2 Illustration

We have identified two graphical syntaxes to represent ontologies: the syntax proposed by *OntoGraph* [11] and the one proposed by *OWLGred* [28]. The *OntoGraph* syntax is the one used in [17]. Unfortunately, it does not allow the representation of some domain model elements such as attributes or cardinalities. For this illustration, we have thus decided to use the *OWLGred* syntax. For readability purposes, we have

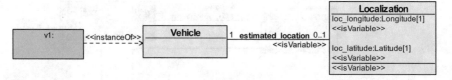

Fig. 9 *localization_component_0*: ontology associated with the root level of the goal diagram of Fig. 1

decided to represent the *isVariable* property only when it is set to *true* and to remove optional characteristics representation.

Figures 9, 10, and 11 represent the domain models associated, respectively, with the root level of the goal diagram of Fig. 1 (*localization_component_0*), with its first refinement level (*localization_component_1*) and with its second one (*localization_component_2*).

5.2.1 Ontology Associated with the Root Level

In ontology *localization_component_0* (Fig. 9), a vehicle is modeled as an instance of **Concept** named Vehicle and its localization is represented through an instance of **Concept** named Localization. Since it is possible to dynamically add or remove vehicle localizations, the property *isVariable* of Localization is set to *true*, which is represented by the stereotype «*isVariable*». Since the system is designed to control a single vehicle, it is not possible to dynamically add new ones. The involved vehicle is thus modeled as an instance of **Individual** named v1 having Vehicle as *type*. Localization is the *domain* of two attributes: the latitude modeled as an instance of **Attribute** named loc_latitude and the longitude modeled as an attribute named loc_longitude. Attribute loc_latitude has, as range, an instance of **CustomDataSet** named Latitude and loc_longitude an instance of **CustomDataSet** named Longitude. Since it is possible to dynamically change the localization of a vehicle, the property *isVariable* of loc_latitude and that of loc_longitude are set to *true*, which is represented by the stereotype «*isVariable*». The association between an individual of Vehicle and an individual of localization is represented through an instance of **Relation** named estimated_location. Its associated domain cardinality has *minCardinality=maxCardinality=1*, and its associated range cardinality has *minCardinality=0* and *maxCardinality=1*.

5.2.2 Ontology Associated with the First Refinement Level

Ontology *localization_component_1* (Fig. 10) has ontology *localization_component_0* (Fig. 9) as parent and defines new concepts and relations. Each reused

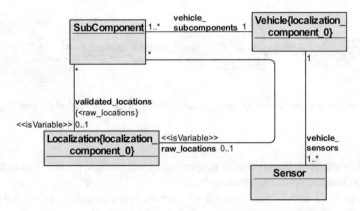

Fig. 10 *localization_component_1*: ontology associated with the first refinement level of the goal diagram of Fig. 1

element is annotated with *localization_component_0*, the parent domain model name. SubComponent, which is an instance of **Concept**, is introduced to represent sub components of a vehicle. Each instance of **Individual** of *type* SubComponent associates the vehicle with a *raw location*. Sensor, which is also an instance of **Concept**, is introduced to represent vehicle sensors used to validate the raw locations. Raw locations which are validated through sensors are called validated locations and are used to compute the vehicle estimated location. Each vehicle has at least one sub component and one sensor.

5.2.3 Ontology Associated with the Second Refinement Level

Ontology *localization_component_2* (Fig. 11) has ontology *localization_component_1* (Fig. 10) as parent. This third abstraction level represents child concepts of SubComponent and Sensor. A subcomponent is either a GPS, represented through an instance of **Concept** named Gps, or a Wi-Fi, represented through an instance of **Concept** named Wifi. A sensor is either an accelerometer, represented through an instance of **Concept** named Accelerometer, or a speed sensor, represented through an instance of **Concept** named SpeedSensor. Finally, v1 is associated to an instance of **Individual** of *type* Gps named g1 and to an instance of **Individual** of *type* Wifi named w1 through vehicle_subcomponents, an instance of **Relation** introduced in *localization_component_1*. It is also associated to a speed sensor called s1 and to an accelerometer called a1.

The constraint *"a GPS is more precise than a Wi-Fi"* is translated into an instance of **Predicate** represented through formula 5: If an instance of **Term**, named x, having Wifi as its *type*, has px as its *precision* and an instance of **Term**, named y, having Gps as its *type*, has py as its *precision*, then $py > px$.

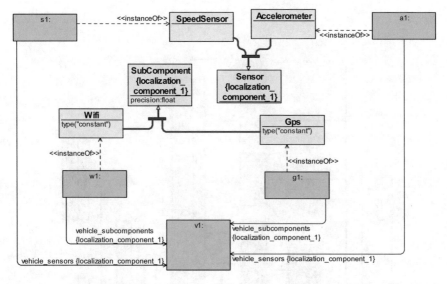

Fig. 11 *localization_component_2*: ontology associated with the second refinement level of the goal diagram of Fig. 1

$$greaterThan(?py, ?px) \leftarrow Wifi(?x) \wedge precision(?x, ?px) \wedge Gps(?y) \wedge precision(?y, ?py)$$

$$(5)$$

6   Conclusion

In this paper, we have first presented the explicitness of the semantics of SysML/KAOS goal models in Event-B. Then, we have drawn up the state of the art related to domain modeling in requirements engineering. After positioning ourselves as to the existing, we have presented our domain modeling approach consisting in representing domain entities and constraints using an ontology modeling language for which a metamodel is defined. The proposal is illustrated with a case study dealing with the specification of a localization component for a *Cycab* vehicle.

Work in progress is aimed at developing mechanisms for the explicitness of the semantics of SysML/KAOS domain models in Event-B. We are also working on integrating the language within the open-source platform *Openflexo* [21] which federates the various contributions of *FORMOSE* project partners [4].

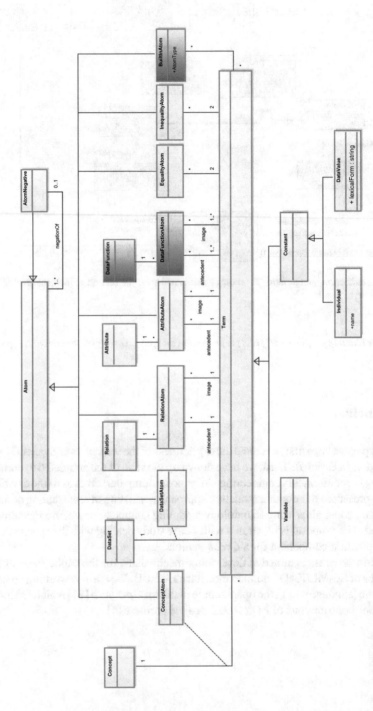

Fig. 12 Fifth part of the metamodel associated with the domain modeling language

Acknowledgements This work is carried out within the framework of the *FORMOSE* project [4] funded by the French National Research Agency (ANR).

References

1. J. Abrial, *Modeling in Event-B - System and Software Engineering* (Cambridge University Press, 2010)
2. Y. Aït Ameur, M. Baron, L. Bellatreche, S. Jean, E. Sardet, Ontologies in engineering: the OntoDB/OntoQL platform. Soft Comput. **21**(2), 369–389 (2017)
3. ANR-06-SETIN-017: TACOS ANR project (2006)
4. ANR-14-CE28-0009: Formose ANR project (2014). http://formose.lacl.fr/
5. Atelier B, *The Industrial Tool to Efficiently Deploy the B Method*. http://www.atelierb.eu/index-en.php (2008). Access date 22.03. 2015
6. D. Bjørner, A. Eir, Compositionality: ontology and mereology of domains, in *Concurrency, Compositionality, and Correctness, Essays in Honor of Willem-Paul de Roever*, Lecture Notes in Computer Science, vol. 5930. Springer (2010), pp. 22–59
7. M. Broy, *Domain Modeling and Domain Engineering: Key Tasks in Requirements Engineering*. (Springer, Berlin, Heidelberg, 2013), pp. 15–30
8. ClearSy: Atelier B: B System (2014). http://ajhurst.org/~ajh/teaching/ClearSy-Industrial_Use_of_B.pdf
9. D. Dermeval, J. Vilela, I.I. Bittencourt, J. Castro, S. Isotani, P. Brito, A. Silva, Applications of ontologies in requirements engineering: a systematic review of the literature. Requir. Eng. **21**(4), 405–437 (2016)
10. V. Devedzic, Knowledge modeling - state of the art. Integr. Comput. Aided Eng. **8**(3), 257–281 (2001)
11. S. Falconer, Protégé - ontograph (2010). http://protegewiki.stanford.edu/wiki/OntoGraf
12. M. Hause et al., The SysML modelling language, in *Fifteenth European Systems Engineering Conference*, vol. 9. Citeseer (2006)
13. M. Kifer, G. Lausen, F-logic: a higher-order language for reasoning about objects, inheritance, and scheme, in *Proceedings of the 1989 ACM SIGMOD* (ACM Press, 1989), pp. 134–146
14. M. Kitamura, R. Hasegawa, H. Kaiya, M. Saeki, An integrated tool for supporting ontology driven requirements elicitation, in *ICSOFT 2007*, Volume SE, ed. by J. Filipe, B. Shishkov, M. Helfert (INSTICC Press, 2007), pp. 73–80
15. R. Laleau, F. Semmak, A. Matoussi, D. Petit, A. Hammad, B. Tatibouet, A first attempt to combine SysML requirements diagrams and B. Innov. Syst. Softw. Eng. **6**(1–2), 47–54 (2010)
16. A. van Lamsweerde, *Requirements Engineering - From System Goals to UML Models to Software Specifications* (Wiley, 2009)
17. A. Mammar, R. Laleau, On the use of domain and system knowledge modeling in goal-based Event-B specifications, in *ISoLA 2016*, Lecture Notes in Computer Science, vol. 9952 (2016), pp. 325–339
18. A. Matoussi, F. Gervais, R. Laleau, A goal-based approach to guide the design of an abstract Event-B specification, in *ICECCS 2011* (IEEE Computer Society, 2011), pp. 139–148
19. W.E. McUmber, B.H.C. Cheng, A general framework for formalizing UML with formal languages, in *ICSE 2001* (IEEE Computer Society, 2001), pp. 433–442
20. T.H. Nguyen, B.Q. Vo, M. Lumpe, J. Grundy, KBRE: a framework for knowledge-based requirements engineering. Softw. Qual. J. **22**(1), 87–119 (2014)
21. Openflexo: Openflexo project (2015). http://www.openflexo.org
22. G. Pierra, The PLIB ontology-based approach to data integration, in *IFIP 18th World Computer Congress*, IFIP, vol. 156 (Kluwer/Springer, 2004), pp. 13–18
23. G. Pierra, Context representation in domain ontologies and its use for semantic integration of data. J. Data Semant. **10**, 174–211 (2008)

24. S. Sekhavat, J.H. Valadez, The Cycab robot: a differentially flat system, in *IROS 2000* (IEEE, 2000), pp. 312–317
25. K. Sengupta, P. Hitzler, Web ontology language (OWL), in *Encyclopedia of Social Network Analysis and Mining* (2014), pp. 2374–2378
26. M. Shibaoka, H. Kaiya, M. Saeki, GOORE: goal-oriented and ontology driven requirements elicitation method, in *ER 2007 Workshops*, Lecture Notes in Computer Science, vol. 4802. (Springer, 2007), pp. 225–234
27. S. Tueno, A. Mammar, R. Laleau, M. Frappier, Event-B expression and validation of translation rules between SysML/KAOS domain models and B system specifications, in *ABZ 2018*, Lecture Notes in Computer Science, vol. 10817 (Springer, 2018), pp. 55–70
28. OWLGrEd home (2017). http://owlgred.lumii.lv/
29. L. Zong-yong, W. Zhi-xu, Z. Ai-hui, X. Yong, The domain ontology and domain rules based requirements model checking. Int. J. Softw. Eng. Appl. **1**(1), 89–100 (2007)

Knowledge Based Modelling

Operations over Lightweight Ontologies and Their Implementation

Marco A. Casanova and Rômulo C. Magalhães

Abstract This chapter first defines a set of operations that create new ontologies, including their constraints, out of other ontologies. The *projection*, *union*, and *deprecation* operations help define new ontologies by reusing fragments of other ontologies, the *intersection* operation constructs the constraints that hold in two ontologies, and the *difference* operation returns the constraints that hold in one ontology, but not in the other. Then, the chapter discusses how to implement the operations for a class of ontologies, called *lightweight ontologies*. The key question is how to concretely construct the constraints of the resulting ontology, which is solved with the help of a structural proof procedure for lightweight ontologies. Lastly, it addresses the question of minimizing the set of constraints of a lightweight ontology.

1 Introduction

We argued elsewhere [12] that certain familiar ontology design problems are profitably addressed by treating ontologies as theories and by defining a set of operations that create new ontologies, including their constraints, out of other ontologies. The *projection*, *union*, and *deprecation* operations help define new ontologies by reusing fragments of known ontologies, the *intersection* operation constructs the constraints that hold in two ontologies, and the *difference* operation returns the constraints that hold in one ontology, but not in the other.

In this chapter, we concentrate on *lightweight ontologies*, that is, ontologies whose constraints are *lightweight inclusions*, which are expressive enough to cover the types of constraints commonly used in conceptual modeling and which are as expressive as the class of inclusions considered in DL-Lite core with arbitrary number restrictions

M. A. Casanova (✉) · R. C. Magalhães
Department of Informatics, Pontifical Catholic University of Rio de Janeiro—PUC-Ri, Rua Marques de S. Vicente, 225, Gávea, 22451900 Rio de Janeiro, Brazil
e-mail: casanova@inf.puc-rio.br

R. C. Magalhães
e-mail: romulo.eng@gmail.com

© Springer Nature Singapore Pte Ltd. 2021
Y. Ait-Ameur et al. (eds.), *Implicit and Explicit Semantics Integration in Proof-Based Developments of Discrete Systems*,
https://doi.org/10.1007/978-981-15-5054-6_4

[3]. We show how to implement projection, union, deprecation, and intersection for this class of ontologies. Difference poses problems, as discussed in the chapter. We single out the question of minimizing a set of lightweight inclusions, which is a step that the implementation of the operations has in common.

The key question about the implementation of the operations is how to concretely construct the constraints of the resulting ontology. Indeed, the implementation of an operation must: (1) return a (finite) set of constraints that characterize the resulting ontology; and (2) guarantee that the constraints of the resulting ontology are of the appropriate class, that is, lightweight inclusions in our case. These are not obvious points, which we solve with the help of a structural proof procedure for lightweight ontologies.

In the context of conceptual modeling, this chapter, therefore, addresses two goals of the present book: *(G1) What are the candidate formal modeling languages and techniques to model such domain knowledge? What are the reasoning capabilities entailed by these modeling languages? (G2) Define composition mechanisms to handle domain knowledge in formal modeling techniques.* The chapter argues that lightweight ontologies are expressive enough to cover certain formal conceptual modeling scenarios (G1) and that the operations introduced provide composition mechanisms (G2) to facilitate the construction of lightweight ontologies.

The paper is organized as follows: Section 2 defines the operations. Section 3 introduces a decision procedure for lightweight inclusions, based on the notion of constraint graphs, and discusses the problem of minimizing a set of lightweight inclusions. Section 4 shows how to compute the operations for lightweight ontologies. Section 5 summarizes related work. Section 6 contains the conclusions.

2   A Formal Framework

2.1   A Brief Review of Basic Concepts

The definition of the operations depends only on the notion of theory, which we introduce in the context of Description Logic (DL) [4].

Briefly, a *vocabulary V* consists of a set of *atomic concepts*, a set of *atomic roles*, and the *bottom concept* $\perp$. A *language* in V is a set of strings, using symbols in V, whose definition depends on the specific variation of Description Logic adopted; the definition of the language typically includes definitions for the set of *concept descriptions in V* and the set of *role descriptions in V*.

An *inclusion in V* is a statement of the form $u \sqsubseteq v$, where u and v both are concept descriptions in V or both are role descriptions in V. We use $u \equiv v$ *(equivalence)* as an abbreviation for the pair of inclusions $u \sqsubseteq v$ and $v \sqsubseteq u$.

An *interpretation s* for V consists of a nonempty set Δ^s, the *domain* of s, whose elements are called *individuals*, and an *interpretation function*, also denoted s, where

$s(\bot) = \emptyset$

$s(A) \subseteq \Delta^s$ for each atomic concept A in V

$s(P) \subseteq \Delta^s \times \Delta^s$ for each atomic role P in V

The function s is extended to role and concept descriptions in V. The exact definition again depends on the specific variation of Description Logic adopted. We use $s(e)$ to indicate the value that s assigns to a concept description or a role description e in V.

Let σ and σ' be two inclusions in V and Σ be a set of inclusions in V. Assume that σ is of the form $u \sqsubseteq v$. We say that

- s *satisfies* σ or s is a *model* of σ, denoted $s \vDash \sigma$, iff $s(u) \subseteq s(v)$.
- s *satisfies* Σ or s is a *model* of Σ, denoted $s \vDash \Sigma$, iff s satisfies all inclusions in Σ.
- σ is *valid*, denoted $\vDash \sigma$, iff any interpretation for V satisfies σ.
- σ and σ' are *tautologically equivalent* iff any model of σ is a model of σ' and vice versa.
- Σ *logically implies* σ, or σ is a *logical consequence* of Σ, denoted $\Sigma \vDash \sigma$, iff any model of Σ satisfies σ.
- Σ is *satisfiable* or *consistent* iff there is a model of Σ.

The *theory* of Σ in V, denoted $\tau[\Sigma]$, is the set of all inclusions in V that are logical consequences of Σ. We say that two sets of inclusions, Γ and Θ, are *equivalent*, denoted $\Gamma \equiv \Theta$, iff $\tau[\Gamma] = \tau[\Theta]$.

Finally, an *ontology* is a pair $O = (V, \Sigma)$ such that V is a finite vocabulary, whose atomic concepts and atomic roles are called *classes* and *properties* of O, respectively, and Σ is a set of inclusions in V, called the *constraints* of O. Two ontologies $O_1 = (V_1, \Sigma_1)$ and $O_2 = (V_2, \Sigma_2)$ are *equivalent*, denoted $O_1 \equiv O_2$, iff Σ_1 and Σ_2 are equivalent.

2.2 Definition of the Ontology Operations

In this section, we introduce a collection of operations over ontologies, whose definition is not restricted to any specific variation of DL.

Definition 1 Let $O_1 = (V_1, \Sigma_1)$ and $O_2 = (V_2, \Sigma_2)$ be two ontologies, W be a subset of V_1, and Ψ be a set of constraints in V_1.

(i) The *projection* of $O_1 = (V_1, \Sigma_1)$ over W, denoted $\pi[W](O_1)$, returns the ontology $O_P = (V_P, \Sigma_P)$, where $V_P = W$ and Σ_P is the subset of the constraints in $\tau[\Sigma_1]$ that use only classes and properties in W.

(ii) The *deprecation* of Ψ from $O_1 = (V_1, \Sigma_1)$, denoted $\delta[\Psi](O_1)$, returns the ontology $O_D = (V_D, \Sigma_D)$, where $V_D = V_1$ and $\Sigma_D = \Sigma_1 - \Psi$.

(iii) The *union* of $O_1 = (V_1, \Sigma_1)$ and $O_2 = (V_2, \Sigma_2)$, denoted $O_1 \cup O_2$, returns the ontology $O_U = (V_U, \Sigma_U)$, where $V_U = V_1 \cup V_2$ and $\Sigma_U = \Sigma_1 \cup \Sigma_2$.

(iv) The *intersection* of $O_1=(V_1,\Sigma_1)$ and $O_2=(V_2,\Sigma_2)$, denoted $O_1 \cap O_2$, returns the ontology $O_N=(V_N,\Sigma_N)$, where $V_N=V_1 \cap V_2$ and $\Sigma_N=\tau[\Sigma_1] \cap \tau[\Sigma_2]$.

(v) The *difference* of $O_1=(V_1,\Sigma_1)$ and $O_2=(V_2,\Sigma_2)$, denoted O_1-O_2, returns the ontology $O_F=(V_F,\Sigma_F)$, where $V_F=V_1$ and $\Sigma_F=\tau[\Sigma_1]-\tau[\Sigma_2]$. □

We refer the reader to [12] for examples of these operations. We observe that the ontology that results from an operation is unique, by definition. However, there might be several ontologies that are equivalent to the resulting ontology. For example, if $O_P=(V_P,\Sigma_P)$ is the projection of O_1 on W, there might be several sets of constraints that are equivalent to the set of constraints in the theory of O_1 that use only terms in W. This simple observation will be helpful in Sect. 4, which addresses how to implement the operations. We also observe that we may generalize union, intersection, and difference by considering a renaming of one or both vocabularies of the ontologies involved and propagating the renaming to the terms that occur in the constraints when comparing the theories. Out of simplicity, we do not consider this extension in the chapter.

We now briefly discuss the conceptual design problems that motivated the definition of the operations. Consider first the problem of designing an ontology to publish data on the Web. Following the Linked Data principles [7], the designer should select known ontologies, as much as possible, to organize the data so that applications "can dereference the URIs that identify vocabulary terms in order to find their definition". We argue that the designer should go further and analyze the constraints of the ontologies from which he is drawing the terms to construct his ontology. To facilitate conceptual design from this perspective, we introduced the *projection, union,* and *deprecation* operations.

Given two ontologies, if the designer wants to know what they have in common, he should create a mapping between their vocabularies and detect which constraints hold in both ontologies after the terms are mapped. The *intersection* operation answers this question. We argued elsewhere [11] that intersection is also useful to address the design of mediated schemas that combine several export schemas in a way that the data exposed by the mediator is always consistent.

Likewise, given two ontologies, if the designer wants to know what holds in one, but not in the other, he should again create a mapping between their vocabularies and detect which constraints hold in the theory of the first ontology, but not in the theory of the second, after the terms are appropriately mapped. The *difference* operation answers this question. Note that a variant of ontology comparison is the problem of analyzing what changed from one version of an ontology to the next.

2.3 Lightweight Description Logic

The procedures that implement the operations, introduced in Sect. 4, assume that the inclusions meet certain restrictions, imposed by a variation of Description Logic, that we call *Lightweight Description Logic,* or *Lightweight DL.*

Lightweight DL is characterized by the following definitions and restrictions on the sets of concept descriptions, role descriptions, and inclusions.

Definition 2 Let V be a vocabulary.

(i) A *lightweight role description* in V is an atomic role P in V or a string of the form P^- (*inverse role*), where P is an atomic role in V.

(ii) A *lightweight basic concept description* in V is the bottom concept $\perp$, an atomic concept in V, or an *at-least restriction* of the form $(\geq n\, p)$, where p is a lightweight role description in V and n is a positive integer.

(iii) A *lightweight concept description* in V is a lightweight basic concept description in V, or a *lightweight negated concept* of the form $\neg e$, where e is a lightweight basic concept description in V.

(iv) A *lightweight inclusion* in V is a string of one of the forms:

- $e \sqsubseteq f$, where e is an atomic concept or an at-least restriction in V and f is the bottom concept $\perp$, an atomic concept in V, or an at-least restriction.
- $e \sqsubseteq \neg f$, where e and f are atomic concepts or at-least restrictions in V. $\square$

Definition 3 Let V be a vocabulary and s be an interpretation for V. The function s is extended to lightweight role and concept descriptions in V as follows (where P is an atomic role, e is a lightweight basic concept description, and p is a lightweight role description):

(i) $s(P^-) = s(P)^-$ (the inverse of $s(P)$)

(ii) $s(\neg e) = \Delta^s - s(e)$ (the complement of $s(e)$ with respect to Δ^s)

(iii) $s(\geq n\, p) = \{I \in \Delta^s /\ card(\{J \in \Delta^s\ /\ (I,J) \in s(p)\}) \geq n\}$ (the set of individuals that $s(p)$ relates to at least n distinct individuals, where $card(S)$ denotes the cardinality of a set S). $\square$

Since lightweight inclusions are a special case of inclusions, the notion of satisfiability, etc., remain as in Sect. 2.1.

Definition 4 An ontology $O = (V, \Sigma)$ is a *lightweight ontology* iff Σ is a set of lightweight inclusions in V. $\square$

We use the following abbreviations, where p is a lightweight role description

- "$\top$" (universal concept) for "$\neg \perp$"
- "$\exists p$" (existential quantification) for "$(\geq 1\, p)$"
- "$(\leq n\, p)$" (at-most restriction) for "$\neg (\geq n + 1\, p)$"

By an *unabbreviated* concept description, we mean a concept description that does not use such abbreviations. Care must be taken to eliminate the abbreviated concept descriptions before checking if an inclusion is indeed a lightweight inclusion. Also, in view of the restrictions in Definition 2(iv), a *lightweight equivalence* $e \equiv f$ is such that e and f both are atomic concepts or at-least restrictions in V.

Let e and f be lightweight basic concept descriptions. Inclusions of the following forms are not lightweight inclusions: $\perp \sqsubseteq f$, $\perp \sqsubseteq \neg f$, $e \sqsubseteq \neg \perp$, $\neg f \sqsubseteq \neg e$, and $\neg e \sqsubseteq f$. However, we note that

(1) $\bot \sqsubseteq f$, $\bot \sqsubseteq \neg f$, and $e \sqsubseteq \neg\bot$ are valid (satisfiable by any interpretation)

(2) $\neg f \sqsubseteq \neg e$ is tautologically equivalent to $e \sqsubseteq f$

Thus, when defining an ontology, inclusions as in (1) can be ignored, since they are vacuous constraints, and $\neg f \sqsubseteq \neg e$ can be replaced by $e \sqsubseteq f$.

We remark that the definitions of $DL - Lite^{N}_{core}$ inclusions [3] and lightweight inclusions differ only in that the latter, but not the former, rules out inclusions of the forms in (1). However, this is semantically immaterial since, given a set Σ of $DL - Lite^{N}_{core}$ inclusions, we can always drop from Σ inclusions as in (1) without affecting the theory of Σ, since these inclusions are valid. On the other hand, inclusions as in (1) would unnecessarily complicate the structural proof procedure introduced in Sect. 3.

Finally, we observe that lightweight inclusions are sufficiently expressive to cover the simplest types of constraints used in conceptual modeling (see Table 1). We refer the reader to [3] for a detailed account of the DL family.

Table 1 Common constraint types used in conceptual modeling

Constraint type	Abbreviated form	Unabbreviated form	Informal semantics
Domain constraint	$\exists P \sqsubseteq C$	$(\geq 1 P) \sqsubseteq C$	Property P has class C as domain, that is, if *(a, b)* is a pair in P, then a is an individual in C
Range constraint	$\exists P^{-} \sqsubseteq C$	$(\geq 1 P^{-}) \sqsubseteq C$	Property P has class C as range, that is, if *(a, b)* is a pair in P, then b is an individual in C
minCardinality constraint		$C \sqsubseteq (\geq k P)$ or $C \sqsubseteq (\geq k P^{-})$	Property P or its inverse P^{-} maps each individual in class C to at least k distinct individuals
maxCardinality constraint	$C \sqsubseteq (\leq k P)$ or $C \sqsubseteq (\leq k P^{-})$	$C \sqsubseteq \neg(\geq k + 1 P)$ or $C \sqsubseteq \neg(\geq k + 1 P^{-})$	Property P or its inverse P^{-} maps each individual in class C to at most k distinct individuals
Subset constraint		$C \sqsubseteq D$	Each individual in C is also in D, that is, class C denotes a subset of class D
Disjointness constraint		$C \sqsubseteq \neg D$	No individual is in both C and D, that is, classes C and D are disjoint

3   Basic Procedures for Lightweight Inclusions

3.1   A Decision Procedure for Lightweight Inclusions

In this section, we review a decision procedure for lightweight inclusions, based on the notion of constraint graphs [11]. We also discuss the problem of minimizing a set of lightweight inclusions, which affects the implementation of the ontology operations. We stress that the concepts introduced in this section refer only to lightweight inclusions. Thus, we often omit explicit reference to this variation of DL, a simplification that the reader must bear in mind.

We say that the *complement* of a basic concept description b is $\neg b$, and vice versa. If e is a basic concept description or the negation of a basic concept description, then $\bar{e}$ denotes the complement of e.

Let Σ be a set of lightweight inclusions and Ω be a set of lightweight concept descriptions.

Definition 5   The labeled graph $g(\Sigma,\Omega) = (\gamma,\delta,\kappa)$ that *captures* Σ and Ω, where κ labels each node with a concept description, is defined as follows:

(i)   For each concept description e that occurs on the right- or left-hand side of an inclusion in Σ, or that occurs in Ω, there is exactly one node in γ labeled with e. If necessary, the set of nodes is augmented with new nodes so that

  (a)   For each atomic concept C in Σ or in Ω, there is exactly one node in γ labeled with C.
  (b)   For each atomic role P in Σ or in Ω, there is exactly one node in γ labeled with $(\geq 1\ P)$ and exactly one node labeled with $(\geq 1\ P^{-})$.

(ii)   If there is a node in γ labeled with a concept description e, then there must be exactly one node in γ labeled with $\bar{e}$.

(iii)   For each inclusion $e \sqsubseteq f$ in Σ, there is an arc (M,N) in δ, where M and N are the nodes labeled with e and f, respectively.

(iv)   If there are nodes M and N in γ labeled with $(\geq m\ p)$ and $(\geq n\ p)$ such that $m < n$, where p is either P or P^{-}, then there is an arc (N,M) in δ. Such arcs are called *tautological arcs*.

(v)   If there is an arc (M,N) in δ such that M and N are labeled with e and f, respectively, then there is an arc (K,L) in δ such that K and L are the nodes labeled with $\bar{f}$ and $\bar{e}$, respectively.

(vi)   These are the only nodes and arcs of $g(\Sigma,\Omega)$.

If Ω is empty, we write $g(\Sigma)$ and say that $g(\Sigma)$ is the graph that *captures* Σ.   □

Definition 6   The *constraint graph* for Σ and Ω is the labeled graph $G(\Sigma,\Omega) = (\eta,\varepsilon,\lambda)$, where λ labels each node with a set of concept descriptions. The graph $G(\Sigma,\Omega)$ is defined by collapsing each strongly connected component of $g(\Sigma,\Omega)$ into a single node, labeled with the set of concept descriptions that previously labeled

Table 2 The constraints of the ontology *APO* (unabbreviated form)

Constraint	Informal specification
$(\geq 1 \text{ foaf:name}) \sqsubseteq \text{foaf:Person}$	The domain of foaf:name is foaf:Person
$(\geq 1 \text{ foaf:name}^{-)} \sqsubseteq \text{xsd:string}$	The range of foaf:name is xsd:string
$(\geq 1 \text{ mo:member\_of}) \sqsubseteq \text{foaf:Person}$	The domain of mo:member_of is foaf:Person
$(\geq 1 \text{ mo:member\_of}^{-)} \sqsubseteq \text{foaf:Group}$	The range of mo:member_of is foaf:Group
mo:MusicArtist $\sqsubseteq$ foaf:Agent	mo:MusicArtist is a subset of foaf:Agent
foaf:Group $\sqsubseteq$ foaf:Agent	foaf:Group is a subset of foaf:Agent
foaf:Organization $\sqsubseteq$ foaf:Agent	foaf:Organization is a subset of foaf:Agent
mo:SoloMusicArtist $\sqsubseteq$ foaf:Person	mo:SoloMusicArtist is a subset of foaf:Person
mo:SoloMusicArtist $\sqsubseteq$ mo:MusicArtist	mo:SoloMusicArtist is a subset of mo:MusicArtist
mo:MusicGroup $\sqsubseteq$ mo:MusicArtist	mo:MusicGroup is a subset of mo:MusicArtist
mo:MusicGroup $\sqsubseteq$ foaf:Group	mo:MusicGroup is a subset of foaf:Group
mo:CorporateBody $\sqsubseteq$ foaf:Organization	mo:CorporateBody is a subset of foaf:Organization
mo:Label $\sqsubseteq$ mo:CorporateBody	mo:Label is a subset of mo:CorporateBody
foaf:Person $\sqsubseteq \neg$foaf:Organization	foaf:Person and foaf:Organization are disjoint

the nodes in the strongly connected component. When Ω is the empty set, we write $G(\Sigma)$ and say that $G(\Sigma)$ is the *constraint graph* for Σ.   ☐

If a node K of $G(\Sigma,\Omega)$ is labeled with e, then $\bar{K}$ denotes the node labeled with $\bar{e}$; we say that K and $\bar{K}$ are *dual nodes* and (M,N) and $(\bar{N}, \bar{M})$ are *dual arcs*.

Example 1 Let "foaf:" refer to the vocabulary of the "Friend-of-a-Friend" ontology, "mo:" to the vocabulary of the "music" ontology, and "xsd:" to the XML schema vocabulary. Consider the *Agent-Person* ontology, $APO = (V_{APO}, \Sigma_{APO})$, where

V_{APO} = {foaf:Agent, foaf:Person, foaf:Group, foaf:Organization, mo:MusicArtist, mo:CorporateBody, mo:SoloMusicArtist, mo:MusicGroup, mo:Label, mo:member_of, foaf:name, xsd:string}

Σ_{APO} = (the set of constraints is shown in Table 2)

Figure 1 depicts the constraint graph $g(\Sigma_{APO})$ for Σ_{APO}. Since $g(\Sigma_{APO})$ has no strongly connected components, $g(\Sigma_{APO})$ and $G(\Sigma_{APO})$ are in fact the same graph. Note that there is a path from the node labeled with mo:Label to the node labeled with $\neg(\geq 1 \text{ mo:member\_of})$, which indicates that mo:Label $\sqsubseteq \neg(\geq 1 \text{ mo:member\_of})$ is a logical consequence of Σ_{APO}. This logical implication would not be captured if we constructed the graph with just the concept descriptions that occur in Σ_{APO}. Hence, it provides an example of why we need Conditions (ii) and (v) in Definition 5.

We use $K \rightarrow M$ to indicate that there is a path in $G(\Sigma,\Omega)$ from K to M. Also, as a convenience, a *path of length 0* is a path consisting of a single node.

Definition 7 Let $G(\Sigma,\Omega) = (\eta,\varepsilon,\lambda)$ be the constraint graph for Σ and Ω.

(i)   We say that a node K of $G(\Sigma,\Omega)$ is a $\perp$-*node of rank 0* iff

(a) K is labeled with $\perp$, or

(b)   There are nodes M and N, not necessarily distinct from K, and a basic concept description b such that M and N are labeled with b and $\neg b$, respectively, and $K \rightarrow M$ and $K \rightarrow N$.

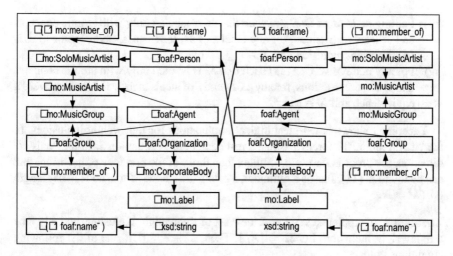

Fig. 1 The constraint graph $G(\Sigma_{APO})$ for Σ_{APO}

(ii) For each positive integer n, we say that a node K of $G(\Sigma,\Omega)$ is a $\perp$-*node of rank n* iff K is not a $\perp$-node of rank m, with $m < n$, and there is a $\perp$-node L of rank $n-1$ such that

    (a) (K,L) is an arc of $G(\Sigma,\Omega)$, or

    (b) L is labeled with $(\geq 1\ P^-)$ and K is labeled with $(\geq 1\ P)$, or

    (c) L is labeled with $(\geq 1\ P)$ and K is labeled with $(\geq 1\ P^-)$.   ☐

    Case (ii-b) captures the fact that, given an interpretation s, if $s((\geq 1\ P^-)) = \emptyset$, then $s(P) = s((\geq 1\ P)) = \emptyset$. Case (ii-c) follows likewise, when $s((\geq 1\ P)) = \emptyset$. In view of these cases, the notion of rank is necessary to avoid a circular definition.

Definition 8 Let $G(\Sigma,\Omega) = (\eta,\varepsilon,\lambda)$ be the constraint graph for Σ and Ω. Let K be a node of $G(\Sigma,\Omega)$. We say that K is a $\perp$-*node* iff K is a $\perp$-node with rank n, for some nonnegative integer n. We say that K is a $\top$-*node* iff $\bar{K}$ is a $\perp$-node.   ☐

    To simplify the procedures described in Sect. 4, we tag the $\perp$-nodes and the $\top$-nodes of a constraint graph with "$\perp$-node" and "$\top$-node", respectively.

Definition 9 The *tagged* constraint graph for Σ and Ω is the constraint graph for Σ and Ω, with the $\perp$-nodes and the $\top$-nodes tagged with "$\perp$-node" and "$\top$-node", respectively.   ☐

    In what follows, let $G(\Sigma,\Omega)$ be the constraint graph for a set Σ of lightweight inclusions and a set Ω of lightweight concept descriptions. We refer the reader to [11] for detailed proofs.

Proposition 3 (Consistency).

(i) For any pair of nodes M and N of $G(\Sigma,\Omega)$ such that M is labeled only with atomic concepts and at-least restrictions, for any label e of M, for any label f of N, if $M \to N$ then $e \sqsubseteq f$ is a lightweight inclusion and $\Sigma \vDash e \sqsubseteq f$.

(iii) For any node M of $G(\Sigma,\Omega)$ such that M is labeled only with atomic concepts and at-least restrictions, for any label e of M, if M is a $\bot$-node, then $e \sqsubseteq \bot$ is a lightweight inclusion and $\Sigma \vDash e \sqsubseteq \bot$.
(iv) For any node M of $G(\Sigma,\Omega)$ such that M is labeled only with atomic concepts and at-least restrictions, for any pair e and f of labels of M, $e \equiv f$ is a lightweight equivalence and $\Sigma \vDash e \equiv f$.

Theorem 1 shows how to test logical implications for lightweight inclusions. In the "if" direction, Theorem 1 is just a restatement of Proposition 3. In the "only if" direction, the proof is far more complex than those of the previous propositions and can be found in [11]. Just as a reminder, a path from a node M to a node N has length 0 iff $M = N$.

Theorem 1 (Completeness). Let Σ be a set of lightweight inclusions and $e \sqsubseteq f$ be a lightweight inclusion. Let $\Omega = \{e,f\}$. Then, $\Sigma \vDash e \sqsubseteq f$ iff one of the following conditions holds:

(i)   The node of $G(\Sigma,\Omega)$ labeled with e is a $\bot$-node, or
(ii)   The node of $G(\Sigma,\Omega)$ labeled with f is a $\top$-node, or
(iii) There is a path in $G(\Sigma,\Omega)$, possibly with length 0, from the node labeled with e to the node labeled with f. □

Theorem 1 leads to a structural decision procedure, **Implies**, to test if a lightweight inclusion $e \sqsubseteq f$ is a logical consequence of a set Σ of lightweight inclusions (see Fig. 2).

3.2 Minimizing the Set of Constraints of a Lightweight Ontology

We briefly discuss the problem of minimizing the set of constraints of a lightweight ontology, which is similar to finding a *minimal equivalent graph (MEG)* of a graph G, defined as a graph H with a minimal set of edges such that the transitive closure of G and H are equal. This problem has a polynomial solution when G is acyclic and is NP-hard for strongly connected graphs [1, 16, 19].

Figure 3 contains all procedures developed to address this question, as well as procedures to construct constraint graphs and to test if a lightweight inclusion is a logical consequence of a set of lightweight inclusions, based on Theorem 1. The **MinimizeGraph** procedure is based on the strategy of finding the MEG of a constraint graph. Step (2) of **MinimizeGraph** can be implemented in polynomial time [1, 16], since H is acyclic (because so is G, by Proposition 2(i)). In view of Propositions 1 and 2, Step (2) considers just the nodes of H labeled only with atomic concepts and at-least restrictions. Furthermore, since Step (2) drops the dual arcs, H satisfies the properties listed in Propositions 1 and 2. The **GenerateConstraints**

ConstructGraph:

Input: a set Σ of lightweight inclusions and an optional lightweight inclusion $e \sqsubseteq f$

Output: the tagged constraint graph $G(\Sigma,\Omega)$

1. Construct the constraint graph $G(\Sigma,\Omega)$ for Σ and $\Omega = \{e,f\}$, using Definition 6.
2. Tag $G(\Sigma,\Omega)$, using Definition 9.
3. Return $G(\Sigma,\Omega)$.

Implies:

Input: a lightweight inclusion $e \sqsubseteq f$ and a set Σ of lightweight inclusions

Output: "True", if $e \sqsubseteq f$ is a logical consequence of a set Σ, and "False", otherwise

1. Call **ConstructGraph** to construct the constraint graph $G(\Sigma,\Omega)$ for Σ and $\Omega = \{e,f\}$.
2. Return "True" if
3. The node of $G(\Sigma,\Omega)$ labeled with e is a $\perp$-node, or
4. The node of $G(\Sigma,\Omega)$ labeled with f is a T-node, or
5. There is a path in $G(\Sigma,\Omega)$, possibly with length 0, from the node labeled with e to the node labeled with f.
6. Return "False", otherwise.

MinimizeGraph:

Input: a tagged constraint graph G

Output: a MEG H of G

1. Initialize H with the same nodes, arcs, labels and tags as G.
2. For each node L of H labeled only with atomic concepts and at-least restrictions,
3.   For each arc (L,M) in H,
4.   For each node N in H, do:
5.   If there are arcs (M,N) and (L,N) in H
6.   such that (L,N) is not a tautological arc,
7.   Drop from H both the arc (L,N) and the arc (N,L) connecting the dual nodes of L and M.

GenerateConstraints:

Input: a tagged constraint graph H

Output: a set of constraints Σ_2

Fig. 2 Basic procedures

1.   Initialize Σ_2 to be the empty set.
2.   Mark all arcs of H as unprocessed.
3.   For each node M of H labeled only with atomic concepts and at-least restrictions, do:
4.   If M is tagged as a "$\perp$-node", then
5.   For each label e of M,
6.   Add to Σ_2 a constraint of the form $e \sqsubseteq \perp$.
7.   If M is not tagged as "$\perp$-node", then
8.   Order the labels of M, creating a list $e_1,...,e_n$, and
9.   Add to Σ_2 the constraints $e_1 \sqsubseteq e_2, e_2 \sqsubseteq e_3,..., e_{n-1} \sqsubseteq e_n$ and $e_n \sqsubseteq e_1$.
10.   For each arc (M,N) of H such that (M,N) is unprocessed, do:
11.   Select a label e of M and a label f of N and
12.   Add to Σ_2 a constraint of the form $e \sqsubseteq f$.
13.   Mark both (M,N) and (N,M) as processed.
14.   Return Σ_2.

MinimizeConstraints:

Input: a set of lightweight constraints Σ_1

Output: an equivalent, minimal set of constraints Σ_2

1.   Call **ConstructGraph** to construct the tagged constraint graph G for Σ_1.
2.   Call **MinimizeGraph** with G to generate H.
3.   Call **GenerateContraints** with H to generate Σ_2.
4.   Return Σ_2.

Fig. 2 (continued)

Projection:

Input: $O_1 = (V_1, \Sigma_1)$ be a lightweight ontology and W be a subset of V_1

Output: $O_P = (W, \Gamma_P)$, a lightweight ontology equivalent to the projection of O_1 on W

(1) Construct $G(\Sigma_1)$, the tagged constraint graph for Σ_1.
(2) Construct $G^*(\Sigma_1)$, the transitive closure of $G(\Sigma_1)$. The nodes of $G^*(\Sigma_1)$ retain all labels and tags as in $G(\Sigma_1)$.
(3) Use $G^*(\Sigma_1)$ to create a graph G_W by discarding all concept descriptions that label nodes of $G^*(\Sigma_1)$ and that involve classes and properties which are not in W; nodes that end up with no labels are discarded, as well as their adjacent arcs. The nodes of G_W retain all tags as in $G^*(\Sigma_1)$.
(4) Call **MinimizeGraph** with G_W to generate H.
(5) Call **GenerateContraints** with H to generate Γ_P.
(6) Return $O_P = (W, \Gamma_P)$.

Fig. 3 Projection Procedure

procedure transforms graph H into a set of constraints Σ_2. Again, in view of Proposition 1 and 2, Step (3) of **GenerateConstraints** considers just the nodes of H labeled only with atomic concepts and at-least restrictions. The **MinimizeConstraints** uses the previous procedure to transform a set of lightweight constraints Σ_1 and output an equivalent, minimal set of constraints Σ_2. The correctness of **MinimizeConstraints** is stated in Theorem 2.

Theorem 2 Let Σ_1 be a set of lightweight constraints and Σ_2 be the result of applying **MinimizeConstraints** to Σ_1. Then, Σ_1 and Σ_2 are equivalent, that is, $\tau[\Sigma_1] = \tau[\Sigma_2]$. □

4 Implementation of the Operations

4.1 A Brief Discussion on the Implementation of the Operations

We start with a few simple observations that impact the implementation of the operations. First, Definition 1 guarantees that each operation is a function, as expected, that is, each operation returns a unique result O_R for each input. However, we consider it acceptable that the implementation of an operation computes an ontology O_E which is equivalent to O_R. Furthermore, if the input ontologies have a finite set of constraints, we require that the implementation returns an ontology that has a finite set of constraints. This may be problematic for projection, intersection, and difference, whose definitions use the theories of the sets of constraints involved, rather than the sets of constraints themselves, as in the definition of deprecation and union. Lastly, the resulting ontology must be lightweight if the input ontologies are.

Let $O_1 = (V_1, \Sigma_1)$ and $O_2 = (V_2, \Sigma_2)$ be two lightweight ontologies, W be a subset of V_1 and Ψ be a set of constraints in V_1. From the perspective of the difficulty of implementation, we may divide the operations into three groups:

Group 1: deprecation and union.

These operations have direct implementations from Definitions 1(ii) and (iii). Given O_1 and Ψ, the **Deprecation** procedure returns the ontology $O_D = (V_D, \Sigma_D)$, where $V_D = V_1$ and Σ_D is the result of minimizing $\Sigma_1 - \Psi$. Given O_1 and O_2, the **Union** procedure returns the ontology $O_U = (V_U, \Sigma_U)$, where $V_U = V_1 \cup V_2$ and Σ_U is the result of minimizing $\Sigma_1 \cup \Sigma_2$. Hence, these procedures are quite simple and will not be further discussed.

Group 2: projection and intersection.

These operations have implementations that depend on Theorem 1. The **Projection** procedure computes the projection of O_1 onto W and follows from Definition 1(i), Theorem 1, and constraint minimization. The **Intersection** procedure likewise follows from Definition 1(iv).

Group 3: difference.

This operation raises difficulties, as discussed in Sect. 4.4.

4.2 Implementation of *Projection*

Let $O_1 = (V_1, \Sigma_1)$ be a lightweight ontology and W be a subset of V_1. Recall that the projection of O_1 over W is the ontology $O_P = (V_P, \Sigma_P)$, where $V_P = W$ and Σ_P is the set of constraints in $\tau[\Sigma_1]$ that use only classes and properties in W.

Procedure **Projection**, shown in Fig. 3, computes Γ_P so that $\tau[\Gamma_P] = \tau[\Sigma_P]$. That is, given any lightweight inclusion $e \sqsubseteq f$ that involves only classes and properties in W, $e \sqsubseteq f$ is a logical consequence of Γ_P iff $e \sqsubseteq f$ is a logical consequence of Σ_1. Note that this does not mean that $e \sqsubseteq f$ is a logical consequence of the subset of Σ_1 whose inclusions involve only classes and properties in W.

The correctness of **Projection** is established in Theorem 3, whose proof follows directly from Theorem 1. In particular, the transitive closure $G^*(\Sigma_1)$, generated in Step (2), is simply a convenient way to capture all paths in $G(\Sigma_1)$ required to apply Condition (iii) of Theorem 1.

Theorem 3 (Correctness of **Projection**). Let $O_1 = (V_1, \Sigma_1)$ be a lightweight ontology and W be a subset of V_1. Let $O_P = (W, \Gamma_P)$ be the ontology that **Projection** returns for O_1 and W. Then, for any lightweight inclusion $e \sqsubseteq f$ that involves only classes and properties in W, we have that $\Sigma_1 \vDash e \sqsubseteq f$ iff $\Gamma_P \vDash e \sqsubseteq f$. □

4.3 Implementation of *Intersection*

Let $O_k = (V_k, \Sigma_k)$, for $k = 1,2$, be two lightweight ontologies. Recall that the intersection of O_1 and O_2 is the ontology $O_N = (V_N, \Sigma_N)$, where $V_N = V_1 \cap V_2$ and $\Sigma_N = \tau[\Sigma_1] \cap \tau[\Sigma_2]$.

Procedure **Intersection**, shown in Fig. 4, computes Γ_N so that $\tau[\Gamma_N] = \tau[\Sigma_N]$. That is, a lightweight inclusion is a logical consequence of Γ_N iff it is a logical consequence of Σ_k, for $k = 1,2$. We now discuss the decisions that lead to the **Intersection** procedure. Recall from Theorem 1 that a lightweight inclusion $e \sqsubseteq f$ is a logical consequence of Σ_k iff there are nodes M and N of $G(\Sigma_k, \Omega)$, with $\Omega = \{e, f\}$, such that

Intersection:

input: $O_1 = (V_1, \Sigma_1)$ and $O_2 = (V_2, \Sigma_2)$, two lightweight ontologies

output: $O_N = (V_1 \cap V_2, \Gamma_N)$, a lightweight ontology equivalent to the intersection of O_1 and O_2.

(1) Construct the closure Δ of Σ_1 and Σ_2 with respect to each other.
(2) Construct $G(\Sigma_1, \Delta)$ and $G(\Sigma_2, \Delta)$, the tagged constraint graphs for Σ_1 and Δ and Σ_2 and Δ, respectively.
(3) Construct a set of constraints Σ_3 as follows (see Table 3):
    (a) Initialize Σ_3 to be the empty set.
    (b) For each node M of $G(\Sigma_1, \Delta)$ tagged with "⊥-node" and
       labeled only with atomic concepts and at-least restrictions,
          for each label e of M, do:
            (i)   If e also labels a node of $G(\Sigma_2, \Delta)$ tagged with "⊥-node", then add $e \sqsubseteq \perp$ to Σ_3.
            (ii)   For each node K of $G(\Sigma_2, \Delta)$ tagged with "⊤-node",
                for each label f of K, add $e \sqsubseteq f$ to Σ_3.
            (iii) For each path of $G(\Sigma_2, \Delta)$, possibly with length 0, from a node labeled
                with e to a node labeled with f, add $e \sqsubseteq f$ to Σ_3.
    (c) For each node M of $G(\Sigma_1, \Delta)$ not tagged with "⊥-node" and
       labeled only with atomic concepts and at-least restrictions,
          for each path in $G(\Sigma_1, \Delta)$ from M to a node N,
            for each label e of M,
              for each label f of N ($f \neq e$, if $M=N$), do:
                (i)   If e also labels a node of $G(\Sigma_2, \Delta)$ tagged with "⊥-node",
                     then add $e \sqsubseteq f$ to Σ_3.
                (ii)   If f also labels a node of $G(\Sigma_2, \Delta)$ tagged with "⊤-node",
                     then add $e \sqsubseteq f$ to Σ_3.
                (iii) If there is a path in $G(\Sigma_2, \Delta)$, possibly with length 0, from a node
                     labeled with e to a node labeled with f, then add $e \sqsubseteq f$ to Σ_3.
(4) Call **MinimizeConstraints** with Σ_3 to generate Γ_N.
(5) Return $O_N = (V_1 \cap V_2, \Gamma_N)$.

Fig. 4 Intersection Procedure

(i)   The node of $G(\Sigma_k, \Omega)$ labeled with e is a $\bot$-node, or
(ii)   The node of $G(\Sigma_k, \Omega)$ labeled with f is a $\top$-node, or
(iii) There is a path in $G(\Sigma_k, \Omega)$, possibly with length 0, from the node labeled with e to the node labeled with f.

Therefore, we must construct Γ_N so that $e \sqsubseteq f$ is a logical consequence of Γ_N iff $e \sqsubseteq f$ satisfies the above conditions with respect to Σ_k, for $k = 1,2$. However, a direct application of Theorem 1 depends on the inclusion $e \sqsubseteq f$ being tested (since the theorem depends on the constraint graph $G(\Sigma, \Omega)$, with $\Omega = \{e, f\}$. We argue that we can simplify the application of Theorem 1 in the context of the intersection operation if we define a set of concept descriptions as follows:

Definition 10 Let Σ_1 and Σ_2 be two sets of lightweight inclusions. The *closure of Σ_1 and Σ_2 with respect to each other* is the set Δ of concept descriptions defined so that a concept description e is in Δ iff, for $k = 1,2$, e occurs in an inclusion of Σ_k but not in an inclusion of Σ_{k+1} (sum is module 2). $\quad\square$

Then, $G(\Sigma_1, \Delta)$ and $G(\Sigma_2, \Delta)$ satisfy the following property.

Proposition 4 Let Σ_1 and Σ_2 be two sets of lightweight inclusions and Δ be the closure of Σ_1 and Σ_2 with respect to each other. Then, for $k = 1,2$, any lightweight inclusion $e \sqsubseteq f$ in Σ_{k+1} (sum is module 2) is a logical consequence of Σ_k iff there are nodes M and N of $G(\Sigma_k, \Delta)$ such that

(i)   The node of $G(\Sigma_k, \Delta)$ labeled with e is a $\bot$-node, or
(ii)   The node of $G(\Sigma_k, \Delta)$ labeled with f is a $\top$-node, or
(iii) There is a path in $G(\Sigma_k, \Delta)$, possibly with length 0, from the node labeled with e to the node labeled with f. $\quad\square$

The final case analysis to compute the intersection operation is summarized in Table 3 and results in a set of lightweight inclusions (Column C of Table 3). Step (3), the core of the **Intersection** procedure, directly captures such case analysis. We decided to create a set of lightweight inclusions, rather than a constraint graph, to clarify the decisions behind the **Intersection** procedure. The actual implementation is optimized and avoids this intermediate step.

Note that we need not consider $\top$-nodes of $G(\Sigma_1, \Delta)$ (or of $G(\Sigma_2, \Delta)$). Indeed, by Proposition 1, there is a $\top$-node of $G(\Sigma_1, \Delta)$ labeled with f iff there is a $\bot$-node labeled with $\bar{f}$. Furthermore, there is a path, possibly with length 0, from a node labeled with e to a node labeled with f iff there is a path, possibly with length 0, from a node labeled with $\bar{f}$ to a node labeled with $\bar{e}$. Therefore, Cases 4, 5, and 6 of Table 8, respectively, reduce to Cases 2, 1, and 3 of Table 3.

Theorem 4 (Correctness of **Intersection**): Let $O_1 = (V_1, \Sigma_1)$ and $O_2 = (V_2, \Sigma_2)$ be two lightweight ontologies. Let Δ be the closure of Σ_1 and Σ_2 with respect to each other. Let $O_N = (V_1 \cap V_2, \Gamma_N)$ be the ontology that **Intersection** returns for O_1 and O_2. Let $e \sqsubseteq f$ be a lightweight inclusion. Then, $\Gamma_N \models e \sqsubseteq f$ iff $\Sigma_1 \models e \sqsubseteq f$ and $\Sigma_2 \models e \sqsubseteq f$. $\quad\square$

Table 3 Case analysis for the intersection operation

Case	(A) Condition on $G(\Sigma_1, \Delta)^{1,2}$	(B) Condition on $G(\Sigma_2, \Delta)$	(C) Inclusion in Σ_3
1	There is a $\bot$-node labeled with e	There is a $\bot$-node labeled with e	$e \sqsubseteq \bot$
2		There is a $\top$-node labeled with f	$e \sqsubseteq f$
3		There is a path, possibly with length 0, from a node labeled with e to a node labeled with f	$e \sqsubseteq f$
4	There is a $\top$-node labeled with f	There is a $\bot$-node labeled with e	$e \sqsubseteq f$
5		There is a $\top$-node labeled with f	$\top \sqsubseteq f$
6		There is a path, possibly with length 0, from a node labeled with e to a node labeled with f	$e \sqsubseteq f$
7	There is a path, possibly with length 0, from a node labeled with e to a node labeled with f	There is a $\bot$-node labeled with e	$e \sqsubseteq f$
8		There is a $\top$-node labeled with f	$e \sqsubseteq f$
9		There is a path, possibly with length 0, from a node labeled with e to a node labeled with f	$e \sqsubseteq f$

Notes (see Definition 2):

(1) e is an atomic concept or an at-least restriction.
(2) f is the bottom concept $\bot$, an atomic concept, a lightweight at-least restriction, a negated atomic concept or a negated at-least restrictions.

4.4 A Note on *Difference*

The problem of creating a procedure to compute the difference between two ontologies, $O_1 = (V_1, \Sigma_1)$ and $O_2 = (V_2, \Sigma_2)$, lies in that it might not be possible to obtain a finite set of inclusions Δ_N in such a way that

(1) $\tau[\Delta_N] = \tau[\Sigma_1] - \tau[\Sigma_2]$

This invalidates the effort to create a procedure to obtain a finite set of inclusions Δ_N satisfying (1), along the lines of those exhibited in Sects. 4.2 and 4.3. This remark puts in doubt the usefulness of a (generic) difference operation. For example, consider the following two sets of inclusions:

(2) $\Sigma_1 = \{e \sqsubseteq g, g \sqsubseteq f\}$

(3) $\Sigma_2 = \{e \sqsubseteq f\}$

Then, ignoring tautologies when computing $\tau[\Sigma_k]$, $k = 1,2$, we have:

(4) $\tau[\Sigma_1] = \{e \sqsubseteq g, g \sqsubseteq f, e \sqsubseteq f\}$

(5) $\tau[\Sigma_2] = \{e \sqsubseteq f\}$

(6) $\Delta_N = \tau[\Sigma_1] - \tau[\Sigma_2] = \{e \sqsubseteq g, g \sqsubseteq f\} = \Sigma_1$

But this definition of Δ_N is not satisfactory since we have

(7) $\tau[\Delta_N] = \tau[\Sigma_1] = \{e \sqsubseteq g, g \sqsubseteq f, e \sqsubseteq f\}$

That is, to compute the difference $\Delta_N = \tau[\Sigma_1] - \tau[\Sigma_2]$, we remove "$e \sqsubseteq f$" from $\tau[\Sigma_1]$, only to get "$e \sqsubseteq f$" back by logical implication from Δ_N. In fact, in this rather obvious example, we cannot obtain a set of inclusions Δ_N such that $\tau[\Delta_N] = \tau[\Sigma_1] - \tau[\Sigma_2]$. Indeed, since the set of inclusions must not logically imply "$e \sqsubseteq f$", the only candidates are

(8) $\Delta_1 = \{e \sqsubseteq g\}$

(9) $\Delta_2 = \{g \sqsubseteq f\}$

In both cases, we have (ignoring tautologies when computing $\tau[\Delta_k]$, $k = 1,2$):

(10) $\tau[\Delta_k] = \Delta_k \subset \tau[\Sigma_1] - \tau[\Sigma_2]$

In view of this discussion, we cannot hope to always compute the exact difference between two ontologies.

5 Selected Related Work

We start with a very brief note about the algebraic specification, which is not the main setting of this chapter, and then relate the concepts and results of this chapter with work on ontology specification, especially ontology modularization.

The idea of theories as the semantics of specifications goes back at least to the early 1980s [8]. This idea was then carried on by the algebraic specification community in the late 1980s. In this context, a rather abstract notion of module and operations on modules were extensively explored (see, for example, [5, 14]). Operations on objects include pullback and amalgamated sum [6], which generalize intersection and union, respectively.

Treating ontologies as theories can be traced back at least to Uschold and Gruninger [24] seminal paper, where they argue that "for any given ontology, the goal is to agree upon a *shared terminology* and *set of constraints* on this terminology". Specifically, in the context of Linked Data, Jain et al. [18] stressed that the design of Linked Data sources must indicate the semantics of the concepts. We argue that such semantics must be expressed as constraints *derived* from those of the underlying ontologies used in the description of the data source, which was the primary motivation for the work reported here.

Ontology modularization is a well-explored notion (see for example the papers in [23]), especially for description logics and the Web Ontology Language. The idea of ontological module extraction is to pull out, from a large ontology, those constraints (or axioms) that are relevant to certain terms (or concepts) of interest, that is, to reuse only those parts that cover all the knowledge about the subset of relevant terms. Module extraction strategies can be *structure-based*, that is, based on the syntactical structure of the constraints and hierarchy of concepts [13, 22, 21] or *logic-based*, that is, based on the theory the ontology constraints define.

In more detail, as defined in [17], given an ontology $O = (V_O, \Sigma_O)$, a *module* $M = (V_M, \Sigma_M)$ is a subset of O, that is, $V_M \subseteq V_O$ and $\Sigma_M \subseteq \Sigma_O$. A module M is *relevant* for a set of terms W iff all consequences of O that can be expressed over W are also consequences of M; in this case, O is said to be a *conservative extension* of M. A stronger property is that every model of M extends to a model of O; in this case, O is said to be a *model conservative extension*.

The notion of S-module introduced in [15] is defined as follows: Let S be a signature and $M = (V_M, \Sigma_M)$ and $O = (V_O, \Sigma_O)$ be two ontologies such that M is a subset of O. Then, M is an *S-module* in O with respect to an ontology language L iff, for every ontology $P = (V_P, \Sigma_P)$ and constraint α expressed in L with $\mathrm{Sig}(P) \cap \mathrm{Sig}(O) \subseteq S$ and $\mathrm{Sig}(\alpha) \subseteq \mathrm{Sig}(P)$, we have $\Sigma_P \cup \Sigma_O \models \alpha$ if and only if $\Sigma_P \cup \Sigma_M \models \alpha$. The authors show that testing if an ontology is an S-module of another ontology, is undecidable for the fragment of OWL DL, that disallows transitive roles, role hierarchies, inverse roles, and cardinality restrictions.

The projection operation, as defined in this chapter, is stricter than the notion of conservative extension and the notion of S-module. Indeed, given an ontology $O = (V_O, \Sigma_O)$ and a set of terms $W \subseteq V_O$, recall that the *projection* of O on W is the ontology $M = (W, \Sigma_M)$ such that, for every constraint α expressed in W, we have that

$\Sigma_O \models \alpha$ if and only if $\Sigma_M \models \alpha$. In this chapter, for lightweight ontologies, we showed how to compute the projection operation in polynomial time (see Fig. 3). We also note that the notion of modularization expressed by the projection operation is logic-based, by Definition 1(i), and yet the implementation described in Sect. 4.2 is structural, in the sense that it explores the constraint graph, which, in turn, captures the structure of the lightweight inclusions of the ontology we want to project.

We went further and considered other operations. The union operation is quite simple to implement, but it can generate redundant constraints. We then showed how to minimize the constraints of the union of two lightweight ontologies. The implementation of the intersection operation for lightweight ontologies, with finite sets of inclusions, described in Sect. 4.3, explores the constraint graphs of the input ontologies to construct the finite set of lightweight inclusions of the resulting ontology. However, the implementation of the intersection is far more intricated than that of projection. The intersection is useful to address the design of mediated schemas that combine several export schemas in a way that the data exposed by the mediator is always consistent [11]. It was this problem that triggered the developments reported in Sect. 3 of this chapter. In fact, previous work by the authors [9] introduced the notion of the open fragment, which is captured by the projection operation, whereas a preliminary version of the union and difference operations was presented in [12].

Finally, Volz et al. [25] proposed a tool that implements an operation similar to the projection by the creation of a database view resulting from query execution. However, this tool does not allow the generation of semantic information captured by the constraints that apply to the vocabulary terms. Ibáñez-García et al. [17] addressed module extraction within the context of DOL—Distributed Ontology Language (DOL). Other tools, such as OAPT [2], addressed the question of partitioning an ontology into modules to facilitate ontology reuse. The *OntologyManagerTab* [20] is a full implementation of the operations described in this chapter as a Protégé plug-in.

6 Conclusions

In this chapter, we defined a set of operations that create new ontologies, including their constraints, out of other ontologies and that can be efficiently implemented for lightweight ontologies. We argued that such ontologies are expressive enough to cover certain formal conceptual modeling scenarios, and that the operations introduced provide composition mechanisms to facilitate the construction of lightweight ontologies.

As future work, we intend to expand the implementation of the operations to cover a more expressive family of ontologies, using the results presented in [10]. We also intend to explore the use of assistant theorem provers to compute approximations of the operations, as in [15]. The challenge, in this case, would be to figure out how a theorem prover would help construct sets of constraints, in much the same way that the decision procedure that explores the constraint graph does.

Acknowledgements We are grateful to the referees of an earlier version of this chapter for urging us to revise the introduction and the related work sections. Also, this work was partly supported by CNPq under grant 302303/2017-0 and by FAPERJ under grant E-26-202.818/2017.

References

1. A.V. Aho, M.R. Garey, J.D. Ullman, The transitive reduction of a directed graph. SIAM J. Comp. **1**(2), 131–137 (1972)
2. Algergawy, A., Babalou, S., Klan, F., Koenig-Ries, B., 2016. OAPT: A tool for ontology analysis and partitioning, in *Proceedings of the 19th International Conference on Extending Database Technology*, March 15–18, 2016, Bordeaux, France (2016), pp. 644–647
3. A. Artale, D. Calvanese, R. Kontchakov, M. Zakharyaschev, The DL-Lite family and relations. J. Artif. Intell. Res. **36**, 1–69 (2009)
4. F. Baader, W. Nutt, Basic description logics, in *The Description Logic Handbook: Theory, Implementation and Applications* (Cambridge University Press, Cambridge, UK, 2003), pp. 43–95
5. J.A. Bergstra, J. Heering, P. Klint, Module algebra. *J. ACM* **37**(2), 335–372 (1990)
6. M. Barr, C. Wells, *Toposes, Triples and Theories*. Grundlehren der mathematischen Wissenschaften, vol. 278 (Springer, New York, 1985)
7. C. Bizer, R. Cyganiak, T. Heath, How to publish Linked Data on the Web (2007), http://www4.wiwiss.fu-berlin.de/bizer/pub/LinkedDataTutorial/
8. R.L. Carvalho, T.S.E. Maibaum, T.H.C. Pequeno, A.A. Pereda, P.A.S. Veloso, A model theoretic approach to the semantics of data types and structures, in *Proceedings of the International Computer Symposium*, Taiwan (1982)
9. M.A. Casanova, K.K. Breitman, A.L. Furtado, V.M.P. Vidal, J.A.F. Macêdo, The role of constraints in linked data, in *Proceedings of the Confederated International Conferences: CoopIS, DOA-SVI, and ODBASE 2011, Part II. LNCS 7045* (Springer, Berlin, 2011), pp. 781–799
10. M.A. Casanova, K.K. Breitman, A.L. Furtado, V.M.P. Vidal, J.A.F. Macêdo, An efficient proof procedure for a family of lightweight database schemas, in *Conquering Complexity* ed by M.G. Hinchey (ed.), . Springer, 2012a, 431–461
11. M.A. Casanova, T. Lauschner, L.A.P.P. Leme, K.K. Breitman, A.L. Furtado, V.M.P. Vidal, Revising the constraints of lightweight mediated schemas. Data Knowl. Eng. **69**(12), 1274–1301 (2010)
12. M.A. Casanova, J.A.F. Macêdo, E. Sacramento, A.M.A. Pinheiro, V.M.P. Vidal, K.K. Breitman, A.L. Furtado, Operations over lightweight ontologies, in *Proceedings of the 11th International Conference on Ontologies, Databases, and Applications of Semantics. LNCS 7566* (Springer, Berlin, 2012b), pp. 646–663
13. P. Doran, V. Tamma, L. Iannone, Ontology module extraction for ontology reuse: An ontology engineering perspective, in *Proceedings of the 16th ACM Conference on Information and Knowledge Management (CIKM '07)* (ACM, New York, NY, USA, 2007), pp. 61–70
14. H. Ehrig, B. Mahr, *Fundamentals of Algebraic Specification 2: Module Specifications and Constraints* (Springer Science & Business Media, Berlin, 2012)
15. B.C. Grau, I. Horrocks, Y. Kazakov, U. Sattler, Extracting modules from ontologies: A logic-based approach, in *Modular Ontologies: Concepts, Theories and Techniques for Knowledge Modularization* (Springer, Berlin, Heidelberg, 2009). ISBN-10 3-642-01906-4
16. H.T. Hsu, An algorithm for finding a minimal equivalent graph of a digraph. J. ACM **22**(1), 11–16 (1975)
17. Y.A. Ibáñez-García, T. Mossakowski, D. Sannella, A. Tarlecki, Modularity of ontologies in an arbitrary institution, in *Logic, Rewriting, and Concurrency—Essays dedicated to José Meseguer on the Occasion of His 65th Birthday*. LNCS 9200 (Springer, Berlin, 2015), pp. 361–379
18. P. Jain, P. Hitzler, P.Z. Yeh, K. Verma, A.P. Sheth, Linked data is merely more data, in *Proceedings of the AAAI Spring Symposium: 'Linked Data Meets Artificial Intelligence'*, pp. 82–86 (2010)
19. S. Khuller, B. Raghavachari, N. Young, Approximating the minimum equivalent digraph. SIAM J. Comp. **24**(4), 859–872 (1995)

20. R.C. Magalhães, M.A. Casanova, B.P. Nunes, G.R. Lopes, On the implementation of an algebra of lightweight ontologies, in *Proceedings of the 21th International Database Engineering & Applications Symposium*, Bristol, England, July 12–14 (2017)
21. Seidenberg, J., Rector, A.L., 2006. Web ontology segmentation: analysis, classification and use. In: Proc.15th Int'l. Conf. on World Wide Web, Edinburgh, UK, 23–26 May 2006, pp. 13–22
22. H. Stuckenschmidt, M. Klein, Structure-based partitioning of large concept hierarchies, in *The Semantic Web*—ISWC 2004. LNCS 3298 (Springer, Berlin, Heidelberg, 2004)
23. H. Stuckenschmidt, C. Parent, S. Spaccapietra, (eds.), *Modular Ontologies: Concepts, Theories and Techniques for Knowledge Modularization*. Theoretical Computer Science and General Issues 5445 (Springer, Berlin, Heidelberg, 2009). ISBN-10: 3-642-01906-4
24. Uschold, M., Gruninger, M., 1996. Ontologies: principles, methods and applications. *The Knowledge Engineering Review*, 11(2), June 1996, 93-136
25. R. Volz, D. Oberle, R. Studer, Views for light-weight web ontologies, in Proceedings of the 7th International Database Engineering and Application Symposium (IDEAS 2003), pp. 160–169 (2003)

Formal Ontological Analysis for Medical Protocols

Neeraj Kumar Singh, Yamine Ait-Ameur, and Dominique Méry

Abstract Clinical guidelines systematically assist practitioners to provide an appropriate health care in specific clinical circumstances. A significant number of guidelines and protocols is lacking in quality. Indeed, ambiguity and incompleteness are likely anomalies in medical practice. In order to find anomalies and to improve the quality of medical protocols, this paper presents a stepwise formal development of a medical protocol. In this development, we define the domain concepts based on ontologies and integrate them with the medical protocol in an explicit way. In this work, we use the Event B language for modelling a domain model using ontologies and capturing the functional behaviour of the medical protocol. Our main contributions are: to use domain-specific knowledge in a system model explicitly; to link a domain model and a system model using an annotation mechanism; and to use a proof-based formal approach to evaluate a medical protocol. An assessment of the proposed approach is given through a case study, relative to a real-life reference protocol (electrocardiogram (ECG) interpretation), which covers a wide variety of protocol characteristics related to different heart conditions.

1 Introduction

Over the past few decades, much research has been done in the area of medical domain to address the growing challenges in the field of biomedical informatics, life sciences,

IMPEX Project (ANR-13-INSE-0001), webpage: http://impex.loria.fr.

N. K. Singh (✉) · Y. Ait-Ameur
ENSEEIHT-INPT/IRIT, University of Toulouse, Toulouse, France
e-mail: neeraj.singh@toulouse-inp.fr

Y. Ait-Ameur
e-mail: yamine.aitameur@toulouse-inp.fr

D. Méry
LORIA, Université de Lorraine, BP 239, Nancy, France
e-mail: mery@loria.fr

© Springer Nature Singapore Pte Ltd. 2021
Y. Ait-Ameur et al. (eds.), *Implicit and Explicit Semantics Integration in Proof-Based Developments of Discrete Systems*,
https://doi.org/10.1007/978-981-15-5054-6_5

pharmacology, neuroscience and clinical research. There are several databases to manage different kinds of biological information. However, these databases face new challenges in terms of increasing amount of data due to the growing number of users. These new challenges are: data are voluminous, unstructured and collected from a variety of incompatible sources; difficult to use and understand the available data, information and knowledge; needs of better techniques and tools to manage the databases; needs of semantical description of biological domain and medical systems; and needs of some sound techniques to meet regulators and certification standards [37].

Mostly, the system development process does not consider the domain knowledge explicitly. However, such knowledge is provided implicitly during the system development by making some assumptions on an environment and some of the past experiences. It is very common that such implicit domain knowledge often shows some contradictory results, which may lead to a system failure state. Integrating domain knowledge into a system model explicitly may improve the quality of the development process. Note that one of the main reasons for not integrating domain knowledge into the system development is the lack of modelling languages. Most of the languages are unable to express environment requirements related to a system [5].

Medical guidelines are "*systematically developed statements to assist practitioners and patients to determine appropriate health care for specific circumstances*" [27, 45]. Medical protocols provide healthcare testimonials and facilitate high standard practices. For developing high-quality protocols, we need regular amendments. Medical bodies worldwide have made efforts for improving existing protocols and their development process. However, these initiatives are not sufficient since they rely on informal methods and they do not apply domain knowledge during the development of protocols [28].

We are concerned with a different approach, namely the quality improvement of medical protocols using formal methods. In order to find anomalies and to improve the quality of medical protocols, this paper presents a stepwise formal development of a medical protocol. The whole development is composed of two different models: domain model and system model. The domain model contains domain concepts based on ontologies [15] and the system model contains the required functional behaviour of a given medical protocol. Note that an annotation mechanism is used to integrate these two models for developing and verifying the medical protocol. Combining these two models allows us to verify some new properties related to the domain knowledge within the enriched design medical protocol. In this work, we use the Event-B language for modelling the domain model and the system protocol model. Our main contributions are:

- to use domain-specific knowledge in a system model explicitly;
- to link a domain model and a system model using an annotation mechanism;
- to use a proof-based formal approach to evaluate a medical protocol;
- to find ambiguity, incompleteness and inconsistency in a medical protocol.

The main goal of this work is to translate an informal description of a medical protocol into a formal language, with the aim of analysing a set of properties. Such kinds of formal verification allow us to expose problematic parts in the protocol by analysing the formal description of the protocol. The current work intends to explore those problems related to the modelling of medical protocols. Moreover, an incremental development of the medical protocol model helps to discover the ambiguous, incomplete or even inconsistent elements in the medical protocol under the explicit domain knowledge. The electrocardiogram (ECG) protocol covers a wide variety of characteristics related to different heart conditions. Formal modelling and verification of the ECG clinical protocols have been carried out as a case study to assess the feasibility of this approach.

The outline of the remainder of the paper is as follows: Section 2 describes the ontology concepts and the modelling framework is presented in Sect. 3. Section 4 presents a modelling methodology. In Sect. 5, we explore the incremental proof-based formal development of the ECG protocol, including domain model. Related work is presented in Sects. 6 and 7 concludes the paper.

2 Ontology

Ontology—"science of being"—is originated in philosophy, which is defined as *"hierarchical structuring of knowledge about concepts by sub-classing them according to their properties and qualities"* [19]. It can also be defined as *"a declarative model of a domain that defines and represents the concepts existing in that domain, their attributes and the relationships between them"* [19, 20]. Ontology provides a description of concepts along with desired relations. The concept plays a very important role in data sharing and knowledge representation. Nowadays ontologies are adopted by almost every area of science and engineering for a common understanding between different user groups. In general, ontologies are classified as (i) Upper ontologies, (ii) General Ontologies (iii) Domain Ontologies and (iv) Application ontologies. All these classes are provided according to a detailed conceptual knowledge. Upper ontologies or top-level ontologies provide a very generic knowledge applicable to various domains. They mainly contain basic notions of objects, relations, events and processes. General ontologies are not dedicated to any specific domain or field. These ontologies represent general knowledge of a large field at an intermediate level without addressing low-level details. Domain ontologies are only applicable to a domain with a specific viewpoint that represent knowledge about a particular field or area of the world. Application ontologies are the specialisation of domain ontologies that are designed for specific tasks.

The prime use of defining or developing ontology is to share knowledge or information with groups, who work in the same domain. The main reasons for developing ontologies are [33, 46]: (1) to share knowledge in the same domain; (2) to reuse existing developed ontologies; (3) to provide an explicit list of domain assumptions;

(4) to separate domain knowledge from operational knowledge; and (5) to perform domain-specific methodical analyses.

2.1 Ontology in Medical Domain

Medicine is a branch of science dealing with the maintenance of health, and the prevention and treatment of diseases. It offers a solid foundation in the core biomedical subjects, such as anatomy, physiology, pharmacology, neuroscience, etc. A medical domain is characterised by abundant knowledge of medical science collected from various sources. It is constantly growing due to new discoveries provided by medical experts and researchers. Most of the existing data are distributed into heterogeneous databases and architecture. They have different implementations and are not compatible with each other. Integrating heterogeneous databases can be a solution to provide a centralised and reliable database, but it is a very costly operation and requires huge resources [37]. Note that most of the individual databases are developed by different research groups for their own purpose that do not follow standard approaches. Data collected from different sources are mainly inconsistent and hard to understand. Therefore, an approach is required to systematically represent medical knowledge that could be used for analysis, clinical practices and supporting the different healthcare activities [12]. Ontology has played a significant role in representing medical knowledge systematically in an independent format to share and reuse across the other biomedical domains. The medical ontology framework provides a common medical concepts, relationships, properties, and axioms related to biomedical, disease, diagnosis, treatment, anatomy, pharmacology, clinical procedure and so on. There are several medical ontologies, such as GALAN [14], OpenCyc [13], WordNet [30], UMLS [12], SNOMED-CT [26], FMA [36] and Gene Ontology [6] developed by researchers, industries and medical centres. In our work, we adopt these ontologies to define the domain concepts related to the selected medical protocol.

3 The Modelling Framework: Event-B

This section describes the essential components of modelling framework. In particular, we will use the Event-B modelling language [3] for modelling a complex system in a progressive way. There are two main components in Event-B: *context* and *machine*. A *context* is a formal static structure that is composed of several other components, such as *carrier sets, constants, axioms* and *theorems*. A *machine* is a formal dynamic structure that is composed of *variables, invariants, theorems, variants* and *events*. A machine and a context can be connected with *sees* relationships.

An Event-B model is characterised by a list of *state variables* possibly modified by a list of *events*. Events play an important role in modelling the functional behaviour of a system. An event is a state transition that contains two main components: *guard* and *action*. A *guard* is a predicate based on the state variables that define a necessary condition for enabling the event. An *action* is also a predicate that allows modifying the state variables when the given guard becomes true. A set of invariants defines required safety properties that must be satisfied by all the defined state variables. There are several proof obligations, such as invariant preservation, nondeterministic action feasibility, guard strengthening in refinements, simulation, variant, well-definedness, that must be checked during the modelling and verification process.

The Event-B modelling language allows us modelling a complex system gradually using refinement. Refinement enables us to introduce more detailed behaviour and the required safety properties by transforming an abstract model to a concrete version. At each refinement step, the events can be refined by (1) keeping the event as it is; (2) splitting an event into several events; or (3) refining by introducing another event to maintain state variables. Note that the refinement always preserves a relation between an abstract model and its corresponding concrete model. The newly generated proof obligations related to refinement ensures that the given abstract model is correctly refined by its concrete version. Note that the refined version of the model always reduces the degree of nondeterminism by strengthening the guards and/or predicates. The modelling framework has a very powerful tool support (Rodin [35]) for project management, model development, conducting proofs, model checking and animation and automatic code generation. There are numerous publications and books available for an introduction to Event-B and its related refinement strategies [3].

4 Modelling Methodology

In this section, we present a modelling methodology, which is described in [5]. Figure 1 depicts a stepwise modelling methodology, which contains the different modelling steps: domain modelling, system modelling, model annotation and model verification. These modelling steps are described as follows:

1. **Domain Modelling.** Domain knowledge plays an important role in making assumptions for a given system. Mostly, the required information related to a domain may be considered hypothetically based on previous experiences and the available domain knowledge. Note that an ontology modelling language can be used to characterise and formally specify a domain knowledge in the form of domain ontology through the definition of concepts, entities, relationships, constraints and rules. In this work for modelling a domain model, we choose the Event-B modelling language [3] to formalise the required domain concepts derived from the domain ontology, which can be described in Event-B context using *sets*, *constants*, *axioms* and *theorems*.

2. **System Modelling.** For developing a safe system considering all the required functionalities is a challenging problem. In order to design a safe system, we can use any formal modelling language to describe a desired behaviour under the

Fig. 1 Four steps modelling
methodology

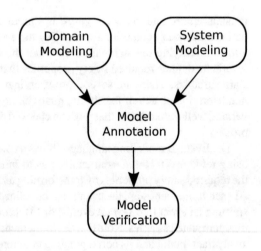

given specification. The selected modelling language and associated verification
approach allows us to check the required behaviour. In this work for modelling a
system model, we also choose the Event-B modelling language [3], which allows
us the progressive development of the system behaviour satisfying the required
safety properties using machines and contexts.

In many cases, we should develop a domain model before developing a system
model so that we can use the domain model during the system model development.
If we do not have any domain model before developing a system model then we
need to introduce the domain-specific information implicitly to design a correct
system model and to check the system model independently. Note that during
the annotation for combining the system model and domain model, we need to
remove the implicit domain-specific concepts.

3. **Model Annotation.** Model annotation is a mechanism that allows us to establish
a relationship between the domain model and the system model by describing the
design model entities and ontology concepts. The annotation mechanism can be
defined independently and can be used to annotate both the system and domain
models. These models can be developed using the same formal notations or different formal notations. An independent annotation mechanism, like a plugin, can
be used to bind the domain model and the system model together. Developing a
new annotation mechanism is beyond the scope of this paper. Note that in our
current work, we have not used any specific annotation mechanism to apply our
approach. In fact, we have used the same modelling language (Event-B [3]) to
design the two models. Thus we have got a free implicit annotation mechanism
(i.e. *see* context relationship) for our purpose used to integrate the domain and system models together. In order to integrate the domain model and system model in
Event-B, we use the domain model as a set of contexts during the development of
the system model for describing the desired properties and functional behaviour.

4. **Model Verification.** This is the last step of the modelling methodology, which can be performed when a system model is annotated with a domain model. The annotated design model is enriched by the domain properties expressed in the ontology. For verifying the annotated model, we should apply the verification in two steps. The first verification must be conducted on the designed system model before annotation (may be no longer correct after annotation) to check the consistency and then the second verification must be conducted on the designed system model after annotation to check the overall consistency considering the domain knowledge. Note that in the second step, the verification also allows us for checking the new emerging properties due to the integration of domain model and system model using annotation mechanism.

5   Case Study: ECG Protocol

An electrocardiogram (EKG or ECG) [10, 24, 44] reflects an electrical activity of the heart that shows depolarization and repolarization of the atria and ventricles. The typical one-cycle ECG is shown in Fig. 2, which is a sequence of different segments and intervals to represent the time evolution of electrical activity in the heart. These sequences are denoted as P-QRS-T-U to show different functionalities of the heart. These sequences are described as follows:

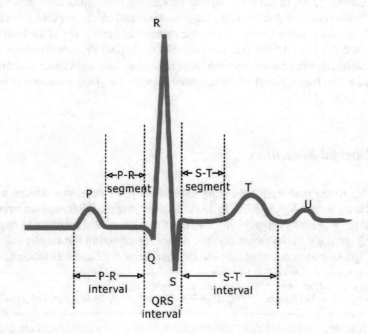

Fig. 2 ECG Deflections

- **P-wave:** It is a small deflection caused by the atrial depolarization before contraction to show an electrical wave propagation from the SA (sinus) node through the atria;
- **PR interval**—It is an interval between the beginning of the P-wave to the beginning of the Q-wave;
- **PR segment**—It is a flat segment between the end of the P-wave and the start of the QRS interval;
- **QRS interval:** It is an interval between the P-wave and T-wave with greater amplitude to show the depolarization of the ventricles;
- **ST interval:** It is an interval between the end of S-wave and the beginning of T-wave;
- **ST-segment**—It is a flat segment starts at the end of the S-wave and finishes at the start of the T-wave;
- **T-wave:** It is a small deflection caused by the ventricular repolarization, whereby the cardiac muscle is prepared for the next cycle of ECG;
- **U-wave:** It is a small deflection immediately following the T-wave due to repolarization of the Purkinje fibres;

For analysing the different heart conditions and possible behaviour, Electrocardiogram (ECG) plays a key role in clinical trials. Medical practitioners heavily depend on the result of ECG interpretation. A series of deflections and wave of the ECG has different characteristics to show the different clinical conditions, which can be used for diagnosis purpose. To our knowledge, there are several databases and ontologies to represent the ECG. In our work, we use the existing ontological definition [25] to define the domain. The domain knowledge encapsulates all the required knowledge in ontology relationship. For describing the conceptual knowledge of the biological process, we use the OBO (Open Biomedical Ontologies) Process Ontology [11], which can be automated using the first-order reasoning. The main OBO relations are classified as the foundational relation, spatial relation, temporal relation and participation relation [11].

5.1 Domain Modelling

According to our four steps modelling methodology, we develop a domain model derived from the ontology of ECG. In this work, we define the OBO relations using the Event-B [3] modelling language then we use these formalised relations to describe the ECG ontology to develop a domain model for capturing the required domain knowledge. In the current work, we use the foundational relations as follows:

$$A \ is\_a \ B = \forall x[instance\_of(x, A) \Rightarrow instance\_of(x, B)]$$
$$A \ part\_of \ B = \forall x[instance\_of(x, A) \Rightarrow \exists y(instance\_of(y, B) \ \& \ x \ PartOf\_Inst \ y)]$$

The first relation states that every instance of class A is an instance of class B and the second relation states that $A \ part\_of \ B$ holds if and only if: for every individual

x, if x instantiates A then there is some individual y such that y instantiates B and x is a part of y. The *instance_of* is a relation between a class instance and a class which it instantiates and the *PartOf_Inst* is a relation between two class instances. The foundational relations, **is_a** and **part_of**, are defined in Event-B context using axioms ($axm1$–$axm5$). $axm2$ and $axm3$ define **is_a** relation and **part_of** relation, respectively. Other axioms ($axm1$, $axm4$ and $axm5$) are used to support the formal definition of the defined relations. Note that these defined OBO relations are used further to define the required domain knowledge for formalising the ECG protocol.

$$axm1 : HAS\_INSTANCES = CLASS \leftrightarrow INSTANCE$$

$$axm2 : IS\_A = \{IsA | IsA \in CLASS \leftrightarrow CLASS \wedge (\forall x, y \cdot (x \in CLASS \wedge y \in CLASS \wedge x \mapsto y \in IsA$$
$$\Leftrightarrow$$
$$union(\{r \cdot r \in HAS\_INSTANCES | ran(\{x\} \lhd r)\})$$
$$\subseteq$$
$$union(\{r \cdot r \in HAS\_INSTANCES | ran(\{y\} \lhd r)\})))\}$$

$$axm3 : PART\_OF = \{PartOf | PartOf \in CLASS \leftrightarrow CLASS \wedge$$
$$(\forall x, y \cdot (x \in CLASS \wedge y \in CLASS \wedge x \mapsto y \in PartOf$$
$$\Leftrightarrow$$
$$\forall p \cdot p \in union(\{r \cdot r \in HAS\_INSTANCES | ran(\{x\} \lhd r)\}) \Rightarrow$$
$$(\exists q \cdot q \in union(\{r \cdot r \in HAS\_INSTANCES | ran(\{y\} \lhd r)\}) \wedge p \mapsto q \in PartOf\_Inst)))\}$$

$$axm4 : PartOf\_Inst \in INSTANCE \leftrightarrow INSTANCE$$

$$axm5 : (\forall p \cdot p \in INSTANCE \Rightarrow p \mapsto p \in PartOf\_Inst) \wedge$$
$$(\forall p, q \cdot p \in INSTANCE \wedge q \in INSTANCE \wedge$$
$$p \mapsto q \in PartOf\_Inst \wedge q \mapsto p \in PartOf\_Inst \Rightarrow p = q) \wedge$$
$$(\forall p, q, r \cdot p \in INSTANCE \wedge q \in INSTANCE \wedge$$
$$r \in INSTANCE \wedge p \mapsto q \in PartOf\_Inst \wedge q \mapsto r \in PartOf\_Inst \Rightarrow$$
$$p \mapsto r \in PartOf\_Inst)$$

In our work, we adopt the existing available work [1, 2, 17, 18, 24] for designing and developing the domain model of ECG. Note that the developed ECG domain model based on existing ontologies contains a very abstract information related to the heart and ECG by hiding the main complexities. It is important to include complex details to consider every aspect of the domain knowledge. For the sake of simplicity, the produced domain model is used only for realising the case study of ECG protocol.

In order to define an ECG domain model, we define several small models based on sub-ontologies. These sub-ontologies are human heart, blood circulation, bioelectric phenomena and ECG, which are depicted in Fig. 3. All these sub-ontologies are connected to each other using dependency relationships. For instance, the sub-ontology of the human heart depends on the other sub-ontologies ECG, bioelectric phenomena and blood circulation.

Human Heart. It is a domain model based on sub-ontology depicted in Fig. 4 to describe a very high-level abstraction of the heart. The heart consists of four chambers: left atrium, right atrium, left ventricle and right ventricle. There is the **part_of** relationships between the heart and four chambers. These relationships indicate that the heart has an only single chamber of each type. For instance, only one left atrium.

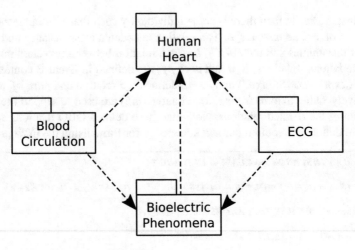

Fig. 3 Overview of a domain model based on ontology

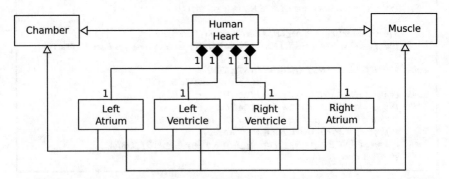

Fig. 4 Human heart domain model based on human heart sub-ontology

There is also the *is_a* relationship between the Chamber and four heart chambers. Similarly, there is the *is_a* relationship between Muscle and four heart chambers.

The human heart domain model is formalised in the Event-B modelling language using a context. This context is an extension of our previous context, which contains the formal description of OBO relations. In this extended context, we use the *is_a* and *part_of* relationships to describe the relational properties between different biological entities according to the Fig. 4. All the possible relationships are defined using axioms (*axm*1–*axm*6). The next axiom (*axm*7) is declared as an enumerated set to define a set of physical units, which can be associated with variables and constants to maintain the physical unit consistency between variables during a calculation. For example, in our case, we define the beat per minute (bpm), centimetre (cm), millimetre (mm), micrometre (mu_m). *axm*8 defines a function to map between physical units and integer numbers that can be associated with any class to describe the class attributes. The next two axioms (*axm*9 and *axm*10) are used to define the heart rate

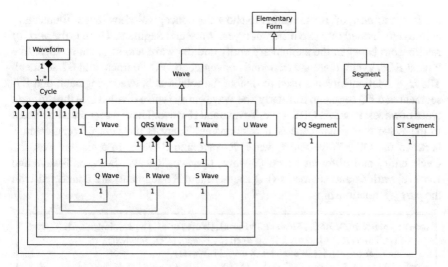

Fig. 5 ECG domain model based on ECG sub-ontology

considering the physical unit *bpm*. In a similar way, the next two axioms (*axm*11 and *axm*12) are used to define the normal heart rate associating with the same unit. The last axiom (*axm*13) defines the abnormal heart rate using the previous definitions of the heart rate and normal heart rate.

$axm1 : partition(CLASS, \{Heart\}, \{Chamber\}, \{Muscle\}, \{LeftAtrium\}, \{RightAtrium\},$
$\qquad \{LeftVentricle\}, \{RightVentricle\})$
$axm2 : \{Heart \mapsto Chamber\} \in IS\_A \wedge \{Heart \mapsto Muscle\} \in IS\_A$
$axm3 : \{LeftAtrium \mapsto Chamber\} \in IS\_A \wedge \{LeftAtrium \mapsto Muscle\} \in IS\_A \wedge$
$\qquad \{LeftAtrium \mapsto Heart\} \in PART\_OF$
$axm4 : \{RightAtrium \mapsto Chamber\} \in IS\_A \wedge \{RightAtrium \mapsto Muscle\} \in IS\_A \wedge$
$\qquad \{RightAtrium \mapsto Heart\} \in PART\_OF$
$axm5 : \{LeftVentricle \mapsto Chamber\} \in IS\_A \wedge \{LeftVentricle \mapsto Muscle\} \in IS\_A \wedge$
$\qquad \{LeftVentricle \mapsto Heart\} \in PART\_OF$
$axm6 : \{RightVentricle \mapsto Chamber\} \in IS\_A \wedge \{RightVentricle \mapsto Muscle\} \in IS\_A \wedge$
$\qquad \{RightVentricle \mapsto Heart\} \in PART\_OF$
$axm7 : partition(UNIT, bpm, mm, cm, mu\_m)$
$axm8 : F\_UNIT \in UNIT \rightarrow \mathbb{P}(\mathbb{Z})$
$axm9 : HEART\_RATE \in \{Heart\} \leftrightarrow F\_UNIT$
$axm10 : HEART\_RATE = \{Heart \mapsto (bpm \mapsto 1 .. 300)\}$
$axm11 : NORMAL\_HEART\_RATE \in \{Heart\} \leftrightarrow F\_UNIT$
$axm12 : NORMAL\_HEART\_RATE = \{Heart \mapsto (bpm \mapsto 60 .. 100)\}$
$axm13 : ABNORMAL\_HEART\_RATE = HEART\_RATE \setminus NORMAL\_HEART\_RATE$

ECG. It is a domain model based on sub-ontology depicted in Fig. 5 to describe a very high-level abstraction of the ECG. As we have previously described that the ECG is a sequence of deflections. All these elementary deflections can be described in different types of waves and segments, which form a typical ECG cycle. All these elementary entities, such as waves, segments and cycle, of ECG are organised using the *is_a* and **part_of** relationships to show the conceptual knowledge. Elementary entities are divided into two different types: Wave and Segment. There is the **part_of** relationship between the elementary entity and the wave and segment entities. In a typical ECG cycle, there are two kinds of segments, PQ Segment and ST-segment. The *is_a* relationships are used to denote the relations between Segment, and ST-segment and PQ segment. Similarly, the Wave entity is also divided into the different types of waves: P- wave, QRS wave, T-wave and U-wave. These waves are also related to the Wave entity using the *is_a* relationship. There is the **part_of** relationships between the QRS Wave and Q-wave, R-wave and S-wave. In a similar way, the Cycle entity and different waves (P-wave, Q-wave, R-wave, S-wave, T-wave and U-wave) entities and segments (PQ segment and ST-segment) are connected with the **part_of** relationship.

$$
\begin{aligned}
&axm1 : partition(CLASS, \{ElementaryForm\}, \{Waveform\}, \{Wave\}, \{Segment\}, \{Cycle\}, \\
&\qquad \{P\_Wave\}, \{QRS\_Wave\}, \{T\_Wave\}, \{U\_Wave\}, \{PQ\_Segment\}, \\
&\qquad \{ST\_Segment\}, \{Q\_Wave\}, \{R\_Wave\}, \{S\_Wave\}) \\
&axm2 : \{Wave \mapsto ElementaryForm\} \in IS\_A \\
&axm3 : \{Segment \mapsto ElementaryForm\} \in IS\_A \\
&axm4 : \{Cycle \mapsto Waveform\} \in PART\_OF \\
&axm5 : \{P\_Wave \mapsto Wave\} \in IS\_A \wedge \{P\_Wave \mapsto Cycle\} \in PART\_OF \\
&axm6 : \{QRS\_Wave \mapsto Wave\} \in IS\_A \\
&axm7 : \{T\_Wave \mapsto Wave\} \in IS\_A \wedge \{T\_Wave \mapsto Cycle\} \in PART\_OF \\
&axm8 : \{U\_Wave \mapsto Wave\} \in IS\_A \wedge \{U\_Wave \mapsto Cycle\} \in PART\_OF \\
&axm9 : \{PQ\_Segment \mapsto Segment\} \in IS\_A \wedge \{PQ\_Segment \mapsto Cycle\} \in PART\_OF \\
&axm10 : \{ST\_Segment \mapsto Segment\} \in IS\_A \wedge \{ST\_Segment \mapsto Cycle\} \in PART\_OF \\
&axm11 : \{Q\_Wave \mapsto QRS\_Wave\} \in PART\_OF \wedge \{Q\_Wave \mapsto Cycle\} \in PART\_OF \\
&axm12 : \{R\_Wave \mapsto QRS\_Wave\} \in PART\_OF \wedge \{R\_Wave \mapsto Cycle\} \in PART\_OF \\
&axm13 : \{S\_Wave \mapsto QRS\_Wave\} \in PART\_OF \wedge \{S\_Wave \mapsto Cycle\} \in PART\_OF \\
&axm14 : partition(LEADS, \{I\}, \{II\}, \{III\}, \{aVR\}, \{aVL\}, \{aVF\}, \{V1\}, \{V2\}, \{V3\}, \\
&\qquad \{V4\}, \{V5\}, \{V6\}) \\
&axm15 : RR\_Int\_equidistant \in \{Cycle\} \times LEADS \rightarrow BOOL \\
&axm16 : PP\_Int\_equidistant \in \{Cycle\} \times LEADS \rightarrow BOOL \\
&axm17 : P\_Positive \in \{P\_Wave\} \times LEADS \rightarrow BOOL \\
&axm18 : PP\_Interval \in \{Cycle\} \times LEADS \rightarrow BOOL \\
&axm19 : RR\_Interval \in \{Cycle\} \times LEADS \rightarrow BOOL
\end{aligned}
$$

The ECG sub-ontology is formalised in the Event-B modelling language using the OBO relationships (*is_a* and **part_of**) according to the Fig. 5. All the possible relationships are defined in axioms (*axm1–axm13*). The next axiom (*axm14*) defines a set of leads (12-leads) as an enumerated set that represents the heart's electrical activity recorded from electrodes on the body surface. The next five axioms (*axm15–axm19*) are defined as functions to characterise the ECG signal. These functions are *RR_Int_equidistant* to show the boolean state of the equidistant of RR interval; *PP_Int_equidistant* to show the boolean state of the equidistant of PP interval; *P_Positive* to show the positive visualisation of the P-wave, *PP_Interval* to represent the PP interval; and *RR_Interval* to represent the RR interval.

5.2 System Modelling

This section describes the second step of our modelling methodology. To design a system model (ECG protocol) using the domain knowledge, we revisit our developed case study of the ECG interpretation protocol [28, 42]. In this case study, our main objective is to utilise the domain knowledge explicitly in the development of ECG protocol. The ECG protocol is formalised to detect possible anomalies in the existing ECG protocol. In this development, we use the Event-B [3] modelling language that allows us to develop the whole complex ECG protocol using a *correct by construction* approach to introduce the detailed clinical properties of the ECG protocol. Figure 7 depicts an incremental formal development of the ECG interpretation protocol. Every refinement level introduces a *diagnosis* criteria for different components of the ECG signal, and each new criterion helps to analyse a particular set of diseases. The whole development of the ECG protocol is summarised below.

5.2.1 Abstract Model (Assessing Rhythm and Rate)

Figure 6 depicts a standard clinical procedure for analysing the ECG protocol abstractly that is taken from [24]. This is a basic procedure that is used by most of the medical practitioners at the initial stage of clinical procedure for analysing the different heart conditions. In this basic procedure, the ECG protocol assesses the *rhythm* and *heart rate* to distinguish between the normal and abnormal heart conditions. We have used this clinical step for modelling the abstract model and the other clinical steps will be introduced progressively in the next refinement levels that are also adopted from [24].

In order to define the static properties, we define *State* and *YesNoState* as enumerated sets in axioms (*axm*1 and *axm*2). These two axioms are further used to define

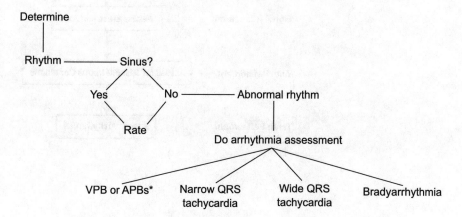

Fig. 6 Basic diagram of assessing rhythm and rate [24]

Fig. 7 Refinements of ECG
protocol

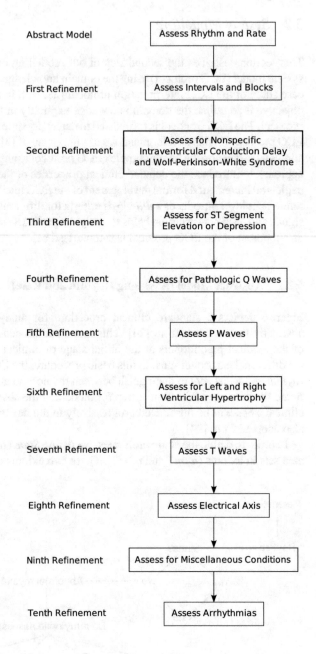

the normal and abnormal states of the heart in $axm3$ and the sinus states of the heart in $axm4$.

$axm1 : partition(State, \{OK\}, \{KO\})$
$axm2 : partition(YesNoState, \{Yes\}, \{No\})$
$axm3 : HState \in \{Heart\} \rightarrow State$
$axm4 : HYNState \in \{Heart\} \rightarrow YesNoState$
$CS1 : ClinicalProp1 = (\lambda x \mapsto y \cdot x = Cycle \land y = P\_Wave \land ((\exists l \cdot l \in \{II, V1, V2\} \land$
　　　$PP\_Int\_equidistant(x \mapsto l) = TRUE \land RR\_Int\_equidistant(x \mapsto l) = TRUE \land$
　　　$RR\_Interval(x \mapsto l) = PP\_Interval(x \mapsto l)) \land P\_Positive(y \mapsto II) = TRUE)|TRUE)$
$CS2 : ClinicalProp2 = (\lambda x \mapsto y \cdot x = Cycle \land y = P\_Wave \land ((\forall l \cdot l \in \{II, V1, V2\} \Rightarrow$
　　　$PP\_Int\_equidistant(x \mapsto l) = FALSE \lor RR\_Int\_equidistant(x \mapsto l) = FALSE \lor$
　　　$RR\_Interval(x \mapsto l) \neq PP\_Interval(x \mapsto l)) \lor P\_Positive(y \mapsto II) = FALSE)|TRUE)$

In our development, we define the clinical steps for analysing the ECG protocol. The first clinical step $CS1$ is defined as a function $ClinicalProp1$. This function has two input parameters $Cycle$ and P_Wave, which are used to assess the ECG signal by the following clinical strategy: there exists equivalent in the PP interval, equivalent in the RR interval, the RR interval and PP interval are equal in leads (II, V1 and V2) and the positive visualisation of P-wave in lead II is $TRUE$. In order to satisfy this property the function $ClinicalProp1$ results as $TRUE$. In a similar way, we define the second clinical step $CS2$ defined as a function $ClinicalProp2$. This function has also two input parameters $Cycle$ and P_Wave, which are used to assess the ECG signal by the following clinical strategy: the PP interval and RR interval are not equidistant, the RR intervals and PP intervals are not equivalent in all leads (II, V1 and V2), or the positive visualisation of P-wave in lead II is $FALSE$. In order to satisfy this property the function $ClinicalProp2$ results as $TRUE$. Note that these clinical properties are defined explicitly to build the domain knowledge of the ECG protocol.

$inv1 : Sinus \in HYNState$
$inv2 : Heart\_Rate \in HEART\_RATE$
$inv3 : Heart\_State \in HState$
$saf1 : P\_Positive(P\_Wave \mapsto II) = FALSE \Rightarrow Sinus = Heart \mapsto No$
$saf2 : Sinus = Heart \mapsto Yes \Rightarrow ClinicalProp1(Cycle \mapsto P\_Wave) = TRUE$
$saf3 : ClinicalProp2(Cycle \mapsto P\_Wave) = TRUE \Rightarrow Sinus = Heart \mapsto No$
$saf4 : Heart\_Rate \in NORMAL\_HEART\_RATE \land Sinus = Heart \mapsto Yes \Rightarrow$
　　　$Heart\_State = Heart \mapsto OK$
$saf5 : Heart\_Rate \in ABNORMAL\_HEART\_RATE \land Sinus = Heart \mapsto Yes \Rightarrow$
　　　$Heart\_State = Heart \mapsto KO$
$saf6 : Heart\_Rate \in NORMAL\_HEART\_RATE \land Sinus = Heart \mapsto No \Rightarrow$
　　　$Heart\_State = Heart \mapsto KO$

To define the initial clinical procedure for assessing the rhythm and heart rate, we define three variables ($inv1$–$inv3$): $Sinus$ to represent the sinus state of the heart; $Heart\_Rate$ to represent the heart rate limit; and $Heart\_State$ to show the normal or abnormal state of the heart. In the abstract model, we provide a list of safety properties using invariants ($saf1$–$saf6$) to verify the required conditions for the ECG interpretation protocol based on analysis of the signal features.

All these invariants are generated from the ECG protocol and extracted from the required documents with the help of medical experts. The first safety property ($saf1$) states that if the positive visualisation of P-wave in lead II is $FALSE$, then there is no sinus rhythm. The next safety property ($saf2$) states that if the sinus is yes then the clinical property $Clinical Prop1$ must be $TRUE$. This clinical property is defined in the context. Similarly, the next safety property ($saf3$) states that if the clinical property $Clinical Prop2$ is $TRUE$ then there is no sinus rhythm. The next two safety properties ($saf4$ and $saf5$) state that if the heart rate belongs to the range of the normal heart rate and sinus rhythm is yes then the heart state is OK, and if the heart rate belongs to the abnormal heart rate and the sinus rhythm is yes then the heart state is KO. The last safety property ($saf6$) states that if the heart rate belongs to the range of the normal heart rate and the sinus rhythm is no then the heart state is KO.

In this abstract model, we define three events *Rhythm_test_TRUE, Rhythm_test_-FALSE* and *Rhythm_test_TRUE_abRate*. The guards of the first event state that the clinical property $Clinical Prop1$ is $TRUE$ in the selected $Cycle$ and P_Wave, and the heart rate belongs to the normal heart rate. The action of this event shows that the sinus rhythm is yes, the current heart rate is assigned and the heart state is OK.

```
EVENT Rhythm_test_TRUE
 ANY rate
 WHEN
   grd1 : Clinical Prop1(Cycle ↦ P_Wave) = TRUE
   grd2 : rate ∈ NORMAL_HEART_RATE
 THEN
   act1 : Sinus := Heart ↦ Yes
   act2 : Heart_Rate := rate
   act3 : Heart_State := Heart ↦ OK
 END
```

In a similar way, the second event is used to assess the ECG to determine that the sinus rhythm is no, the current heart rate is assigned and the heart state is KO. The guards of this event state that the clinical property $Clinical Prop2$ is $TRUE$ in the selected $Cycle$ and P_Wave, and the heart rate belongs to the heart rate.

```
EVENT Rhythm_test_FALSE
 ANY rate
 WHEN
   grd1 : Clinical Prop2(Cycle ↦ P_Wave) = TRUE
   grd2 : rate ∈ HEART_RATE
 THEN
   act1 : Sinus := Heart ↦ No
   act2 : Heart_Rate := rate
   act3 : Heart_State := Heart ↦ KO
 END
```

The last event also represents the ECG assessment for determining the sinus rhythm is *Yes* and the heart state is *KO* in the case of an abnormal heart rate. The guards of this event state that the clinical property $ClinicalProp1$ is $TRUE$ in the selected $Cycle$ and P_Wave, and the heart rate belongs to the abnormal heart rate.

```
EVENT Rhythm_test_TRUE_abRate
  ANY rate
  WHEN
    grd1 : ClinicalProp1(Cycle ↦ P_Wave) = TRUE
    grd2 : rate ∈ ABNORMAL_HEART_RATE
  THEN
    act1 : Sinus := Heart ↦ Yes
    act2 : Heart_Rate := rate
    act3 : Heart_State := Heart ↦ KO
END
```

5.2.2 An Overview of Refinement

This section describes an overview of the progressive development of the ECG protocol by defining new properties and introducing new recommended clinical practices to identify the possible heart diseases. Note that all the refinement steps correspond to the standard analyses steps of the ECG protocol [24]. Due to limited space, we present only a summary of each refinement development to understand the overall development process. A detailed formal development of the ECG protocol is available in the technical report [29].

- **First Refinement (Intervals and blocks).** To classify different types of heart diseases, this refinement introduces a set of intervals and blocks. In particular, the PR interval and QRS interval are introduced to characterise the ECG signal. These intervals play an important role to assess the RBBB (Right Bundle Branch Block) and LBBB (Left Bundle Branch Block). In this development, we introduce new events for assessing the RBBB and LBBB through carefully analysing the QRS complex signal, and assessing the first degree AV block using the PR interval.

- **Second Refinement (Nonspecific intraventricular conduction delay and Wolff–Parkinson–White syndrome).** The second refinement step is used to introduce the clinical analysis steps for the nonspecific intraventricular conduction delay (IVCD) and Wolff–Parkinson–White (WPW) syndrome. The WPW syndrome may mimic as an inferior MI, which is further analysed in the next refinements. According to the standard clinical process if the WPW syndrome, RBBB or LBBB is not detected during the clinical process then it indicates the presence of the nonspecific intraventricular conduction delay (IVCD).

- **Third Refinement (ST-segment elevation or depression).** In this refinement, we introduce the ST-segment to analyse the ST-segments elevation or depression by defining the textual criteria, which is given in [24]. According to the clinical analysis step, it is necessary to assess the ST-segment before assessing the T-waves, QT-interval, electrical axis and hypertrophy because the diagnosis of acute MI or ischemia is vital and depends on the careful assessment of the ST-segment. A list of events is introduced to assess the ST-segment elevation, detection of troponin or CK-MB and acute inferior or anterior MI.

- **Fourth Refinement (Q-wave).** This refinement is used to introduce the Q-wave for assessing the ECG signal. The introduction of Q-wave allows us to characterise the different clinical conditions, such as normal Q-wave assessment, abnormal Q-wave assessment for inferior MI (IMI) and anterolateral MI (AMI). In addition, we also introduce the R-wave to analyse the normal and abnormal pathological conditions of the R-wave together with the Q-wave.

- **Fifth Refinement (P-wave).** This refinement introduces clinical assessment steps for analysing the P-wave to detect possible diseases due to an abnormality in the P-wave and atrial hypertrophy in ECG.

- **Sixth Refinement (Left and right ventricular hypertrophy).** In this refinement, we introduce the clinical step for assessing the Left Ventricular Hypertrophy (LVH) and Right Ventricular Hypertrophy (RVH). According to the clinical step, the LVH and RVH do not require to determine if any bundle branch block (RBBB or LBBB) is present. Thus, it is necessary to exclude the possible clinical assessment for the LBBB and RBBB, which are described in the refinement 2 and refinement 3.

- **Seventh Refinement (T-wave).** In this refinement, we introduce the T-wave to analyse the changing pattern of T-wave in ECG signal collected from 12-leads. The T-wave changes are usually nonspecific, but the T-wave inversion associated with other ST-segments indicates the myocardial ischemia, posterior MI, Hyperkalemia and pulmonary embolism.

- **Eighth Refinement (Electrical Axis).** During the clinical assessment of the ECG, the electrical axis plays an important role to determine the different and correct positions of leads for detecting a desired quality of the ECG signal. According to the ECG protocol, there are two main criteria. First, if the leads I and aVF are upright then the axis is normal. Second, the axis is perpendicular to the lead with the most equiphasic or smallest QRS deflection. The left-axis deviation and commonly associated left anterior fascicular block are always visible in the ECG.

- **Ninth Refinement (Miscellaneous conditions).** After several steps of clinical analysis, there are still several diseases which group together. To distinguish each disease at this level is very difficult due to the ambiguous nature of the clinical protocol and the associated properties. In this refinement level, we introduce the QT-interval and the required clinical steps for assessing the QT-interval. Moreover, this refinement also determines that if the electronic pacing is required using a pacemaker then there is no need to further assess the ECG signal. Otherwise, this refinement allows grouping of multiple miscellaneous conditions of the ECG for further clinical analysis.

- **Tenth Refinement (Arrhythmias).** This is the last refinement of the ECG interpretation protocol, in which we introduce different types of tachyarrhythmias. In particular, we introduce clinical steps for determining the narrow complex tachycardia and the wide complex tachycardia to assess the different kinds of heart diseases.

5.3   Model Annotation

In our selected case study, we have developed the domain model and the ECG protocol as the system model. Both these models are formalised in Event-B. The domain model is described according to the ontology descriptions (see Sect. 5.1) and the ECG protocol is developed using the refinement approach by describing the intended clinical steps and the required properties. The domain model and system model are linked together with the annotation mechanism (i.e. *see* context relationship). In fact, this step is performed by using the ECG domain model, defined in context, in the ECG interoperation protocol for describing the stepwise clinical protocol. This annotation mechanism provides a specific relationship between the ECG protocol and ECG ontology concepts. For example, in the abstract model of the medical protocol, a variable $Heart\_Rate$ is defined as the type of $HEART\_RATE$, which is described in the domain model of the heart. Similarly, the two clinical assessment properties ($CS1$ and $CS2$) are defined as the functions $ClinicalProp1$ and $ClinicalProp2$ in the context $C0$ to use in the process of assessment of the ECG protocol.

Figure 8 depicts annotation relations between the domain model and ECG protocol. Note that the different arrow lines are used to show the use of ontology concepts of the ECG domain knowledge in the ECG interpretation protocol. Each arrow is linked with a specific refinement level, which uses required domain knowledge defined in the ECG domain model. Note that this annotation mechanism allows us to link the domain model and the system model explicitly.

5.4   Model Verification

This section describes the proof statistics of the generated proof obligations of the developed ECG interpretation protocol using stepwise refinement by considering the domain knowledge in the form of ontology and ECG protocol as a system model. In this development, we use the Event-B modelling language, which supports the *consistency checking* and *refinement checking*. The *consistency checking* guarantees that all the events of the model preserve all the given invariants. The *refinement checking* allows checking the correct refinement relation between progressively developed models. Table 1 shows the proof statistics of the revisited formal development of the ECG protocol. To guarantee the correctness of the system behaviour, we provide a list of safety properties in the incremental refinements. This development results in 592 (100%) POs, in which 401 (68%) POs are proved automatically, and the remaining 191 (32%) POs are proved interactively using the Rodin prover and SMT solver. These interactive proof obligations are mainly related to the refinement and complex mathematical expressions, which are simplified through interaction, providing additional information for assisting the Rodin prover. Some of the proofs are quite simple that is achieved by simplifying the predicates. According to the Table 1 in this new development, the proof efforts have been decreased significantly

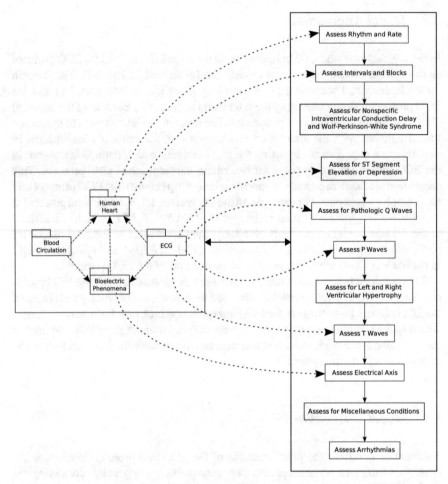

Fig. 8 Annotation between domain model (ECG) and system model (ECG interpretation protocol)

compared with the previous development of the ECG protocol [28, 42]. The proposed modelling approach restructures the system model by integrating the domain model explicitly. Note that the domain model has been developed progressively by analysing the domain-specific ontology and previously developed the system model. For instance, in the development of ECG interpretation protocol, we use the existing model and the ECG ontology to design the domain model. Note that several elements, such as constants, axioms, variables, invariants and functions, are removed/redefined in the system model and domain model. The modelling and integration of system model and domain model reduce the overall system complexity, proof efforts and improves the model consistency. For example, the clinical properties (CS1 and CS2) are defined once in the context model using the domain concepts. These properties have been used later in the ECG system model to define safety properties (see $saf\,2$

Table 1 Proof statistics

Model	Total number of POs	Automatic Proof	Interactive Proof
Abstract model	45	22(49%)	23(51%)
First refinement	55	38(70%)	17(30%)
Second refinement	43	33(77%)	10(23%)
Third refinement	50	39(78%)	11(22%)
Fourth refinement	57	36(63%)	21(37%)
Fifth refinement	39	29(75%)	10(25%)
Sixth refinement	36	24(67%)	12(33%)
Seventh refinement	128	77(60%)	51(40%)
Eighth refinement	57	35(62%)	22(38%)
Ninth refinement	15	12(80%)	3(20%)
Tenth refinement	67	56(84%)	11(16%)
Total	592	401(68%)	191(32%)

and $saf3$) and guards ($grd1$ in all three events). Once these properties are proved (CS1 and CS2) then these are used automatically in the process of proof automation for discharging the other POs. This indicates that the new development applying domain knowledge explicitly in the system development has improved the modelling processes and proof strategies.

5.5 Anomalies

In this work, we have discovered several anomalies in the ECG interpretation protocol that are categorised into three main groups: *ambiguity*, *inconsistency* and *incompleteness*. Ambiguity is a well known anomaly that can represent more than one possible meaning of a fact, which causes possible confusion in a decision. For example, in our work, we encountered a problem to determine whether the terms "ST-depression" and "ST-elevation" have the same meaning or not. Inconsistency anomaly always leads to a conflicting result or different decision for similar data/input. For instance, in our work we found an inconsistency in form of applicable conditions, which state that the given conditions are applicable for both "male" and "female" subjects, however, elsewhere in the protocol it is advised that the given conditions are not applicable for "female". Incompleteness anomaly is related to either missing piece of information or insufficient information in the original document. For instance, the original protocol contains "normal variant" factors to be considered for assessing T-wave. However, the meaning of "normal variant" is not defined in the protocol. Note that we have not listed all anomalies. In our work, we have identified these anomalies which may help for improving the quality of medical protocols.

6 Related Work

The use of ontology in software engineering for designing a complex system has great interest by several researchers to consider the domain knowledge explicitly. In [4, 5, 43], authors proposed a new approach for handling domain knowledge in design models. In this work, the domain models are developed using ontologies that can be used further during the system development applying the annotation mechanism. Hacid et al. [21, 22] have used the similar approach to develop a domain model based on ontologies for developing a system model using stepwise development. In [31], authors proposed a generic approach for integrating domain descriptions formalised by ontologies into an Event-B development process.

From the last decade, several pioneering works have been done to develop and analyse the medical guidelines and protocols based on expert's requirements. Protocol representation languages, such as Asbru [39, 45], EON [32], PROforma [16] and others [34, 47] are used to represent a formal semantics of guidelines and medical protocols. The main objective of all these languages is to provide some standardisation and to improve the clinical practices by modifying the existing or outdated medical protocols. A detailed survey of different techniques and tools related to the clinical guidelines is described in [23]. The simplification and verification of the clinical guidelines using decision-table are presented in [40, 41] to guarantee the properties of completeness and consistency.

Jonathan et al. [38] proposed interactive formal verification for finding a bug and to improve the quality of medical protocols or guidelines. Simon et al. [9] used the Asbru modelling language and temporal logic to represent the medical protocols and then model checking approach was used for checking consistency and error detection. A European project, Protocure [7], developed the techniques and tools for improving the medical protocols by identifying anomalies like ambiguity and incompleteness in medical guidelines and protocols by using formal methods. This project used the Asbru [45] modelling language for describing the medical protocols and KIV interactive theorem prover [8] was used for formal proof of the medical protocol. Méry et al. [28, 29, 42] proposed a new approach for developing a complex medical protocol using a *correct by construction* approach in Event-B. Note that this case study is revisited in this work by developing the domain model and system model separately, and then these models are linked through an annotation approach (*see* context relationship) according to the four steps modelling methodology [5].

7 Conclusion

We have presented an approach to the development of medical protocols or guidelines using a *correct by construction* method that explicitly represents domain knowledge. Considering domain knowledge in the system development can be an excellent way for determining the confidence we have that a system model is safe, secure and effective by respecting all the required domain properties, used physical units and possible relations. In this work, we have presented a stepwise formal development of the medical protocol. The development model contains two different models: domain model and system model. The domain model describes the domain concepts based on ontologies [15] and the system model describes the functional behaviour of the medical protocol. The domain knowledge has been described in Event-B context using ontology relations to capture the functional behaviour of the medical protocols. The medical protocol is developed as a system model to assess the clinical protocol. Note that both the domain model and system model are linked through annotations, in which the system model uses all axioms and theorems defined in the domain ontology model. The main objective of this work is to check the consistency of a clinical medical protocol using a refinement based development that integrates domain knowledge explicitly. Moreover, the same approach can be used for developing any other medical protocols. In this work, we have used an ECG medical protocol and conducted a systematical analysis to verify that the formalisation complies with certain medically relevant protocol properties. This approach allows us to identify possible anomalies and improve the quality of medical protocols.

References

1. http://aber-owl.net/ontology/ECG
2. https://bioportal.bioontology.org/ontologies/ECG
3. J.-R. Abrial, *Modeling in Event-B - System and Software Engineering* (Cambridge University Press, Cambridge, 2010)
4. Y. Ait-Ameur, J.P. Gibson, D. Méry, On implicit and explicit semantics: integration issues in proof-based development of systems (Springer, Berlin, 2014), pp. 604–618
5. Y. Ait-Ameur, D. Méry, Making explicit domain knowledge in formal system development. Sci. Comput. Program. **121**(C), 100–127 (2016)
6. M. Ashburner, C.A. Ball, J.A. Blake, D. Botstein, H. Butler, J.M. Cherry, A.P. Davis, K. Dolinski, S.S. Dwight, J.T. Eppig, M.A. Harris, D.P. Hill, L. Issel-Tarver, A. Kasarskis, S. Lewis, J.C. Matese, J.E. Richardson, M. Ringwald, G.M. Rubin, G. Sherlock, Gene ontology: tool for the unification of biology. Nat. Genet. **25**(1), 25–29, 5 (2000)
7. M. Balser, O. Coltell, J. Van Croonenborg, C. Duelli, F. Van Harmelen, A. Jovell, P. Lucas, M. Marcos, S. Miksch, W. Reif, K. Rosenbr, A. Seyfang, A.T. Teije, Protocure: supporting the development of medical protocols through formal methods, in *SCPG-04, Studies in Health Technology and Informatics*, vol. 101 (IOS Press, Amsterdam, 2004), pp. 103–108
8. M. Balser, W. Reif, G. Schellhorn, K. Stenzel, Kiv 3.0 for provably correct systems, in *Proceedings of the International Workshop on Current Trends in Applied Formal Method: Applied Formal Methods, FM-Trends 98*, London, UK (Springer, 1999), pp. 330–337
9. S. Bäumler, M. Balser, A. Dunets, W. Reif, J. Schmitt, Verification of medical guidelines by model checking - a case study, in *Model Checking Software*, ed. by A. Valmari. LNCS (Springer, Berlin, 2006), pp. 219–233
10. V.N. Batcharov, A.B. de Luna, M. Malik, The morphology of the Electrocardiogram, in *The ESC Textbook of Cardiovascular Medicine* (Blackwell Publishing Ltd., Oxford, 2006)
11. T. Bittner, M. Donnelly, Logical properties of foundational relations in bio-ontologies. Artif. Intell. Med. **39**(3), 197–216 (2007)
12. O. Bodenreider, The unified medical language system (UMLS): integrating biomedical terminology. Nucl. Acids Res. **32**(Database-Issue), 267–270 (2004)

13. O. Bodenreider, A. Burgun, Biomedical ontologies (Springer US, Boston, 2005), pp. 211–236
14. C. Doulaverakis, G. Nikolaidis, A. Kleontas, I. Kompatsiaris, Galenowl: ontology-based drug recommendations discovery. J. Biomed. Semant. **3**(1), 14 (2012)
15. D. Fensel, *Ontologies: A Silver Bullet for Knowledge Management and Electronic Commerce*, 2nd edn. (Springer-Verlag New York, Inc, Secaucus, 2003)
16. J. Fox, N. Johns, A. Rahmanzadeh, Disseminating medical knowledge: the proforma approach. Artif. Intell. Med. **14**(1–2), 157–182 (1998)
17. B. Gonçalves, G. Guizzardi, J.G. Pereira Filho, An electrocardiogram (ECG) domain ontology (2007)
18. B. Gonçalves, G. Guizzardi, J.G. Pereira Filho, Using an ECG reference ontology for semantic interoperability of ECG data. J. Biomed. Inform. **44**(1), 126–136 (2011)
19. T.R. Gruber, Toward principles for the design of ontologies used for knowledge sharing. Int. J. Hum.-Comput. Stud. **43**(5–6), 907–928 (1995)
20. N. Guarino, *Formal Ontology in Information Systems: Proceedings of the 1st International Conference, June 6–8, 1998, Trento, Italy*, 1st edn. (IOS Press, Amsterdam, 1998)
21. K. Hacid, Y. Ait-Ameur, Annotation of engineering models by references to domain ontologies (Springer International Publishing, Cham, 2016), pp. 234–244
22. K. Hacid, Y. Ait-Ameur, Strengthening MDE and formal design models by references to domain ontologies. A model annotation based approach (Springer International Publishing, Cham, 2016), pp. 340–357
23. D. Isern, A. Moreno, Computer-based execution of clinical guidelines: a review. Int. J. Med. Inform. **77**(12), 787–808 (2008)
24. M.G. Khan, *Rapid ECG Interpretation* (Humana Press, Totowa, 2008)
25. P. Lambrix, H. Tan, V. Jakoniene, L. Strömbäck, Biological ontologies (Springer US, Boston, 2007), pp. 85–99
26. D. Lee, R. Cornet, F. Lau, N. de Keizer, A survey of SNOMED CT implementations. J. Biomed. Inform. **46**(1), 87–96 (2013)
27. K.N. Lohr, M.J. Field, *Clinical Practice Guidelines: Directions for a New Program/Committee to Advise the Public Health Service on Clinical Practice Guidelines, United States and Institute of Medicine* (National Academy Press, Washington, 1990)
28. D. Méry, N.K. Singh, Medical protocol diagnosis using formal methods, in *Foundations of Health Informatics Engineering and Systems - First International Symposium, FHIES* (2011), pp. 1–20
29. D. Méry, N.K. Singh, Technical report on interpretation of the electrocardiogram (ECG) signal using formal methods. Technical Report (2011), https://hal.inria.fr/inria-00584177/file/TechRepoECG2011.pdf
30. G.A. Miller, Wordnet: a lexical database for english. Commun. ACM **38**(11), 39–41 (1995)
31. L. Mohand-Oussaid, I. Ait-Sadoune, Formal modelling of domain constraints in Event-B (Springer International Publishing, 2017)
32. M.A. Musen, S.W. Tu, A.K. Das, Y. Shahar, EON: a component-based approach to automation of protocol-directed therapy (1996)
33. N.F. Noy, D.L. McGuinness, Ontology development 101: a guide to creating your first ontology. Technical Report (Stanford University, 2001)
34. M. Peleg, S. Tu, J. Bury, P. Ciccarese, J. Fox, R.A. Greenes, S. Miksch, S. Quaglini, A. Seyfang, E.H. Shortliffe, M. Stefanelli et al., Comparing computer-interpretable guideline models: a case-study approach. J. Am. Med. Inform. Assoc. **10**, 2003 (2003)
35. Project RODIN, Rigorous open development environment for complex systems (2004), http://rodin-b-sharp.sourceforge.net/
36. C. Rosse, J.L.V. Mejino, The foundational model of anatomy ontology (Springer London, London, 2008), pp. 59–117
37. R.K. Saripalle, Current status of ontologies in biomedical and clinical informatics. Int. J. Sci. Inf. (2010)
38. J. Schmitt, A. Hoffmann, M. Balser, W. Reif, M. Marcos, Interactive verification of medical guidelines, in *FM 2006: Formal Methods*, ed. by J. Misra, T. Nipkow, E. Sekerinski. LNCS (Springer, Berlin, 2006), pp. 32–47

39. Y. Shahar, S. Miksch, P. Johnson, The ASGAARD project: a task-specific framework for the application and critiquing of time-oriented clinical guidelines. Artif. Intell. Med. 29–51 (1998)
40. R.N. Shiffman, Representation of clinical practice guidelines in conventional and augmented decision tables. J. Am. Med. Inform. Assoc. **4**(5), 382–393 (1997)
41. R.N. Shiffman, R.A. Greenes, Improving clinical guidelines with logic and decision-table techniques: application to hepatitis immunization recommendations. Med. Decis. Mak. **14**(3), 245–254 (1994)
42. N.K. Singh, *Using Event-B for Critical Device Software Systems* (Springer-Verlag GmbH, Berlin, 2013)
43. N.K. Singh, Y. Aït Ameur, D. Méry, Formal ontology driven model refactoring, in *23rd International Conference on Engineering of Complex Computer Systems, ICECCS 2018, Melbourne, Australia, December 12–14, 2018* (2018), pp. 136–145
44. Societe française cardiologie, J.-Y. Artigou, J.-J. Monsuez, *Cardiologie et maladies vasculaires* (Elsevier Masson, Paris, 2006)
45. A. Ten Teije, M. Marcos, M. Balser, J. van Croonenborg, C. Duelli, F. van Harmelen, P. Lucas, S. Miksch, W. Reif, K. Rosenbrand, A. Seyfang, Improving medical protocols by formal methods. Artif. Intell. Med. **36**(3), 193–209 (2006)
46. P.E. van der Vet, N.J.I. Mars, Bottom-up construction of ontologies. IEEE Trans. Knowl. Data Eng. **10**(4), 513–526 (1998)
47. D. Wang, M. Peleg, S.W. Tu, A.A. Boxwala, R.A. Greenes, V.L. Patel, E.H. Shortliffe, Representation primitives, process models and patient data in computer-interpretable clinical practice guidelines: a literature review of guideline representation models. Int. J. Med. Inform. **68**(1–3), 59–70 (2002)

Deriving Implicit Security Requirements in Safety-Explicit Formal Development of Control Systems

Inna Vistbakka and Elena Troubitsyna

Abstract Nowadays, safety-critical control systems are becoming increasingly open and interconnected. Therefore, while engineering a safety-critical system, we should guarantee that the system safety is not jeopardised by the security attacks. However, often the security requirements are not uncovered until the late design stages. Hence, there is a clear need for the modelling techniques that enable a formal reasoning about safety and security interdependencies at the early stages of the system development. In this work, we present a formal approach that allows the designers to uncover the implicit security requirements that are implied by the explicit system-level safety goals. We rely on modelling and refinement in Event-B to systematically uncover mutual interdependencies between safety and security and derive the constraints that should be imposed on the system to guarantee its safety in the presence of accidental and malicious faults.

1 Introduction

Nowadays, safety-critical control systems are becoming increasingly open and inter-connected. Increasing reliance on networking not only offers a variety of business and technological benefits but also introduces security threats. Exploiting security vulner-abilities might result in a loss of control and situation awareness directly threatening safety. Therefore, while engineering a safety-critical system, we should guarantee

I. Vistbakka (✉)
Åbo Akademi University, Turku, Finland
e-mail: inna.vistbakka@abo.fi

E. Troubitsyna
KTH, Stockholm, Sweden
e-mail: elenatro@kth.se

© Springer Nature Singapore Pte Ltd. 2021   109
Y. Ait-Ameur et al. (eds.), *Implicit and Explicit Semantics Integration*
in Proof-Based Developments of Discrete Systems,
https://doi.org/10.1007/978-981-15-5054-6_6

that the system is protected not only against the random faults but also the malicious ones, i.e. the security attacks.

However, often the security requirements are not uncovered until the late design stages. Moreover, safety and security goals might result in the orthogonal functional requirements that are hard to resolve at the implementation level. Hence, there is a clear need for the modelling techniques that enable a formal reasoning about safety and security interdependencies at the early stages of the system development. In this work, we present a formal approach that allows the designers to uncover the implicit security requirements that are implied by the explicit system-level safety goals.

We rely on modelling and refinement in Event-B [1] to systematically uncover mutual interdependencies between safety and security and derive the constraints that should be imposed on the system to guarantee its safety in the presence of accidental and malicious faults. Event-B is a rigorous approach to correct-by-construction system development by refinement. Development starts from an abstract specification that models the most essential system functionality and then the abstract model is transformed into a detailed specification. While refining the system model, we can explicitly represent both nominal and failure behaviour of the system components, as well as define the mechanisms for error detection and recovery. We can also explicitly represent the effect of security vulnerabilities such as tampering, spoofing and denial-of-service (DOS) attacks and analyse their impact on system safety.

In our formal development, we adopt the systems approach, i.e. specify controlling software together with the relevant behaviour of its environment—sensors, actuators and controlled process. The security failures are modelled by their effect on the system—altering or blocking messages sent over the communication channels. The proposed approach is illustrated by a case study—a battery charging system of an electric vehicle. We believe that the proposed approach facilitates an integration of the security consideration into the safety-driven design of control systems. It allows us to capture the dynamic nature of safety and security interplay, i.e. analyse the impact of deploying the security mechanisms on safety assurance and vice versa.

2   Systems View on Safety and Security Interdependencies

In his seminal work [14], Parnas has introduced the four-variable model as an abstract model of a computer-based control system. The four-variable model [14], shown in Fig. 1 defines the dependencies between the controlled physical process, input/output devices, and controlling software. The goal of the system is to control a certain physical process. The input device—a sensor—monitors the state of the physical process by measuring the value of the controlled parameter. Such a measurement is taken as an input by the controller. Upon receiving the corresponding input, the controller computes the output—the state of the actuator. The actuator affects the behaviour of the controlled physical process, i.e. it changes the value of the monitored variable to achieve the desired behaviour.

Fig. 1 The four-variable model

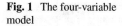

Fig. 2 Generic architecture of a control system

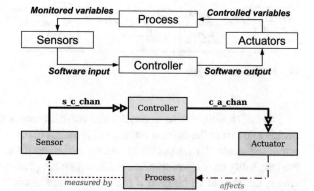

An important requirement, which is usually imposed on the majority of control systems, is to guarantee safety—a freedom of the accidents that might cause a loss of human lives or environmental damage [12]. The four-variable model allows us to derive the behaviour of controlling software that is acceptable from the safety point of view. Thus, we should ensure that under the given definitions of the physical process and behaviour of the input/output devices, the controlling software does not generate the output that puts the monitored variable into the hazardous state.

By applying the four-variable model, we derive two main types of requirements that should be implemented to guarantee safety. The first type is the fault tolerance requirements. Since both the input and output devices can be unreliable, to cope with their failures, either the system should contain redundancy or the controller should be able to put the system in a failsafe state. Moreover, the controller should cater to the imprecision of the sensor. The second type of the requirements is correctness. We should guarantee that the output, which the controller computes, preserves the safety boundaries of the monitored variable.

Nowadays safety-critical systems are increasingly relying on the networked technologies. In this work, we argue that the four-variable model should be extended to take into account the impact of malicious attacks to which the communication channels might be vulnerable. In Fig. 2, we present our proposal to extend the four-variable model to define a generic architecture of a networked control system.

We assume that the state of the controlled physical process is defined by the (physical) variable p_real. The value of p_real is measured by the sensor. The sensor can be a physical device, i.e. a hardware component that converts the physical value of p_real into its digital representation p_sen. However, it can be also a logical sensor— a module of a controlling program that computes an estimate of p_real based on some other measurements of the controlled process. In either case, the sensor corresponds to the input device in the Parnas's model.

In general, sensing is remote, i.e. the measured value p_sen is transmitted to the input of the controller. Since the transmission channel between the sensor and the controller s-c-$chan$ might be untrusted, i.e. it might be a subject of security attack, the value that is received by the controller—p_est might be different from p_sen.

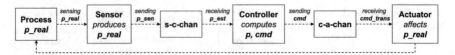

Fig. 3 Data control cycle

The controlling software checks the reasonableness of the received *p_est* and decide to use it as the current estimate of *p_real*, i.e. *p:=p_est* or ignore it. The value *p* that the controller adopts as its current estimate of the process state should pass the feasibility check, i.e. should coincide with the predicted value and the freshness check, i.e. should be ignored if the transmission channel is blocked due to a DOS attack. If the controller ignores the received value *p_est*, it uses the last good value of *p_est* and the maximal variation of the process dynamics to compute *p*.

The value of *p* is then used to calculate the next state of the actuator—the physical device that affects the controlled process, i.e. causes the changes in the value of *p_real*. The command from the controller to the actuator is transmitted over a network. In a similar way, the transmission channel from the controller to the actuator *c-a-chan* might be attacked. Hence, the command *cmd_trans* received by the actuator might be different from the command *cmd* computed by the controller.

The behaviour of the system is cyclic. At each cycle, the sensor measures *p_real*, produces *p_sen* and sends it over the *s-c-chan* channel to the controller. The controller receives *p_est*, checks it and computes *p* and the actuator command *cmd*. After that it sends *cmd* over the *c-a-chan*. The actuator receives the *cmd_trans* command and applies it, which should result in the desirable change of the process state. Figure 3 illustrates a generation and transmission of data within each cycle.

In this work, we focus on the failsafe systems, i.e. consider the control systems that can be put into a safe nonoperational state to preclude an occurrence of a safety failure [12]. Often system safety is defined over the parameters of the controlled physical process. For instance, in our generic control system, we can define safety as the following predicate

$$Safety = p\_real \leq safe\_threshold \lor failsafe{=}TRUE.$$

It means that the controlled process should be kept within the safety boundaries while the system is operational; otherwise, a safe shutdown should be executed.

Design of any system relies on certain assumptions and properties of the domain. In case of a safety-critical software-intensive control system, the aim of the design is to construct controlling software, which is under the given assumptions and properties guarantees safety, i.e. allows us to proof the following judgement:

$$(\textbf{\textit{ASM, DOM, SW}}) \vdash Safety,$$

where *ASM, DOM* and *SW* stand for assumptions, domain and controlling software properties, correspondingly. Below we define these three types of properties that suffice to proof *Safety* for our generic control system

ASM

A1. $p\_sen = p\_real \pm \Delta_1$
A2. $p = p\_sen \land \Delta_2 \land \Delta_2 = k\Delta_3$
A3. $(failsafe = FALSE \land cmd\_trans = cmd) \lor failsafe = TRUE$

DOM

D1. $cmd = incr \Rightarrow p\_real(t + 1) \geq p\_real(t)$ for any t, while the system is operational
D2. $cmd = decr \Rightarrow p\_real(t + 1) < p\_real(t)$ for any t, while the system is operational
D3. $max|(p\_real(t + 1) - p\_real(t))| = \Delta_3$
D4. $failsafe=TRUE \Rightarrow p\_real(t + 1) \leq p\_real(t)$ for any t, while the system is shutdown

SW

S1. $p\_est + \Sigma_{i=1}^{3}\Delta_i \geq safe\_threshold \land failsafe=FALSE \Rightarrow cmd = decr$

Straightforward logical calculations allow us to prove

$$(A1, ... , A3, D1, ... , D4, S1) \vdash Safety.$$

Now we discuss these assumptions and link them with safety and security requirements. The assumption *A1* means that the sensor measurements are sufficiently precise and unprecision is bounded. It implies a safety requirement: sensor should have high reliability. Here Δ_1 is the maximal imprecision value for the sensor.

The assumption *A2* states that the controller always adopts a measurement of the value of the process parameter that either coincides with p_sen, i.e. $k = 0$, or is calculated on the basis of the last good value and Δ_3—the maximal possible increase of the value p_real per cycle ($\Delta_2 = k\Delta_3$, where k is the number of cycles). This assumption implies both safety and security requirements. Firstly, we should guarantee that the channel *s-c-chan* is tamper resistant and the sensor is spoofing resistant. Secondly, we should ensure that the controlling software checks the validity of the input parameter and ignores it if the check fails. The assumption *A2* also implies that, in case of DOS attack on the channel *s-c-chan*, the system continues to function for some time by relying on the last good value. The assumption *A3* implies that if a failure or an attack on the channel *c-a-chan* is detected then the system is shut down. It means that the system should have some (possibly non-programmable) way to execute a shutdown in case the channel *c-a-chan* becomes unreliable.

The domain properties define certain axioms about the physical environment of the system and their interdependencies. The property *D1* states that an execution of the command *incr* results in the increase of the value p_real. The property *D2* is similar to *D1*. The domain property *D3* states that the maximal possible increase of p_real per cycle is known and bounded. *D4* stipulates that when the system is put in the failsafe state, the value of the physical parameter does not increase.

Finally, the software property *S1* corresponds to the safety invariant that controller should maintain: the controller issues the command *decr* to the actuator if at the next cycle the safe threshold can be exceeded.

Our system-level analysis has demonstrated that both safety and security aspects are critical for fulfilling the system-level goal of ensuring safety. Hence, both these aspects should be explicitly addressed during the system development. It is easy to observe, that we had to define a large number of requirements even for a generic high-level system architecture. To facilitate a systematic requirements derivation, we propose to employ formal development framework Event-B.

3   Modelling and Refinement in Event-B

Event-B [1] is a state-based framework that promotes the correct-by-construction approach to system development and formal verification by theorem proving. In Event-B, a system model is specified using the notion of an *abstract state machine* (ASM). ASM encapsulates the model state, represented as a collection of variables, and defines operations on the state, i.e. it describes the dynamic behaviour of a modelled system. A machine also has an accompanying component, called *context*, which includes user-defined sets, constants and their properties given as axioms.

The dynamic behaviour of the system is defined by a set of atomic *events*. Generally, an event has the following form:

$$e \mathrel{\widehat{=}} \textbf{any } a \textbf{ where } G_e \textbf{ then } R_e \textbf{ end},$$

where e is the event's name, a is the list of local variables, the *guard* G_e is a predicate over the local variables of the event and the state variables of the system. The body of an event is defined by a *multiple* (possibly nondeterministic) assignment over the system variables. The guard defines the conditions under which the event is *enabled*, i.e. its body can be executed. If several events are enabled at the same time, any of them can be chosen for execution nondeterministically.

Event-B employs a top-down refinement-based approach to system development. Development typically starts from an abstract specification that nondeterministically models the most essential functional requirements. In a sequence of refinement steps, we gradually reduce nondeterminism and introduce detailed design decisions. In particular, we can add new events, split events, as well as replace abstract variables by their concrete counterparts, i.e. perform *data refinement*. When data refinement is performed, we should define *gluing invariants* as a part of the invariants of the refined machine. They define the relationship between the abstract and concrete variables. The proof of data refinement is often supported by supplying *witnesses*—the concrete values for the replaced abstract variables and parameters. Witnesses are specified in the event clause **with**.

The consistency of Event-B models, i.e. verification of well-formedness, invariant preservation and correctness of refinement steps, is demonstrated by discharging a number of verification conditions—proof obligations. The Rodin platform [19] provides an automated support for formal modelling and verification in Event-B. It automatically generates the required proof obligations and attempts to discharge (prove) them automatically. It also provides a support for an interactive proving.

4 Generic Development of a Control System

In this section, we present a generic methodology for the refinement-based development of control systems that facilitates identifying implicit security requirements that should be fulfilled to satisfy the safety goals.

4.1 Abstract Specification: Overall System Behaviour

In the initial Event-B model, we introduce an abstract representation of the system architecture with the explicit definition of the communication channels as defined in Fig. 2. The abstract model (given in Fig. 4)—the machine ControlSystem_SS_m0—represents the overall behaviour of the system as an interleaving between the events modelling the phases of the *control cycle* defined in Sect. 2.

We define a variable *phase* $\in$ *PHASES* to designate the current stage of the control cycle. The set *PHASES={PROC, SEN, TO_CONTR, CONTR, TO_ACTUA, ACTUA}* contains the constants denoting the corresponding stages of the control cycle, while *phase* is used to enforce the predefined cyclic execution order of events

Process $\rightarrow$ Sensor $\rightarrow$ S_C_Chan $\rightarrow$ Controller $\rightarrow$ C_A_Chan $\rightarrow$ Actuator $\rightarrow$ Process $\rightarrow$ ...

Process event models the behaviour of the monitored physical process. Sensor event models the behaviour of the sensor, while Controller event specifies the behaviour of the controlling software. The events S_C_Chan and C_A_Chan model communication channels. Finally, Actuator event models the behaviour of the actuator.

In the initial model, we also abstractly specify an occurrence of faults. The event FailureDetection models a failure detection by nondeterministically assigning the variable *failure* the value *TRUE* or *FALSE*. In our further refinement steps, we will distinguish between the criticality of failures, and hence, will execute shutdown (enable the event FailSafe) less often.

```
Process ≙
    when phase=PROC then  phase:=SEN end          C_A_Chan ≙
Sensor ≙                                              when phase=TO_ACTUA ∧ failure=FALSE
    when phase=SEN ∧ failure=FALSE                    then  phase:=ACTUA end
    then  phase:=TO_CONTR end                     Actuator ≙
S_C_Chan ≙ ...                                        when phase=ACTUA ∧ failure=FALSE
    when phase=TO_CONTR ∧ failure=FALSE               then  phase:=PROC end
    then  phase:=CONTR end                        FailureDetection ≙ ...
Controller ≙                                      FailSafe ≙ ...
    when phase=CONTR ∧ failure=FALSE              end
    then  phase:=TO_ACTUA end
```

Fig. 4 Events of the machine ControlSystem_SS_m0

Let us note that in our initial specification, we have not yet formulated safety as a model invariant. Since the initial model defines only the control flow, we do not have the sufficiently detailed 'knowledge' to define and prove the desired safety property. This goal will be achieved via the refinement process.

4.2 First Refinement: Introducing Model Data

Our first refinement step aims at augmenting the abstract model with the explicit specification of the variables that are updated by the components and transmitted via the communication channels within the control cycle. At this refinement step, we also elaborate on the model of the controlled process, i.e. define the behaviour of the physical process characterised by the variable p_real. We also model the dependencies between the actuator state and the expected range of p_real value.

In the dynamic part of the Event-B model—the machine ControlSystem_SS_m1—we introduce four variables that explicitly represent the variable p_real and its 'perception' at each stage of the controlled process

- p_real—the current physical value defining the state of the process;
- p_sen—the value of the physical variable measured by the sensor. It can be affected by the sensor imprecision or failures;
- p_est—the value of the sensor measurement received by the controller as an input. It can be affected by the security attacks (hacking or spoofing);
- p—the value adopted by the controller as the current estimate of the process variable value.

Moreover, we also introduce a representation of the controller output—the command issued to the actuator—and corresponding command received by the actuator

- cmd—the controller output;
- cmd_trans—the command received by the actuator. It can be affected by security attacks (hacking or spoofing) on the transmission channel.

The type of the variables cmd and cmd_trans is defined as the enumerated set COMMANDS that contains the elements {INCR, DECR, ND, DOS}. The constants INCR and DECR stand for increasing and decreasing commands to the actuator, ND—for the initialisation command. The constant DOS is an abstract representation of a DOS attack or channel failure. It models the fact that the actuator does not receive any (fresh) command at the current control cycle.

Finally, we define a variable act_state representing the current state of the actuator. The actuator can be either switched off (consequently, $act\_state=OFF$) or switched on ($act\_state=ON$). For simplicity, we assume that when the actuator is switched on then the value of p_real is increasing and when it is switched off, it is decreasing correspondingly.

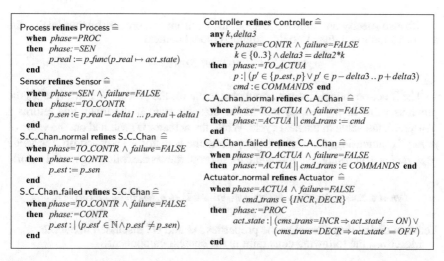

Fig. 5 Events of the machine ControlSystem_SS_m1

The event **Sensor** models the behaviour of the sensor by assigning the variable *p_sen* any value from the range *p_real - delta1 .. p_real + delta1*, where *delta1* is the maximal imprecision value for the sensor introduced as a constant in the model context. Hence, we model the sensor imprecision defined in our assumption *A1*. The event **SensorFailure** models the sensor failure. In case of failure, the sensor produces the reading that is out of the expected range.

The abstract event **S_C_Chan** models the transmission of the sensor reading (*p_sen* value) to the controller. Since the transmission channel *s-c-chan* might be a subject of a security attack, the input *p_est* received by the controller might differ from the *p_sen* value. We refine the event **S_C_Chan** by two concrete ones: modelling the normal and abnormal transmissions.

The event **Controller** is refined to abstractly specify the behaviour of the controller. We model the procedure of computing the current estimate *p*. The controller either accepts the current input or relies on the last good value, or calculates a new value *p* on the basis of the last good value and the maximal possible increase per cycle. The computed value of *p* is used to calculate the output—the next state of the actuator, i.e. update the variable *cmd*. Similar to the **S_C_Chan** event, we refine the abstract event **C_A_Chan** by two events modelling successful and failed command transmission from the controller to the actuator (Fig. 5).

Upon receiving the command from the controller, the actuator changes its state accordingly. This behaviour is modelled by the event **Actuator**. The actuator behaviour preserves the following invariants, postulating that the actuator is switched on when the increasing command is received, and vice versa

$$phase = PROC \land cmd\_trans = INCR \Rightarrow act\_state = ON,$$
$$phase = PROC \land cmd\_trans = DECR \Rightarrow act\_state = OFF.$$

We also specify our knowledge about the environment process by introducing the abstract function p_func into the extended model context:

$$p\_func \in \mathbb{N} \times ACTUATOR\_STATES \rightarrow \mathbb{N}.$$

The function models the next predicted value for the process p_real. It takes the previous value of p_real, as well as the actuator state as the input, and returns a new predicted value in the next cycle. While the actuator is switched on, the value of p_real is increasing. Correspondingly, while the actuator is switched off, the value of p_real is decreasing. We formulate these propertiesas the following model axiom:

$$\forall n \cdot n \in \mathbb{N} \Rightarrow p\_func(n \mapsto INCR) \geq n, \ \forall n \cdot n \in \mathbb{N} \Rightarrow p\_func(n \mapsto DECR) < n.$$

Let us note that they formalise the properties **D1** and **D2** discussed in Sect. 2.
Moreover, the following constraint in the context component

$$\forall n \cdot n \in 0 .. p\_max + delta1 \Rightarrow p\_func(n \mapsto INCR) \leq safe\_threshold.$$

requires that, if the process state is currently in the safe range $[0..p\_max + delta1]$, it cannot exceed the critical range within the next cycle, i.e. the safety gap between p_max and $safe\_threshold$ is sufficiently large.

4.3   Second Refinement: Specifying Controller Logic

Our second refinement step aims at introducing a detailed specification of the behaviour of the controlling software. Namely, we refine the abstract event Controller to represent its different alternatives that depend on the received input.

The output of the controller—the next state of the actuator—depends on the value of p adopted by the controller as the current estimate of the process state. Upon receiving the input—p_est—the controller checks its reasonableness. If the check is successful then p obtains the value of p_est. Then the controller proceeds by checks whether p exceeds p_max or is in the safe range $[0..p\_max]$. These alternatives are modelled by the events Controller_normal_DECR (see Fig. 6) and Controller_normal, correspondingly.

If the input does not pass the reasonableness check, the controller calculates the value of the process parameter using the last good input value and the maximal possible increase of the value p_real per cycle $delta2$. Then, the controller checks whether p exceeds p_max and computes the output. These alternatives are covered by the events Controller_retry_DECR and Controller_retry, correspondingly. Here the variable *retry* is introduced to model the number of retries (cycles) before the failure is considered to be a permanent and system is shutdown.

```
Controller_normal_DECR refines Controller ≙       Controller_retry_DECR refines Controller ≙
  where phase=CONTR ∧ failure=FALSE                  any p_new, delta3
        p_est = p_func(p ↦ act_state)                where phase=CONTR ∧ failure=FALSE ∧
        p_est > p_max                                      p_est ≠ p_func(p ↦ act_state) ∧
  with  k=0                                                retry <= 2 ∧ delta3 = (retry+1)delta2
        delta3=0                                           p_new ∈ p-delta3...p+delta3 ∧ p_est > p_max
  then  phase:=TO_ACTUA                              with  k=retry+1
        p := p_est                                   then  phase:=TO_ACTUA
        cmd := DECR                                        p := p_new
        retry := 0                                         cmd := DECR
  end                                                      retry := retry+1
                                                     end
```

Fig. 6 Some events of the machine ControlSystem_SS_m2

The behaviour of the controller preserves the following invariants:

$$phase = TO\_ACTUA \land p > p\_max \Rightarrow cmd=DECR,$$
$$phase = TO\_ACTUA \land p \in 0..p\_max \Rightarrow cmd=INCR \lor cmd=DECR.$$

They postulate that the controller issues the command *DECR* if the parameter *p* is approaching the critically high value (*safe_threshold*). If the controlled parameter is within the safety region then the controller output might be either *DECR* or *INCR*.

4.4   Third Refinement: Attack Modelling

In a networked control system, the communication channels are used to transmit the sensor data to the controller, as well as the commands issued by the controller to the actuator. However, such channels could be possibly vulnerable to the security attacks. The goal of our third refinement step is to introduce into the Event-B model an abstract representation of the attacks and the system reaction on them.

To achieve this, we add several new events and variables into the refined system specification as shown in Fig. 7. Firstly, we introduce a new event Attack_S_C_Chan to model a possible attack on the channel *s-c-chan*. The attack can happen anytime while transmitting the sensed data to the controller. The variable *attack_s_c* ∈ *BOOL* indicates whether the system is under attack. If the event Attack_S_C_Chan is triggered, the value of *attack_s_c* becomes *TRUE*, otherwise it equals to *FALSE*. Let us note that the event Attack_S_C_Chan is merely an abstraction introduced to represent the results of the security monitoring. In general, a security monitoring relies on the anomaly detection including checks of well-formedness of data packets, deviations in response time or periodicity, etc. (An implementation of the security monitoring mechanisms is out of the scope of this paper.) Similarly to modelling the attack on the channel *s-c-chan*, we introduce the event Attack_S_A_Chan to model a possible attack on the channel *c-a-chan*.

Events ...

Attack_S_C_Chan $\widehat{=}$
  when phase=TO_CONTR ∧ attack_s_c=FALSE
  then attack_s_c:=TRUE **end**
S_C_Chan_normal **refines** S_C_Chan_normal $\widehat{=}$
  when ... ∧ attack_s_c=FALSE
  ...
Attack_C_A_Chan $\widehat{=}$
  when phase=TO_ACTUA ∧ attack_c_a=FALSE
  then attack_c_a:=TRUE **end**

C_A_Chan_failure ...
  when phase=TO_ACTUA ∧ ... ∧ attack_c_a=TRUE
  then phase:=ACTUA
      cmd := DOS
end
C_A_Chan_AttackDetection **refines** FailureDetection
  when ... ∧ failure=FALSE ∧ cmd=DOS
  then failure:=TRUE **end**
...
end

Fig. 7 Some events of the machine ControlSystem_SS_m3

We define the additional guards in the events S_C_Chan_normal and C_A_Chan_normal to ensure that the events are enabled only if no attacks have been detected, i.e. *attack_s_c*=FALSE or *attack_c_a*=FALSE, correspondingly.

According to the assumption *A2*, in case of an attack on the channel *s-c-chan*, the system continues to function for some time: the controller relies on the last good input received. If an attack on the channel *c-a-chan* has occurred then the controller output would differ from the command received by the actuator. In case of the DOS attack (or in general a channel unavailability), the actuator would not receive any command at all. For simplicity, we model it by assigning the constant DOS to the *cmd_trans* variable. (In practice, the DOS attack detection is implemented by checking the timestamp of the received packet.) Safety cannot be ensured if an attack on the channel *c-a-chan* is detected and hence the system should be shut down.

We formulate the following properties as the model invariants and prove them

$$attack\_s\_c=FALSE \land phase=CONTR \Rightarrow p\_est=p\_sen,$$
$$attack\_c\_a=FALSE \land phase=ACTUA \Rightarrow cmd=cmd\_trans.$$

The properties describe the effect of the attacks on the controller input and output.

As a result of this refinement step, we arrive at a sufficiently detailed specification to define and prove the following safety invariant: $p\_real \in 0 .. safe\_threshold$.

Next, we will demonstrate how the proposed generic refinement process can be applied to develop a case study—a battery charging system.

5   Case Study: The Battery Charging System

In this section, we present our use case—a battery charging system and in the next section, we will demonstrate how to develop a detailed specification of this system by refinement and uncover the mutual interdependencies between safety and security requirements through the process of formal development. In our development we will rely on the generic development presented in Sect. 4.

5.1 Case Study Description

Our case study is a battery charging system (BCS) of an electric car. Charging of the car battery is initiated when the vehicle gets connected to an external charging unit [20]. Figure 8 shows the main components of the system: the battery module, the battery management system, the CAN bus, the charging station (with the associated charging interface and the external charging unit). When the charging station detects that an electrical vehicle got connected to its external charging unit, it starts the charging procedure. While charging, the battery management system (BMS)—the controlling software of the system, monitors the measurements received from the battery and issues the signal to the charging station to continue or stop charging.

The communication between the BMS and the charging station goes through the CAN bus. The system behaviour is cyclic: at each cycle, the charging station receives the command from BMS to continue or stop charging. Correspondingly, it either continues or stops to supply the energy to the battery of the car.

The main hazard associated with the system is overcharging of the car's battery, which might result in an explosion. Therefore, the safety goal of the system is to avoid overcharging. In case the system cannot reliably assess the current battery charge or stop charging using the programmable means, a safe shutdown should be executed. Hence, the system architecture should have a reliable mechanism for controlling the charging procedure and, in case of hazardous deviations, be able to abort charging, i.e. make a transition to the failsafe state.

The top-level safety goal of BCS is to keep a battery level parameter within the predefined boundaries. Let bl_real correspond to the real physical value of such a parameter. The system safety property can be formulated as

$$0 \leq bl\_real \leq bl\_max\_crit,$$

where 0 and bl_max_crit denote the lowest and highest boundaries correspondingly. The safety goal is achieved by changing the state of the charging unit that supplies an electricity to the battery.

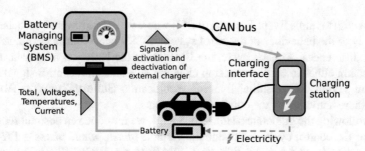

Fig. 8 Architecture of battery charging system

BCS is a typical example of a *control system* discussed in Sect. 2. Indeed, the BMS acts as a *controller*, the charging station (with its associated charger unit)—as an *actuator* and the battery unit as the *process* that the system controls. The battery level parameter can be directly measured by the *sensor* of BMS or computed on the basis of the alternative measurements obtained from the battery. At each cycle, BMS assesses the battery level and sends the corresponding control command.

The charging station and in-car CAN bus are linked by the corresponding communication channel that could be possibly vulnerable to the security attacks. In particular, the attacker can use the in-car charging interface as an entry point by compromising the external charger interface or tampering with the communication between the interfaces to inject a malicious content into the CAN bus. Therefore, while reasoning about the behaviour of such a system, we should also reason about the impact of security threats on its safety. The analysis presented in Sect. 2 shows that safety cannot be guaranteed when the controller-actuator channel is attacked. Therefore, the BCS should include an additional hardware component that should be installed in the car to break the charging circuit if the battery charge level becomes dangerously high. Such a non-programmable switch can override the commands from the controller and put the system in the failsafe state to guarantee safety.

Next, we present an abstract Event-B specification of BCS.

6   Event-B Development of the Battery Charging System

In this section, we will present a formal Event-B development of the described BCS. Our development would rely on the generic development presented in Sect. 4. Since the developed models are fairly large, we only highlight the most important modelling solutions for the development.

6.1   Initial Specification

We start with a simple Event-B model of BCS where we define its essential behaviour. Similarly to the behaviour of a control system, BCS's behaviour is cyclic, yet with several differences. At each cycle, the BMS reads the battery sensor data, makes the decision either to continue or stop charging and sends the control signal to the charging station. In the initial Event-B specification BatteryCharging_Abs, we model these activities by the corresponding events.

By following the guidelines defined in Sect. 4, we introduce an abstract representation of the control cycle and define the variable *phase*, where *phase* $\in$ *PHASES*. Here the set *PHASES* = {BAT, EST, BMS, TRANSM, CHARGST, FIN}. The variable *phase* is used to enforce the predefined cyclic execution of the data flow

Battery $\rightarrow$ BMS_estim $\rightarrow$ BMS_act $\rightarrow$ CAN_bus $\rightarrow$ ChargStation $\rightarrow$ Battery $\rightarrow$ ...

Battery event models the changes of the battery parameter *bl_real* while charging. **BMS_estim** event models the BMS estimation of this parameter (that is defined by *bl* variable). **BMS_act** event specifies the BMS actions (i.e. sending the signal to continue or stop charging) and **CAN_bus** event models transmission of the corresponding command to the charging station. Finally, **ChargStation** event models the required actions from the charging station upon receiving the signal from BMS.

In addition to modelling the control cycle, we also define the event **Connect** that represents the beginning of the charging procedure (i.e. when a vehicle connects to the charging station) and the event **ChargingComplete** representing its completion.

6.2 The First Refinement

In our first refinement step, we focus on modelling properties related to the safety. In particular, we define the safety restrictions imposed on the charging procedure. The procedure begins when a cable is connected. Only after it has been started, the cable is electrified. If the cable is disconnected while charging, the vehicle system and charging station will detect the change and discontinue the power supply to the cable at their respective ends.

We introduce a variable *cable_connected* $\in BOOL$, which has the value $FALSE$ when the cable is connected, and $TRUE$ otherwise. The variable *cable_connected* is updated in the events **Connect** and **ChargingComplete** modelling the beginning and the progress of the charging procedure, respectively. Moreover, to model the state of the energy supplying equipment, we define a variable *cable_electrified* $\in$ $EQUIPMENT\_STATE$, where $EQUIPMENT\_STATE = \{OFF, ON\}$.

The status of charging is modelled by the variable *status*. It can obtain any of three possible values from the set $STATUSES=\{IDLE, CHARGING, CHARGED\}$. When the external charger unit is not connected to the vehicle, *status* has the value *IDLE*. The variable *status* obtains the value *CHARGING* if charging is in progress and the value *CHARGED* when charging has been recently stopped.

The interdependencies between the variables are defined as following invariants (Fig. 9):

$$cable\_connected=FALSE \Rightarrow status=IDLE, \quad status=CHARGED \Rightarrow cable\_electrified=OFF,$$
$$cable\_electrified=ON \wedge failure=FALSE \Rightarrow cable\_connected=TRUE, ...$$

The abstract event **ChargingStation** is refined by three events: **ChargingStation_init, ChargingStation_cont** and **ChargingStation_stop**. They model the beginning of charging (i.e. supplying energy), its continuation and completion, respectively. If the cable is disconnected while charging, to detect this change and stop the power supply to the cable, we define the event **UnsafeUnconnect** event.

Connect $\hat{=}$ **refines** Connect	ChargingComplete $\hat{=}$ **refines** ChargingComplete
where *phase=INIT* $\wedge$ *failure = FALSE* $\wedge$	**where** *phase=FIN* $\wedge$ *failure = FALSE* $\wedge$
cable_connected = FALSE	*status=CHARGED* $\wedge$ *cable_electrified=OFF*
then *phase := BAT*	**then** *status := IDLE*
cable_connected := TRUE	*cable_connected := FALSE*
status := CHARGING	*phase := INIT*
end	**end**

Fig. 9 Some events of the machine BatteryCharging_Ref1

6.3   The Second Refinement

Our second refinement step aims at introducing a detailed specification of the BMS behaviour. We define the control algorithm. The controller calculates the commands to be sent to the charging station using the current estimate of the battery level. At this refinement step, we also elaborate on the dynamics of the controlled process, i.e. define the changes in the real battery level *bl_real* and model different cases of the charging station behaviour. Let us note that this refinement step combines two corresponding refinement steps of our generic development described in Sect. 4.

At each control cycle, the controller receives the current estimate of the battery level from the sensor. The controller checks whether the battery is still not fully charged and it is safe to continue to charge it or charging should be stopped. The decision to continue to charge can be made only if the controller verifies that the battery level at the end of the next cycle will still be in the safe range $[0 \dots bl\_max\_crit]$.

At this development step, we refine the event **BMS_estim** modelling the estimation of the current value of battery parameter by BMS. Consequently, the variable *bl* gets any value from the range *bl_real - bl_delta .. bl_real + bl_delta*, where *bl_delta* is the maximal imprecision value for the battery sensor.

We also specify our knowledge about the process of the battery charging by introducing the following abstract function into the model context: $bl\_fnc \in \mathbb{N} \rightarrow \mathbb{N}$. The function models the next predicted value for the battery level parameter bl_real. It takes the previous value of bl_real and returns its predicted value in the next cycle. Obviously, while the battery is charging, its battery level is increasing, as rendered by the following axiom defined in the context **BatteryCharging_c1**: $\forall n \cdot n \in \mathbb{N} \Rightarrow n < bl\_fnc(n)$. Moreover, the following constraint in the context

$$\forall n \cdot n \in 0 \dots bl\_max + bl\_delta \Rightarrow bl\_fnc(n) \leq bl\_max\_crit$$

defines that, if the battery level is currently in the safe range it cannot exceed the critical range within the next cycle, i.e. the safety gap between *bl_max* and *bl_max_crit* is sufficiently large.

BMS_continue ≙ refines BMS_control
where *phase=BMS ∧ failsafe=FALSE ∧ status=CHARGING ∧*
         bl < bl_max ∧ sensor_reading=OK
then *phase := TRANSM*
         signal:=SUPPLY
end
	CAN_bus ≙ refines CAN_bus
	where *phase=TRANSM ∧ failure=FALSE ∧*
	         status=CHARGING
	then *phase := CHARGST*
	         bus_out:=signal
	end

Fig. 10 Some events of the machine BatteryCharging_Ref2

We also refine the abstract event **BMS_control** to represent the reaction of the controller (BMS) on different values of the input: stopping the charge if the monitored parameter exceeding *bl_max* and continuing it if the monitored parameter is in the safe range *[0..bl_max)*. Note that the monitored value *bl* that BMS relies on here is different from the actual value of the physical process (*bl_real*) updated by the event **Battery**.

The variable *signal* is used to model the control commands issued by the BMS to continue or abort the charging. The abstract constants *SUPPLY* and *STOP* correspond to the external charger being switched on and off. Upon receiving the command, the charging station activates or deactivates the external charger unit (Fig. 10).

The BMS behaviour preserves the following invariants:

$$phase = TRANSM \land bl \geq bl\_max \Rightarrow signal=STOP$$
$$phase = TRANSM \land bl < bl\_max \Rightarrow signal=SUPPLY$$

Indeed, BMS issues the signal to stop charging when the parameter *bl* is approaching the critically high value (*bl_max_crit*), and continue otherwise. To give the system a time to react, BMS sends the stopping command to the charging station whenever the estimated value *bl* reaches the predefined value *bl_max*.

In this refinement step, we have elaborated on the control algorithm and the model of the controlled physical process. However, we have abstracted away from modelling the fact that the charging station reads the signal from the CAN bus. Such an abstraction allows us to further refine the communication model and explicitly define the impact of the security attacks on the system behaviour.

6.4   The Third Refinement

In the architecture of BCS, the CAN bus represents the communication channel in the in-car system. This component is used to transmit the signal issued by the BMS to the charging station. However, such a channel could be possibly vulnerable to the security attacks. The attacker can use the in-car charging interface as an entry point by compromising the external charger interface or tampering with the communications between the interfaces to inject malicious content into the CAN bus. Therefore, the goal of our third refinement step is to incorporate into the model architecture a certain mechanism that would allow the system to transmit the signal in a secure way. The

Invariants ... $(phase = CHARG \wedge bl \geq bl\_max \Rightarrow bus\_out=STOP) \wedge$ ...

$(attack = FALSE \wedge phase = CHARG \wedge bus\_out=STOP \Rightarrow signal=STOP) \wedge$ ...

Events   ChargingStation_stop $\widehat{=}$ **refines** ChargingStation_stop

...   **where** $phase=CHARG \wedge charg\_in = STOP \wedge$

SecurityGateway_no_attack $failure = FALSE$

  where $phase=CHARGST \wedge failure=FALSE \wedge$ **with** $sg=STOP$

    $charg\_in=S0 \wedge attack=FALSE \wedge gateway=FALSE$ **then** $status := CHARGED \parallel phase := FIN$

  then $charg\_in := bus\_out \parallel gateway = TRUE$ $charg\_in := S0 \parallel cable\_electrified := OFF$

end   **end**

Fig. 11 Some events of the machine BatteryCharging_Ref3

possible solution here is to add a new component—security gateway—between the CAN bus and the external charging unit. In general, such a security gateway could control the network access according to predefined security policies and can also inspect the packet content to detect intruder attacks and anomalies.

Some events of the refined model obtained by following the corresponding step of our generic development are presented in Fig. 11. We introduce a new event **Attack** to model a possible attack on the system. The attack can happen anytime while transmitting the signal to the charging interface.

Secondly, we introduce two events SecurityGateway_no_attack and SecurityGateway_attack and a new variable $charg\_in$ that specifies the input buffer of the charging interface. It might obtain the values from the set of possible signals, i.e. $charg\_in \in SIGNALS$. If no attack occurred then the transmission results in copying the signal from the output buffer of the CAN bus (bus_out variable) to the input buffer $charg\_in$ of the charging interface. If a security failure occurred (e.g. the system has been under attack) then output signal would differ from the sent signal. For the sake of simplicity, we consider here a DOS attack. Therefore, the input buffer of the charging interface will get DOS value. This behaviour is modelled by the events SecurityGateway_no_attack and SecurityGateway_attack presented in Fig. 11. In case of the DOS attack, the system will detect this (modelled by the event **AttackDetection**) and make a transition to a failsafe state.

However, while adding security protection to the system architecture, a security gateway might introduce latency into communication between the CAN bus and the charging station, and, in turn, increase the reaction time of charging unit. Thus, a careful analysis should be performed while choosing a suitable value bl_max to ensure the following constraint:

$$bl\_max + bl\_delta + max\_increase \leq bl\_max\_crit,$$

where $max\_increase$ is the maximal increase of the better level value peer cycle.

The specification obtained at this step is sufficiently detailed and allows us to prove the desired safety property as a model invariant: $bl\_real \in 0 .. bl\_max\_crit$.

6.5 Discussion

In the case study, we have followed the generic development process described in Sect. 4. We had to introduce some adjustments while specifying BCS to cater to the behaviour pertaining to this particular system. However, the deviations from the generic development have been minor and hence, we believe that the proposed generic development pattern have been successfully validated. In our future work, we are planning to elaborate on our approach and make it still more generic. In particular, we are planning to formulate the proposed design solutions as separate Event-B patterns. In the context of formal development in Event-B, patterns represent generic modelling solutions that can be reused in similar developments via instantiation.

There are several observations, which we made as a result of the case study. In our modelling, we have adopted an implicit discrete model of time. Namely, we define the abstract function representing the change in the dynamics of the controlled process, as well as the constraints relating the components behaviour in the successive iterations. Such an approach is based on our previous experience in modelling control systems [10, 11]. To enable verification of real-time properties, we can rely on the approach proposed by Iliasov et al. [5] allowing to map Event-B specification into UPPAAL. An alternative approach would be to support the explicit reasoning about the continuous system behaviour, e.g. as proposed by Babin et al. [2].

To support reasoning about safety-security interplay, we have to explicitly model the impact of accidental and malicious faults on the system behaviour, i.e. introduce in our specification an explicit representation of failure modes. As a result, the complexity of the specification can significantly increase. To address this issue, we can rely on the modularisation approach [8], which supports compositional reasoning and specification patterns [7]. Moreover, we can also extend the work proposed to integrate the results of FMEA into the formal models [18].

Our approach has already demonstrated a good scalability and facilitates the development of complex safety-critical control systems. However, to cope with the complexity of a formal specification, which explicitly integrates the failure behaviour, we can employ such an architectural mechanism as the mode-based reasoning, as proposed, e.g. in [6]. We can distinguish between the normal operational mode, the degraded mode caused by the accidental component failures, as well as the attacked and failsafe modes. By defining and verifying such a high-level mode logic, we can facilitate a structured analysis of the complex failure behaviour.

7 Related Work and Conclusions

Research investigating safety and security interaction has received a significant attention. It has been recognised that there is a need for the approaches facilitating an integrated analysis of safety and security [16, 27–30]. This problem has been addressed by several techniques demonstrating how to adapt conventional techniques for analysing safety risks (e.g. FMECA, fault trees) to perform a security-informed safety analysis [4, 21]. The techniques aim at providing the engineers with a structured way to discover and analyse security vulnerabilities that have safety implications. Since the use of such techniques facilitates a systematic analysis of failure modes and results in discovering important safety and security requirements, the proposed approaches can provide a valuable input for our modelling.

There are several works that address the formal analysis of safety/security requirements interactions [9, 17]. Majority of these works demonstrate also how to find conflicts between them. A typical scenario used to demonstrate this is a contradiction between the access control rules and safety measure. In our approach, we treat the problem of safety-security interplay at a more detailed level, i.e. we analyse the system architecture, investigate the impact of security failures on safe implementation of system functions. Such an approach allows us to analyse the dynamic nature of safety-security interactions. The work [17] presents ongoing work on a method to co-engineering of security and safety requirements. Specifically, the paper illustrates how Goal-Oriented Requirements Engineering can support co-engineering to address the safety and security dimensions in cyber-physical systems.

The distributed MILS approach [3] employs a number of advanced modelling techniques to create a platform for a formal architectural analysis of safety and security. The approach supports a powerful analysis of the properties of the data flow using model checking and facilitates derivation of security contracts. Since our approach enables incremental construction of complex distributed architectures, it would be interesting to combine these techniques to support an integrated safety-security analysis throughout the entire formal model-based system development.

An important aspect of demonstrating system safety is its quantitative evaluation. The foundations of the quantitative probabilistic reasoning about safety using formal specifications was established in [13, 22, 26]. This work has been further extended to enable probabilistic assessment of safety and reliability using Event-B specifications [24]. It would be interesting to quantitatively assess the impact of accidental and malicious faults on safety.

The work proposed by Parnas [15] is also based on the four-variable model proposed by Parnas. The authors show how this model is used in the development of safety-critical systems in industry and helps to clarify the behaviours of, and the boundaries between, the environment, sensors, actuators and software. Similarly, in our work four-variable model allows us to derive the behaviour of controlling software that is acceptable from the safety point of view. However, we further employ formal modelling technique to uncover mutual interdependencies between safety and security.

In our work, we have assumed that the hazards associated with the system has been already identified and correspondingly, focused on modelling system behaviour guaranteeing hazard avoidance. Our work can be complemented with the approaches proposed in [23, 25], which address hazard identification and elicitation of safety requirements.

In this work, we have proposed a formal approach enabling derivation of implicit security requirements from system safety goals. The proposed approach allows us in a systematic disciplined manner to derive the constraints that should be imposed on the system to guarantee its safety even in the presence of the security attacks. Our approach has relied on modelling and refinement in Event-B. While specifying the system, we have followed the systems approach, i.e. modelled the controlling software together with its environment. Such an approach has allowed us to systematically derive the constraints that should be imposed on components, communication channels and software to guarantee safety in the presence of accidental (due to the component failures) and security failures. A distinctive feature of our approach is a support for the integrated consideration of safety and security.

The approach presented in this work generalises the results of our experience with formal refinement-based development in the Event-B conducted in the context of verification of safety-critical control system. The results have demonstrated that the formal development significantly facilitates derivation of safety and security requirements. We have also observed that the integrated safety-security modelling in Event-B could be facilitated by the use of external tools supporting constraint solving and continuous behaviour simulation. Such an integration would be interesting to investigate in our future work.

References

1. J.R. Abrial, *Modeling in Event-B* (Cambridge University Press, Cambridge, 2010)
2. G. Babi, Y.A. Ameur, N.K. Singh, M. Pantel, A system substitution mechanism for hybrid systems in event-B. ICFEM **2016**, 106–121 (2016)
3. A. Cimatti, R. DeLong, D. Marcantonio, S. Tonetta, Combining MILS with contract-based design for safety and security requirements, in *SAFECOMP 2015 Workshops*. LNCS, vol. 9338 (Springer, Berlin, 2015), pp. 264–276
4. I.N. Fovino, M. Masera, A.D. Cian, Integrating cyber attacks within fault trees. Rel. Eng. Sys. Safety **94**(9), 1394–1402 (2009)
5. A. Iliasov, A. Romanovsky, L. Laibinis, E. Troubitsyna, T. Latvala, Augmenting event-B modelling with real-time verification, in *FormSERA 2012* (IEEE, Piscataway, 2012), pp. 51–57
6. A. Iliasov, E. Troubitsyna, L. Laibinis, I. Romanovsky, K. Varpaaniemi, D. Ilic, T. Latvala, Developing mode-rich satellite software by refinement in event-B. Sci. Comput. Program. **78**(7), 884–905 (2013)
7. A. Iliasov, E. Troubitsyna, L. Laibinis, A. Romanovsky, Patterns for refinement automation, in *FMCO 2009*. LNCS, vol. 6286 (Springer, Berlin, 2010), pp. 70–88
8. A. Iliasov, E. Troubitsyna, L. Laibinis, A. Romanovsky, K. Varpaaniemi, D. Ilic, T. Latvala, Supporting reuse in event-B development: Modularisation approach, in *ABZ 2010* (Springer, Berlin, 2010), pp. 174–188
9. S. Kriaa, M. Bouissou, F. Colin, Y. Halgand, L. Piètre-Cambacédès, Safety and security interactions modeling using the BDMP formalism: case study of a PipeliLeve, in *SAFECOMP 2014*. LNCS, vol. 8666 (Springer, Berlin, 2014), pp. 326–341
10. L. Laibinis, E. Troubitsyna, Fault tolerance in a layered architecture: A general specification pattern in B, in *SEFM 2004* (IEEE Computer Society, Washington, D.C., 2004), pp. 346–355
11. L. Laibinis, E. Troubitsyna, Refinement of fault tolerant control systems in B, in *Proceedings of the SAFECOMP 2004*, vol. 3219 (Springer, Berlin, 2004), pp. 254–268
12. N.G. Leveson, *Safeware: System Safety and Computers* (Addison-Wesley, Boston, 1995)

13. A. McIver, C. Morgan, E. Troubitsyna, The probabilistic steam boiler: A case study in prob-abilistic data refinement, in *Proceedings of the International Refinement Workshop* (Springer, Berlin, 1998), pp. 250–265
14. D.L. Parnas, J. Madey, Functional documents for computer systems. Sci. Comput. Program. **25**, 41–61 (1995)
15. L.M. Patcas, M. Lawford, T. Maibaum, Implementability of requirements in the four-variable model. Sci. Comput. Program. **111**, 339–362 (2015)
16. S. Paul, L. Rioux, Over 20 years of research into cybersecurity and safety engineering: A short bibliography, in *Safety and Security Engineering VI* (WIT Press, Southampton, 2015), p. 335
17. C. Ponsard, G. Dallons, P. Massone, Goal-oriented co-engineering of security and safety requirements in cyber-physical systems, in *SAFECOMP 2016 Workshops DECS* (Springer International Publishing, Berlin, 2016), pp. 334–345
18. Y. Prokhorova, L. Laibinis, E. Troubitsyna, K. Varpaaniemi, T. Latvala, Derivation and for-mal verification of a mode logic for layered control systems, in *18th Asia Pacific Software Engineering Conference, APSEC 2011* (IEEE Computer Society, Washington, D.C., 2011), pp. 49–56
19. Rodin: Event-B Platform. http://www.event-b.org/
20. C. Schmittner, Z. Ma, P. Puschner, Limitation and improvement of STPA-Sec for safety and security co-analysis, in *SAFECOMP Workshops 2016*. LNCS, vol. 9923 (Springer, Berlin, 2016), pp. 195–209
21. C. Schmittner, Z. Ma, P. Smith, FMVEA for safety and security analysis of intelligent and cooperative vehicles, in *SAFECOMP Workshops 2014*. LNCS, vol. 8696 (Springer, Berlin, 2014), pp. 282–288
22. K. Sere, E. Troubitsyna, Probabilities in action systems, in *Proceedings of the 8th Nordic Workshop on Programming Theory*, pp. 373–387 (1996)
23. K. Sere, E. Troubitsyna, Safety analysis in formal specification, in *FM'99 - Proceedings, Volume II*. LNCS, vol. 1709 (Springer, Berlin, 1999), pp. 1564–1583
24. A. Tarasyuk, E. Troubitsyna, L. Laibinis, Integrating stochastic reasoning into event-b devel-opment. Formal Asp. Comput. **27**(1), 53–77 (2015)
25. E. Troubitsyna, Elicitation and specification of safety requirements, in *The Third International Conference on Systems, ICONS 2008* (IEEE Computer Society, Washington, D.C., 2008), pp. 202–207
26. E. Troubitsyna, *Stepwise Development of Dependable Systems*. Technical Report (Turku Centre for Computer Science, 2000)
27. Troubitsyna, E., Laibinis, L., Pereverzeva, I., Kuismin, T., Ilic, D., Latvala, T.: Towards Security-Explicit Formal Modelling of Safety-Critical Systems. In: SAFECOMP 2016, Pro-ceedings. LNCS, vol. 9922, pp. 213–225. Springer (2016)
28. E. Troubitsyna, L. Laibinis, I. Pereverzeva, T. Kuismin, D. Ilic, T. Latvala, Towards security-explicit formal modelling of safety-critical systems, in *SAFECOMP 2016*. LNCS, vol. 9922 (Springer, Berlin, 2016), pp. 213–225
29. I. Vistbakka, E. Troubitsyna, T. Kuismin, T. Latvala, Co-engineering safety and security in industrial control systems: A formal outlook, in *SERENE 2017, Proceedings*. LNCS, vol. 10479 (Springer, Berlin, 2017), pp. 96–114
30. W. Young, N.G. Leveson, An integrated approach to safety and security based on systems theory. Commun. ACM **57**(2), 31–35 (2014)

Towards an Integration of Probabilistic and Knowledge-Based Data Analysis Using Probabilistic Knowledge Patterns

Klaus-Dieter Schewe and Qing Wang

Abstract Knowledge patterns combine rules defined by definite clauses with conditions specifying when the rules are applicable, and conditions specifying when the application of the rules is not permitted. In combination with an extensional ground database the semantics of a set of knowledge patterns is defined by an inflationary fixed-point. Originally, knowledge patterns have been introduced in connection with the problem of record linkage emphasising intensional identity predicates, thus in the fixed-point model equivalence relations on object identifiers are obtained. Known failures in such a classification give rise to minimal changes to the conditions of some knowledge patterns. However, the fixed-point semantics is not restricted to equivalence relations. In this paper we define an extension to probabilistic knowledge patterns, where the rules become clauses in probabilistic logic. Using maximum entropy semantics for the probabilistic logic the fixed-point construction can be extended resulting in a probabilistic model, i.e. distributions for the randomised relations. It is expected that statistical approaches to data analysis can be interpreted in the context of probabilistic knowledge patterns such that learning of knowledge patterns can be enabled, while the advantages of knowledge patterns with respect to provenance can be preserved.

1 Introduction

Knowledge patterns have been introduced in [27] and further elaborated in [28] as a logical approach to entity resolution (aka record linkage), i.e. the problem to determine, whether two entity representations in a database or dataset refer to the same real-world object or not. In a nutshell, a knowledge pattern combines rules

K.-D. Schewe
Laboratory for Client-Centric Cloud Computing, Linz, Austria
e-mail: kdschewe@acm.org

Q. Wang (✉)
Research School of Computer Science, The Australian National University, Canberra, Australia
e-mail: qing.wang@anu.edu.au

© Springer Nature Singapore Pte Ltd. 2021
Y. Ait-Ameur et al. (eds.), *Implicit and Explicit Semantics Integration in Proof-Based Developments of Discrete Systems*,
https://doi.org/10.1007/978-981-15-5054-6_7

131

defined by definite clauses (aka Horn clauses) with conditions specifying when the rules are applicable, and conditions specifying when the application of the rules is not permitted. Such conditions are formalised by positive and negative entries in a pattern relation. In combination with an extensional ground database the semantics of a set of knowledge patterns is defined by an inflationary fixed-point.

In the context of entity resolution applications, knowledge patterns exploit explicit knowledge, whereas common similarity-based methods do not formulate such explicit knowledge. Instead, they rely on the implicit probabilistic distribution that underlies the given data and exploits this to compute similarity measures that can be used for clustering. Thus, knowledge patterns provide the advantage to enable the explicit capture of application knowledge, whereas similarity-based methods provide strength with respect to uncertainty. Integrating both provides an example for the integration of explicit and implicit semantics, which characterises these two directions—this links our work to the general theme of the NII Shonan workshop.

Knowledge patterns complement the various similarity-based methods for entity resolution (see [9] for a survey). For instance, the probabilistic theory of entity resolution developed in [14] exploits probabilistic decision rules that are optimal when the comparison attributes are independent. In [33] a cost-based Bayesian decision model is developed, which focuses on an optimal solution for the matching of database records in the presence of inconsistencies, errors or missing values. In a similar direction goes the ruled-based approach in [10], which emphasises similarity joins. Statistical machine learning methods have been applied to the entity resolution problem such as supervised learning and active learning [2, 26].

From a theoretical angle knowledge patterns have been investigated in [29] emphasising minimality, containment and optimisation. It was shown that containment is decidable, which follows from the decidability of the containment problem for conjunctive queries. It was also shown that in a minimal knowledge pattern none of the records in the pattern relation can be omitted without changing the result of the fixed-point construction. Furthermore, optimised knowledge patterns result from a three-step process and optimisation is proven to be NP-complete. In [35] a method was developed that can capture the provenance of a record linkage decision, which thus allows to infer minimal changes to exceptions in case a derived decision turns out to be incorrect.

As statements with exceptions are conceptually very different from probabilistic statements, none of the similarity-based methods can be used to substitute a set of knowledge patterns nor the other way round. Therefore, an integration of these fundamentally different methods provides several advantages, if uncertainty can be adequately captured and provenance-based reasoning can be preserved. In this article we make a first attempt to integrate probability distributions into knowledge patterns. In doing so we also remove the technical restriction of knowledge patterns to the application context of entity resolution. So far in the fixed-point model equivalence relations on object identifiers are obtained, but the theory is not restricted to equivalence relations.

For the integration of probabilities we can build on the long tradition of probabilistic logics in artificial intelligence [3, 7, 8, 13, 15, 20, 24]. Of particular interests for an extension of the knowledge pattern approach are Bayesian networks [15] which exploit directed graphs with propositions as vertices subject to local Markov property, which have been generalised to probabilistic conditional logic [21] and Bayesian logic programs [19], and Markov logics that combines first-order logic and probabilistic models by attaching weights to first-order logic formulae [12, 25]. Markov logic has already been studied for entity resolution in [31], though not in connection with explicit default rules. We will exploit relational probabilistic conditional logic [5, 16–18], which is the probabilistic logic with a format of rules that is closest to the one used in knowledge patterns.

Relational probabilistic conditional logic extends probabilistic conditional logic to the relational case and is grounded in an inference mechanism based on the principle of maximum entropy [4, 6, 32]. Whenever dependence conditions for a probability distribution are given there may be many distributions satisfying these dependencies. However, it is known that if among these possible choices the distribution with the maximum entropy is chosen, then no additional assumptions will be made [23]. For instance, if nothing is known about a distribution, the one with the maximum entropy is the equal distribution. Furthermore, it has been shown that the distribution with the maximum entropy is unique [22].

There are three locations in the theory of knowledge patterns, where probabilities can be brought in:

1. As the rule part of a knowledge pattern is a Horn clause, it can be easily replaced by a relational probabilistic conditional [17], i.e. add a probability $p \in [0, 1]$ to the clause to express that the implication holds with the probability p.
2. The positive entries in a pattern relation can be interpreted to have probability 1, while the negative ones have probability 0. So more generally we could assign a probability in [0, 1] to every record in a pattern relation. However, as the rule part in our intended probabilistic extension already contains a probability p, we will only investigate the restricted case that positive records will give rise to rule instantiations with the same probability p, whereas negative records give rise to instantiations with probability 0.
3. The ground database that is decisive for the definition of the fixed-point semantics can be generalised to a probabilistic database, where each record carries a probability. By default records that do not appear in the database will then have the probability 0, and any two records in the ground database will be assumed to be probabilistically independent.

On these grounds we generalise the fixed-point semantics of knowledge patterns to probabilistic knowledge patterns. Following the work by Kern-Isberner et al. we first address the interpretation of rules together with a grounding that is defined by a probabilistic ground database. For this we investigate the aggregate semantics [17]. In addition, the restrictions defined by the pattern relation are taking into account. In general, such a grounding defines constraints for the extension of the probability distribution in the form of linear equations. In case of probabilistic independence these equations only involve instances of the head predicates as variables.

Second, the maximum entropy principle defines a unique solution under these constraints, i.e. an extension of the probabilistic database. This follows from variation theory using Lagrange multipliers [11]. In case of non-recursive rules it is then straightforward to define again a fixed-point semantics, as an extension of the ground database for a particular head predicate in one step of the iteration will not be affected anymore in the following steps. In the recursive case, however, more care is needed, as a follow-on iteration step may produce constraints that contradict the solution obtained by the maximum entropy principle. This problem will be addressed by accumulating the constraints rather than solving the linear optimisation problem for the maximum entropy. The maximum entropy will thus only be used in a final step.

The remainder of this paper will be organised as follows. In Sect. 2 we present a brief, yet more detailed review of relational knowledge patterns following mainly previous work in [28, 29, 35]. Then Sect. 3 contains the novel core contribution of this paper, the definition of probabilistic knowledge patterns and their fixed-point semantics grounded in relational probabilistic conditional logic with maximum entropy as defined in previous work by Kern-Isberner et al. [17]. We conclude with a brief summary and outlook in Sect. 4.

2   Relational Knowledge Patterns

In this section we present the definition of knowledge patterns [28, 29]. However, we slightly deviate from our previous work, which permitted only intensional predicates that were bound to equivalence relations. We emphasise that properties of equivalence relations could be captured by separate rules. In order to highlight the extension in our notations we call these knowledge patterns *relational* in order to emphasise that there semantics is defined on top of relational databases.

We then review the results of our previous investigation of knowledge patterns concerning their optimisation [29] and provenance-based revision of inferences [35]. These results, however, have only been proven in the context of entity resolution, i.e. they hold only, if all intensional predicates correspond to equivalence relations. The generalisation of the results to the general relational case is an open research problem.

2.1   Syntax and Fixed-Point Semantics

Let $\mathbf{R} = \{R_1, \ldots, R_n\}$ be a relational database schema, i.e. a set of relation symbols, where each $R_i \in \mathbf{R}$ is associated with a set $\{A_1, \ldots, A_m\} \subseteq \mathbf{A}$ of attributes. Let $\mathbf{D}$ be a set of domains. Each attribute $A \in \mathbf{A}$ is associated with a domain $dom(A) \in \mathbf{D}$. One of the domains in $\mathbf{D}$, say D_0, is a countable set of identifiers. Let $\mathbf{E} = \{E_1, \ldots, E_k\}$ be another relational database schema, disjoint from $\mathbf{R}$.

A *tuple t* over a relation symbol R assigns to each attribute A_i a value $t(A_i) \in dom(A_i)$. We write $t = (A_1 : t(A_1), \ldots, A_n : t(A_n))$ or simply $t = (t(A_1), \ldots, t(A_n))$ when the order of the attributes is fixed. A finite set of tuples over R is called a *relation over R*. A *database (instance) I* over the schema **R** is a family of relations $I(R_i)$ over R_i, indexed by **R**, i.e. $I = \{I(R_i)\}_{R_i \in \mathbf{R}}$. Naturally, these definitions extend to **E** or any other relational database schema.

Example 1 Let us consider a relational database schema **R** with the relation symbols PERSON, AUTHORSHIP and PUBLICATION. The attributes associated with the relational symbols are as follows:

- PERSON = {ID, Author, Email, Affiliation};
- AUTHORSHIP = {AuthID, PubID, Order};
- PUBLICATION = {PubID, Title, Year, DOI}.

A database instance over **R** is represented in the usual way by the tables in Fig. 1.

Let $\mathbf{S} = \mathbf{R} \cup \mathbf{E}$. Following the usual convention of Datalog [1] we usually call the predicates $R_i \in \mathbf{S}$ *extensional* and the predicates $E_j \in \mathbf{S}$ *intensional* to emphasise their different usage in the rule part of knowledge patterns. An *atom* is a term $\psi = R(t_1, \ldots, t_n)$, where $R \in \mathbf{S}$ and each t_i is either a variable x_i or a constant $v_i \in dom(A_i)$. Let $var(\psi)$ denote the set of variables in the atom ψ.

PERSON			
ID	Name	Email	Affiliation
i_1	S. Lee	sl@gmail.com	Faculty of Medicine, University of Otago
i_2	Susan Lee	sl@massey.ac.nz	Massey University
i_3	S. Maneth		Faculty of Medicine, University of Otago
i_4	S. A. Lee	sl@otago.ac.nz	University of Otago
i_5	S. Lee	sl@massey.ac.nz	Massey University
i_6	Ben Williams	ben@otago.ac.nz	University of Otago
i_7	Ben Williams		Massey University
i_8	A. Timu	ali@acm.org	
i_9	A. Timu	atimu@stanford.edu	Stanford University

AUTHORSHIP		
AuthID	PubID	Order
i_2	b_1	1
i_6	b_1	2
i_8	b_1	3
i_4	b_2	1
i_3	b_3	1
i_5	b_3	2
i_7	b_3	3
i_9	b_3	4

PUBLICATION			
PubID	Title	Year	DOI
b_1	An effective World Wide Web image ...	2001	10.1177/...
b_2	Irrigation in Modern Agriculture	1994	10.4321/...
b_3	Algebraic Topology	2011	10.154/...

Fig. 1 A database instance for the schema $\mathbf{R} = \{\text{PERSON, AUTHORSHIP, PUBLICATION}\}$

Originally, in the theory of knowledge patterns [29] each $E_j \in \mathbf{E}$ is associated with exactly two attributes $\{A_1^i, A_2^i\} \subseteq \mathbf{A}$ with $dom(A_k^i) = D_0$ for each A_k^i ($k = 1, 2$). Furthermore, in any valid database instance I over $\mathbf{S}$ each intensional predicate E_j always corresponds to an equivalence relation.

Definition 1 A *(relational) knowledge pattern* P over $\mathbf{S}$ is a pair (ℓ, r) comprising

- a Horn clause ℓ of the form $\psi(\mathbf{x}) \leftarrow \varphi(\mathbf{x}, \mathbf{y})$, where the head is an atom with an intensional predicate $E_j \in \mathbf{E}$ and variables $\mathbf{x}$ and the body is a conjunction of atoms with predicates in $\mathbf{S}$ and variables $\mathbf{x} \cup \mathbf{y}$, and
- a *pattern relation* r of the arity $n + 1$, where n attributes that are in 1-1 correspondence to the variables $\mathbf{x} \cup \mathbf{y}$ in φ, and the $(n + 1)$st attribute is a *sign attribute* A^* with domain $\{+, -\}$. A special value $\lambda \notin \bigcup_{D_i \in \mathbf{D}} D_i$ may appear as value for any attribute in r, except for the sign attribute A^*.

Example 2 Let us extend Example 1 by some relational knowledge patterns.

First define a name collaboration pattern NC $= \langle \varphi_1, r_1 \rangle$: Two persons are identical if they have the same name and there is another person whose name is not "Stephen Smith", whose affiliation is the same with one of the two persons and is "Faculty of Medicine, University of Otago", and who has co-authored a paper with the other person. Assume that a person named "Stephen Smith" at "Faculty of Medicine, University of Otago" has a colleague and a collaborator who are two different people with the same name.

$\varphi_1 :=$ IDENTICAL$(x, y) \leftarrow$
　　　PERSON$(x, z_6, x_3, x_4) \wedge$ AUTHORSHIP$(x, z_1, x_1) \wedge$ AUTHORSHIP$(z', z_1, x_2) \wedge$
　　　PERSON$(y, z_6, y_3, y_4) \wedge$ IDENTITY$(z, z') \wedge$ PERSON(z, z_3, z_4, y_4)

$r_1 :=$

A_{y_4}	A_{z_3}	...	A^*
Faculty of Medicine, University of Otago	λ	λ	+
λ	Stephen Smith	λ	-

Next define a co-authorship pattern CA$=\langle \varphi_2, r_2 \rangle$: Two persons are identical if they have the same name and both have co-authored with at least one other author who is not a person named "Sue Lee" with the email address "sl@otago.ac.nz". We assume that the person named "Sue Lee" with the email address "sl@otago.ac.nz" has co-authored papers with two different persons who have the same name.

$\varphi_2 :=$ IDENTICAL$(x, y) \leftarrow$
　　　PERSON$(x, z_6, x_3, x_4) \wedge$ AUTHORSHIP$(x, z_1, x_1) \wedge$ AUTHORSHIP$(z', z_1, x_2) \wedge$
　　　PERSON$(y, z_6, y_3, y_4) \wedge$ AUTHORSHIP$(y, z_2, y_1) \wedge$ AUTHORSHIP$(z, z_2, y_2) \wedge$
　　　IDENTITY$(z, z') \wedge$ PERSON(z, z_3, z_4, z_5)

$r_2 :=$

A_z	$A_{z'}$	A_{z_1}	A_{z_2}	A_{z_3}	A_{z_4}	A_{z_5}	...	A^*
λ	λ	λ	λ	λ	λ	λ	λ	+
λ	λ	λ	λ	Sue Lee	sl@otago.ac.nz	λ	λ	−

Definition 2 A *(relational) knowledge model* is a finite, nonempty set **P** of relational knowledge patterns over $\mathbf{S} = \mathbf{R} \cup \mathbf{E}$.

Note that in the original case dedicated to entity resolution the requirement that each intensional predicate E_j should correspond to an equivalence relation gives rise to implicit knowledge patterns capturing the rules for reflexivity, symmetry and transitivity:

Reflexivity pattern. For reflexivity, we may take Horn clauses $E_j(x, x) \leftarrow R_i(\ldots, x, \ldots)$ and a pattern relation with a single tuple $(\lambda, \ldots, \lambda, +)$. That is, for each extensional predicate R_i and each attribute with domain D_0 that corresponds to the attributes in E_j we obtain such a reflexivity pattern, which enforces that every identifier appearing in a database will become equivalent to itself.

Symmetry pattern. For symmetry, we may take a Horn clause $E_j(y, x) \leftarrow E_j(x, y)$ and a pattern relation with a single tuple $(\lambda, \lambda, +)$.

Transitivity pattern. For transitivity, we may take a Horn clause $E_j(x, z) \leftarrow E_j(x, y) \wedge E_j(y, z)$ and a pattern relation with a single tuple $(\lambda, \lambda, \lambda, +)$.

In general, in relational knowledge models such specific patterns must be specified explicitly.

Let r^+ and r^- refer to the subset of tuples in r which have $+$ and $-$ in their attribute values of A^*, respectively. Each tuple $t \in r$ defines an instantiation of the rule ℓ

- For $t(A_i) = v_i \neq \lambda$, replace the variable x_i corresponding to the attribute A_i by v_i.
- For $t(A_i) = \lambda$, do not replace x_i.

We denote this instantiation by $\Phi_t = \psi(\bar{\mathbf{x}}) \leftarrow \varphi(\bar{\mathbf{x}}, \bar{\mathbf{y}})$, where for each $x_i \in \mathbf{x} \cup \mathbf{y}$, depending on the tuple $t \in r$, we either have $x_i = v_i$ or $x_i = x_i$. Let $\mathbf{z}$ refers to the sequence of variables in $\varphi(\bar{\mathbf{x}}, \bar{\mathbf{y}})$. Then for a database instance I over $\mathbf{R} \cup \mathbf{E}$ we define the *interpretation* $\Phi_t(I)$ of Φ_t as

$$\{\mu(\psi(\bar{\mathbf{x}})) \mid \mu \text{ is a valuation of the variables } \mathbf{z} \text{ and } \varphi(\bar{\mathbf{x}}, \bar{\mathbf{y}}) \text{ is true in } I \text{ under } \mu\}.$$

The *interpretation* $P(I)$ of a relational knowledge pattern $P = (\ell, r)$ with $\ell = \psi(\mathbf{x}) \leftarrow \varphi(\mathbf{x}, \mathbf{y})$ over I is thus defined as

$$P(I) = \bigcup_{t \in r^+} \Phi_t(I) - \bigcup_{t \in r^-} \Phi_t(I).$$

Example 3 For the co-authorship pattern CA in Example 2 we obtain two instantiations of the rule:

$\varphi^{+}_{(2,1)} :=$ IDENTICAL$(x, y) \leftarrow$

   PERSON$(x, z_6, x_3, x_4) \wedge$ AUTHORSHIP$(x, z_1, x_1) \wedge$ AUTHORSHIP$(z', z_1, x_2) \wedge$
   PERSON$(y, z_6, y_3, y_4) \wedge$ AUTHORSHIP$(y, z_2, y_1) \wedge$ AUTHORSHIP$(z, z_2, y_2) \wedge$
   IDENTITY$(z, z') \wedge$ PERSON(z, z_3, z_4, z_5)

$\varphi^{-}_{(2,2)} :=$ IDENTICAL$(x, y) \leftarrow$

   PERSON$(x, z_6, x_3, x_4) \wedge$ AUTHORSHIP$(x, z_1, x_1) \wedge$ AUTHORSHIP$(z', z_1, x_2) \wedge$
   PERSON$(y, z_6, y_3, y_4) \wedge$ AUTHORSHIP$(y, z_2, y_1) \wedge$ AUTHORSHIP$(z, z_2, y_2) \wedge$
   IDENTITY$(z, z') \wedge$ PERSON$(z,$ "Sue Lee", "sl@otago.ac.nz"$, z_5)$

and the following rule with IDENTITY at both the head and the body.

$$\text{IDENTITY}(x, y) \leftarrow q^{+}_{(2,1)}(x, y) \wedge \neg q^{-}_{(2,2)}(x, y)$$

For a relational knowledge model **P** of relational knowledge patterns over **S** the semantics over a database instance I of **R** is defined by an inflationary fixed-point analogous to Datalog [1]. For this let $J_0 = I$, which is extended to **S** by $J_0(E_j) = \emptyset$. Then for $n > 0$ define

$$J_n(E_j) = J_{n-1}(E_j) \cup \bigcup_{P \in \mathbf{P}_{E_j}} P(J_{n-1}),$$

where $\mathbf{P}_{E_j}$ is the set of all patterns in **P** with head predicate E_j and $P(J_{n-1})$ is the interpretation of P over J_{n-1} as defined above. Note that for extensional predicates $R_i \in \mathbf{R}$ we always preserve $J_n(R_i) = I(R_i)$, as extensional predicates never appear in the head of Horn clause in a knowledge pattern.

As for each $E_j \in \mathbf{E}$ the sequence $\{J_n(E_j)\}_{n \geq 0}$ is increasing monotonically, the limit exists, and we obtain the *inflationary fixed-point* J_∞ with

$$J_\infty(E_j) = \bigcup_{n \geq 0} J_n(E_j) \quad \text{and} \quad J_\infty(R_i) = I(R_i).$$

Finally note that for the application case of entity resolution, the implicit assumption made in [29] that each intensional predicate E_j corresponds to an equivalence relation yields the same inflationary fixed-points for a given extensional database instance I as the explicit definition of the reflexivity, symmetry and transitivity patterns as elements of **P**.

2.2 Minimising Redundancy in Knowledge Patterns

The complexity of the inflationary fixed-point construction depends heavily on the number of knowledge patterns and accordingly the number of instantiations defined by tuples in the pattern relation of each of them. Therefore, the theoretical investigation of knowledge patterns for entity resolution emphasised redundancy among different patterns, as well as different instantiations within single patterns [29]. We have redundancy between two knowledge patterns P_1 and P_2, i.e. one knowledge pattern P_2 *contains* another one P_1, denoted by $P_1 \sqsubseteq P_2$, iff on any database instance I we always obtain $P_1(I) \subseteq P_2(I)$. For a single knowledge pattern (ℓ, r) we are seeking that the number of tuples in r is *minimal*, and for knowledge patterns with the same rule it is desirable to *optimise* the total number of tuples in their pattern relations. Minimality, containment and optimisation were thus the focus of the research in [29]. However, all these results refer to the case of treating intensional predicates in $\mathbf{E}$ as equivalence relations.

2.2.1 Minimisation

For the construction of the (inflationary) fixed-point it is desirable to minimise the application of the immediate consequence operator. As for a single knowledge pattern $P = (\ell, r)$ this depends strongly on the number of tuples in r, the number of tuples in $r^+ = \{t \in r \mid t.A^* = +\}$ and $r^- = \{t \in r \mid t.A^* = -\}$ should be minimal.

We define $v \preceq \lambda$ for every $v \in D_i \cup \{\lambda\}$. Then we define the notions of subsumption, upward-subsumption and downward-subsumption as follows:

Definition 3 Let t_1 and t_2 be two tuples which are associated with the same set $\mathbf{A}'$ of attributes including A^*. Then

- t_1 *subsumes* t_2 (denoted as $t_2 \sqsubseteq t_1$) if $t_2(A) \preceq t_1(A)$ holds for each attribute $A \in \mathbf{A}'$;
- t_1 *upward-subsumes* t_2 (denoted as $t_2 \sqsubseteq_\uparrow t_1$) if $t_2(A) \preceq t_1(A)$ holds for each non-sign attribute $A \in \mathbf{A}' - \{A^*\}$, $t_1(A^*) = -$ and $t_2(A^*) = +$;
- t_1 *downward-subsumes* t_2 (denoted as $t_2 \sqsubseteq_\downarrow t_1$) if $t_2(A) \preceq t_1(A)$ holds for each non-sign attribute $A \in \mathbf{A}' - \{A^*\}$, $t_1(A^*) = +$ and $t_2(A^*) = -$.

Then we define P to be *minimal* iff for $t_2 \not\sqsubseteq t_1$ and $t_2 \not\sqsubseteq_\uparrow t_1$ hold for any two tuples t_1, t_2 in r.

Example 4 Consider again the name collaboration pattern (φ_1, r_1) from Example 2 with the rule

$\varphi_1 := \text{IDENTICAL}(x, y) \leftarrow$

$\quad \text{PERSON}(x, z_6, x_3, x_4) \wedge \text{AUTHORSHIP}(x, z_1, x_1) \wedge \text{AUTHORSHIP}(z', z_1, x_2) \wedge$

$\quad \text{PERSON}(y, z_6, y_3, y_4) \wedge \text{IDENTITY}(z, z') \wedge \text{PERSON}(z, z_3, z_4, y_4).$

However, change the pattern relation to become

A_{y4}	A_{z3}	...	A^*
λ	λ	λ	+
Faculty of Medicine, University of Otago	λ	λ	+
λ	Stephen Smith	λ	-
University of Otago	Stephen Smith	λ	-

If these tuples are denoted $t_1, \ldots t_4$, then we have that t_1 subsumes t_2 (i.e. $t_2 \sqsubseteq t_1$). Removing t_2 from the pattern relation does not change the result for any fixed-point construction that involves this knowledge pattern. The same holds for t_4, as we have $t_4 \sqsubseteq t_3$.

The following result from [29] states that, for a minimal knowledge pattern P, none of the tuples can be omitted without changing the result of the immediate consequence operator for at least one database instance.

Proposition 1 *Let $P = \langle \ell, r \rangle$ be a pattern and r be minimal. Then the following holds:*

1. *$\varphi_i \not\sqsubseteq \varphi_j$ holds for any two different φ_i and φ_j from Σ_P^+.*
2. *$\varphi_i \not\sqsubseteq \varphi_j$ holds for any two different φ_i and φ_j from Σ_P^-.*
3. *$\varphi_i \not\sqsubseteq \varphi_j$ holds for any $\varphi_i \in \Sigma_P^+$ and $\varphi_j \in \Sigma_P^-$.*

Therefore, if P is a minimal knowledge pattern, then for any different tuples $t_1, t_2 \in r^+$ and $t_1', t_2' \in r^-$ we have

$$P_{t_1} \not\sqsubseteq P_{t_2} \quad P_{t_1'} \not\sqsubseteq P_{t_2'} \quad \text{and} \quad P_{t_1} \not\sqsubseteq P_{t_1'},$$

where P_t indicates a knowledge pattern that has the same rule as P but only has one tuple t in the pattern relation, i.e., $P_t(I) = \Phi_t(I)$ holds for every database instance I.

2.2.2 Containment of Knowledge Patterns

If $P_1(I) \subseteq P_2(I)$ holds for every database instance I, i.e. the knowledge pattern P_1 is contained in the knowledge pattern P_2, naturally, P_1 could be removed from a set **P** of knowledge patterns containing also P_2, as it would not contribute anything to the inflationary fixed-point.

With the following result from [29] the containment problem of knowledge patterns can be reduced to the containment problem of conjunctive queries, and thus is decidable:

Proposition 2 *For two knowledge patterns $P_i = (\ell_i, r_i^+, r_i^-)$ $(i = 1, 2)$ we have $P_1 \subseteq P_2$ iff*

1. *for each $t_1 \in r_1^+$, there exists a $t_2 \in r_2^+$ with $P_{t_1} \subseteq P_{t_2}$, and*
2. *for each $t_1 \in r_1^+$ and $t_2' \in r_2^-$, there exists a $t_1' \in r_1^-$ with $P_{t_1} \cap P_{t_2'} \subseteq P_{t_1} \cap P_{t_1'}$.*

Example 5 Consider the following patterns $P_1 = (\ell_1, r_1)$ and $P_2 = (\ell_2, r_2)$.

$\ell_1 = R(x, y) \leftarrow R_1(x, z_1) \wedge R_2(z_2, y, z_3)$

$r_1 =$

	A_{z_1}	A_{z_2}	A_{z_3}	A^*	
	a	a	λ	$+$	t_1
	b	b	λ	$+$	t_2
	λ	λ	b	$-$	t_3

$\ell_2 = R(x, y) \leftarrow R_1(x, z_1) \wedge R_2(z_1, y, z_3)$

$r_2 =$

	A_{z_1}	A_{z_3}	A^*	
	λ	λ	$+$	t_1
	λ	b	$-$	t_2

To check whether $P_1 \sqsubseteq P_2$ holds according to Proposition 2 we need to check the conditions (1) $\varphi_1^{t_1} \sqsubseteq \varphi_2^{t_1}$, (2) $\varphi_1^{t_3} \wedge \varphi_1^{t_1} \sqsubseteq \varphi_2^{t_2} \wedge \varphi_1^{t_1}$, (3) $\varphi_1^{t_2} \sqsubseteq \varphi_2^{t_1}$, and (4) $\varphi_1^{t_3} \wedge \varphi_1^{t_2} \sqsubseteq \varphi_2^{t_2} \wedge \varphi_1^{t_2}$, where $\varphi_1^{t_1} = \exists z_3.R_1(x, a) \wedge R_2(a, y, z_3)$, $\varphi_1^{t_2} = \exists z_3.R_1(x, b) \wedge R_2(b, y, z_3)$, $\varphi_1^{t_3} = \exists z_1, z_2.R_1(x, z_1) \wedge R_2(z_2, y, b)$, $\varphi_2^{t_1} = \exists z_1, z_3.R_1(x, z_1) \wedge R_2(z_1, y, z_3)$ and $\varphi_2^{t_2} = \exists z_1.R_1(x, z_1) \wedge R_2(z_1, y, b)$ hold.

2.2.3 Optimisation of a Set of Knowledge Patterns

Finally, optimisation aims to reduce the number of tuples in pattern relations in cases, where neither a tuple can be simply removed to obtain minimality, nor two knowledge patterns contain each other.

Therefore, we call $\mathbf{P}'$ a *positive optimisation* of a set $\mathbf{P}$ of knowledge patterns iff $\mathbf{P}$ and $\mathbf{P}'$ are equivalent, the total number of tuples in positive pattern relation $r_i'^+$ of $\mathbf{P}'$ does not exceed the total number of tuples in positive pattern relation r_i^+ of $\mathbf{P}$, and there is no other knowledge pattern $\mathbf{P}''$ also satisfying these properties with a strictly lower number of tuples in positive pattern relation than $\mathbf{P}'$.

Analogously, we call $\mathbf{P}'$ a *negative optimisation* of a set $\mathbf{P}$ of knowledge patterns iff $\mathbf{P}$ and $\mathbf{P}'$ are equivalent, the total number of tuples in negative pattern relation $r_i'^-$ of $\mathbf{P}'$ does not exceed the total number of tuples in negative pattern relation r_i^- of $\mathbf{P}$, and there is no other knowledge pattern $\mathbf{P}''$ also satisfying these properties with a strictly smaller number of tuples in negative pattern relation than $\mathbf{P}'$.

An *optimisation* of $\mathbf{P}$ is a positive optimisation of $\mathbf{P}$ that cannot be further optimised positively nor negatively. Now let us concentrate on knowledge patterns with the same rule. We define the *intersection* $t_1 \curlywedge t_2$ of two tuples by building the minimum on each attribute with respect to $\preceq$. If this is not possible, the intersection is not defined. Analogously, we define the *union* $t_1 \curlyvee t_2$ of two tuples by building the supremum on each attribute with respect to $\preceq$. Then an optimisation of $\mathbf{P}$ can be obtained by applying the following three steps:

Normalisation. For $r^+ = \{t_1, \ldots, t_n\}$, we replace P by n knowledge patterns $P_i = (\ell, \{t_i\} \cup r^-)$, where $i = 1, \ldots, n$.

Elimination. First, for $t_1 \in r_1^-$ and $t_2 \in r_2^+$ with $t_1 \sqsubseteq_\downarrow t_2$, we replace t_1 by $\{t_1 \curlywedge t_3 \mid t_3 \in r_2^-\}$, if $r_2^- \neq \emptyset$, or otherwise omit t_1. Then, we remove redundant knowledge patterns based on containment of the knowledge patterns.

Composition. We define two patterns P_1 and P_2 to be *compatible* iff $r_1^- \curlywedge r_2^+ \sqsubseteq r_2^-$ and $r_1^+ \curlywedge r_2^- \sqsubseteq r_1^-$ hold. Furthermore, if $P_1, \ldots, P_n$ are pairwise compatible, then $t_1 \in r_i^-$ and $t_2 \in r_j^-$ are *mergeable* iff for all tuples $t \in r_k^+$ there exists a tuple

$t^- \in r_k^-$ with $t \curlywedge (t_1 \curlyvee t_2) \sqsubseteq t^-$. Then we decompose the set $\mathbf{P}$ of knowledge patterns into sets of pairwise compatible knowledge patterns and for each such subset build a single knowledge pattern taking the union of the positive pattern relations, and the union of negative pattern relations, such that mergeable tuples t_1, t_2 are replaced by $t_1 \curlyvee t_2$. We can choose the decomposition and the mergeable tuples in such a way that the resulting negative pattern relation has a minimum number of tuples, which is an **NP**-complete optimisation problem.

A sophisticated example for the optimisation of a set of relational knowledge patterns has been given in [29, Sect. 5].

2.3   Provenance-Based Pattern Revision

The inferences based on knowledge patterns are de facto a default reasoning approach, and the applicability conditions and exceptions represented by pattern relations merely reflect partial knowledge about an application. Thus, in reality knowledge patterns will hardly be perfect. In particular, the inference may lead to two identifiers id_1 and id_2 being identified as referring to the same entity, but at the same time new information may violate this inference.

Therefore, in [35] we developed a method based on an *entity resolution index* (ERI), which captures the provenance information of record linkage decisions, and thus enables the analysis of inconsistencies, e.g. why inconsistencies did occur and how they relate to the linkage process. In case a linkage decision turns out to be incorrect, the method permits the inference of minimal changes to the exceptions and can thus significantly reduce human efforts in identifying inconsistent knowledge.

An *entity resolution index* (ERI) is a data structure, in which each index entry has the form (entity e, ER tree t_e), where t_e keeps track of the inferences on matching decisions that are relevant to the entity e. More specifically, an *ER tree t_e* is a binary tree in which each node represents a linkage decision, together with a labelling function θ such that: (1) θ assigns to each leaf a distinct label, which represents a record; (2) θ also assigns to each edge a label which represents the record used in the inference for a linkage decision. This indexing technique ERI allows us not only to store the provenance information of inferences but also to analyse inconsistent inferences. Whenever an inconsistency is found, relevant inferences on linkage decisions can be pinpointed, providing a unified view to understand such an inconsistency. In doing so, the provenance information serves as a ground on which inconsistent inferences can be identified in a meaningful and efficient way.

Two kinds of constraints, so-called *must-link* and *cannot-link* constraints [34] are considered in [35]. A *must-link constraint* $k_1 \simeq k_2$ means that two records k_1 and k_2 must be matched to the same entity, and a *cannot-link constraint* $k_1 \not\simeq k_2$ means that two records k_1 and k_2 cannot be matched to the same entity. Both must-link and cannot-link constraints are instance-level constraints [34], and symmetric in the sense that if $k_1 \simeq k_2$ (resp. $k_1 \not\simeq k_2$) is satisfied, then $k_2 \simeq k_1$ (resp. $k_2 \not\simeq k_1$) is also satisfied. Although must-link and cannot-link constraints look simple, they serve as the building blocks of expressing the integrity of inferences for record linkage. We developed algorithms for two important operations—merge and split—which can eliminate inconsistent inferences violating must-link and cannot-link constraints, respectively. The algorithm for the split operation traverses an ER tree to identify and remove erroneous inferences on matches in order to eliminate inconsistencies. Such traversals downward and upward an ER tree are both efficient. A generic strategy, called CSM, was developed to improve the effectiveness of identifying inconsistencies by taking into account how must-link and cannot-link constraints interact based on the provenance information represented by ERI.

3 Probabilistic Knowledge Patterns

Let $\mathscr{S}$ be a relational database schema, i.e. a finite set of relation symbols $R_1, \ldots, R_k$, where each R_i is associated with an arity n_i. As in Sect. 2 we assume that $\mathscr{S}$ is partitioned into an extensional subschema $\mathscr{R}$ and an intensional subschema $\mathscr{Q}$, i.e. $\mathscr{S} = \mathscr{R} \cup \mathscr{Q}$. As usual, a (relational) database db over $\mathscr{S}$ assigns to each $R_i \in \mathscr{S}$ a finite relation $db(R_i)$ of arity n_i, i.e. a finite set of n_i-tuples.[1]

Definition 4 A *probabilistic database db* over $\mathscr{S}$ is a family $\{db(R_i)\}_{R_i \in \mathscr{S}}$ indexed by $\mathscr{S}$, where each $db(R_i)$ is a finite relation of arity n_i, and each tuple $r \in db(R_i)$ is assigned a probability $p_r \in [0, 1]$.

By default, if an n_i-tuple r does not appear in $db(R_i)$, its probability p_r is 0. A relational database instance over $\mathscr{S}$ appears as a special case of this definition with all probabilities of tuples in the instance being 1.

Each n_i-tuple $r = (v_1, \ldots, v_{n_i})$ defines an event $ev = R_i(v_1, \ldots, v_{n_i})$ with probability $p(ev) = p_r$. Then the probability of the event $\neg R_i(v_1, \ldots, v_{n_i})$ is $1 - p_r$. Given a probabilistic database db over the extensional schema $\mathscr{R}$, we asume all the events defined by it to be probabilistically independent, i.e.

$$p(R_i(v_1, \ldots, v_{n_i}) \wedge R_j(w_1, \ldots, w_{n_j})) = p(R_i(v_1, \ldots, v_{n_i})) \cdot p(R_j(w_1, \ldots, w_{n_j})).$$

Example 6 Let us turn the relational database schema **R** from Example 1 into a probabilistic one. For this we keep the relation symbols PERSON, AUTHORSHIP and PUBLICATION and the associated attributes as they are. We can then reuse the instance from Fig. 1 by assigning probabilities to all tuples. This results in the probabilistic database shown in Fig. 2.

---

[1]Throughout this section we dispense with any consideration of types, say, e.g. that the j'th component in a tuple in $db(R_i)$ must be an integer or a character string, etc. Types can be easily added, but they are not relevant for the development of our theory.

PERSON				
ID	Name	Email	Affiliation	p
i_1	S. Lee	sl@gmail.com	Faculty of Medicine, University of Otago	0.8
i_2	Susan Lee	sl@massey.ac.nz	Massey University	1
i_3	S. Maneth		Faculty of Medicine, University of Otago	0.4
i_4	S. A. Lee	sl@otago.ac.nz	University of Otago	0.7
i_5	S. Lee	sl@massey.ac.nz	Massey University	0.9
i_6	Ben Williams	ben@otago.ac.nz	University of Otago	1
i_7	Ben Williams		Massey University	0.6
i_8	A. Timu	ali@acm.org		1
i_9	A. Timu	atimu@stanford.edu	Stanford University	0.7

AUTHORSHIP			
AuthID	PubID	Order	p
i_2	b_1	1	1
i_6	b_1	2	0.8
i_8	b_1	3	0.2
i_4	b_2	1	1
i_3	b_3	1	0.9
i_5	b_3	2	1
i_7	b_3	3	1
i_9	b_3	4	0.4

PUBLICATION				
PubID	Title	Year	DOI	p
b_1	An effective World Wide Web ...	2001	10.1177/...	0.7
b_2	Irrigation in Modern Agriculture	1994	10.4321/...	0.9
b_3	Algebraic Topology	2011	10.154/...	0.7

Fig. 2 A probabilistic database instance for the schema {PERSON, AUTHORSHIP, PUBLICATION}

3.1 Syntax and Interpretation of Probabilistic Knowledge Patterns

As before an *atom* is a term $\psi = R(t_1, \ldots, t_n)$, where $R \in \mathbf{S}$ and each t_i is either a variable x_i or a constant v_i. If $R \in \mathcal{Q}$ holds (or $R \in \mathcal{R}$, respectively), the atom is called *intensional* (or *extensional*, respectively). We use $var(\psi)$ denote the set of variables in the atom ψ, and extend this notation such that $var(\varphi)$ denotes the set of variables in a formula φ that is composed from such atoms.

Definition 5 A *relational probabilistic conditional* takes the form $\psi(\mathbf{x}) \leftarrow \varphi(\mathbf{x}, \mathbf{y})$ $[\varsigma]$, where $\psi(\mathbf{x})$ is an intensional atom with variables $\mathbf{x}$, $\varphi(\mathbf{x}, \mathbf{y})$ is a conjunction of atoms with variables $\mathbf{x} \cup \mathbf{y}$, and $\varsigma \in [0, 1]$ is a probability.

That is, a relational probabilistic conditional is in essence a Horn clause, but its validity is constrained by a probability.

Definition 6 A *probabilistic knowledge pattern* P over $\mathbf{S}$ is a pair (π, ρ) comprising

- a *pattern rule* π, which takes the form of a relational probabilistic conditional $\psi(\mathbf{x}) \leftarrow \varphi(\mathbf{x}, \mathbf{y})[\varsigma]$ over $\mathscr{S}$, and
- a *pattern relation* ρ of the arity $n + 1$, where the first n position are in 1-1 correspondence to the variables $\mathbf{x} \cup \mathbf{y}$ in φ, and the $(n + 1)$st position can only take

values from $\{+, -\}$. A special variable λ may appear as value in the first n positions in any tuple $t \in \rho$.

We call the probability ς in the pattern rule the *pattern probability* of P.

Definition 7 A *probabilistic knowledge model* is a finite, non-empty set $\mathbf{P}$ of probabilistic knowledge patterns over $\mathbf{S} = \mathbf{R} \cup \mathbf{Q}$.

Example 7 Let us extend Example 6 by some probabilistic knowledge patterns. For this we simply turn the relational knowledge patterns from Example 2 into probabilistic ones.

The name collaboration pattern NC expressed that two persons are identical if they have the same name and there is another person whose name is not "Stephen Smith", whose affiliation is the same with one of the two persons and is at "Faculty of Medicine, University of Otago", and who has co-authored a paper with the other person. Assume that a person named "Stephen Smith" at "Faculty of Medicine, University of Otago" has a colleague and a collaborator who are two different people with the same name.

Now assume that this statement only holds with a probability of 0.9. Then the rule φ_1 becomes

$$\varphi_1 := \text{IDENTICAL}(x, y) \leftarrow$$
$$\text{PERSON}(x, z_6, x_3, x_4) \wedge \text{AUTHORSHIP}(x, z_1, x_1) \wedge \text{AUTHORSHIP}(z', z_1, x_2) \wedge$$
$$\text{PERSON}(y, z_6, y_3, y_4) \wedge \text{IDENTITY}(z, z') \wedge \text{PERSON}(z, z_3, z_4, y_4)[0.9]$$

The pattern relation r_1 remains unchanged. We can make a similar amendment to the co-authorship pattern CA. These little changes have of course an impact on the semantics of knowledge models.

In the following we show how the fixed-point semantics that we defined for relational knowledge models can be generalised to probabilistic knowledge models. The key difference is that in relational knowledge patterns, if the instantiation of the rule body is satisfied in a database instance, the instantiation of the rule head can be added to the instance. In the probabilistic case, however, we have to take the probabilities into account as well, which leads to constraints for a probability distribution over the ground atoms. Using the principle of maximum entropy [22] we can exploit that there exists a unique distribution—the one with maximum entropy—that does not make any additional assumptions.

As in the relational case, each tuple $t \in \rho$ in the pattern relation defines an instantiation of the pattern rule π:

- For $t_i = v_i \neq \lambda$ replace the variable x_i by the constant v_i.
- For $t_i = \lambda$ do not replace x_i.
- For $t_{n+1} = -$ replace the pattern probability ς by 0.

We denote this instantiation by $\Phi_t = \psi(\bar{\mathbf{x}}) \leftarrow \varphi(\bar{\mathbf{x}}, \bar{\mathbf{y}})[\varsigma]$, where for each $x_i \in \mathbf{x} \cup \mathbf{y}$ depending on the tuple $t \in r$ we either have $x_i = v_i$ or $x_i = x_i$, and ς is either the pattern probability or 0. Let $\mathbf{z}$ refers to the sequence of variables in $\varphi(\bar{\mathbf{x}}, \bar{\mathbf{y}})$.

In order to interpret these rule instantiations and thus the probabilistic knowledge patterns take a probabilistic database instance db over $\mathscr{S}$. We will use db as a grounding for the relational probabilistic conditions resulting from the instantiations. While the theory of relational probabilistic conditionals provides different interpretations, we will concentrate here only on the *aggregate semantics*.

So let μ be a valuation of the variables $\mathbf{z}$, and let $\Phi_{t,\mu}$ denote the grounding $\mu(\psi(\bar{\mathbf{x}})) \leftarrow \mu(\varphi(\bar{\mathbf{x}}, \bar{\mathbf{y}}))[\varsigma]$ of the relational probabilistic conditional Φ_t. We consider only those valuations μ with constants that appear in at least one tuple in the probabilistic database db with positive probability. We call such a valuation μ a *db-valuation*. Let $\mathscr{G}$ denote the set of all *db*-valuations for Φ_t. Then the *validity* of Φ_t under the aggregate semantics for the probability distribution p is defined by

$$
p \models_{agg} \Phi_t \quad \Leftrightarrow \quad \frac{\sum\limits_{\mu \in \mathscr{G}} p(\mu(\psi(\bar{\mathbf{x}})) \wedge \mu(\varphi(\bar{\mathbf{x}}, \bar{\mathbf{y}})))}{\sum\limits_{\mu \in \mathscr{G}} p(\mu(\varphi(\bar{\mathbf{x}}, \bar{\mathbf{y}})))} = \varsigma, \tag{1}
$$

where ς is the probability in the instantiation Φ_t.

Definition 8 The *interpretation* $P(db)$ of a knowledge pattern $P = (\pi, \rho)$ on $\mathscr{S}$ with respect to a probabilistic database db over $\mathscr{S}$ is the set $E(P)$ of all equations (1) defined by all instantiations Φ_t with $t \in \rho$.

Note that this definition of the interpretation $P(db)$ does not yet define the probability distribution p such that validity $p \models_{agg} \Phi_t$ holds for all $t \in \rho$. In general, there may be many different solutions for p satisfying all the equations in $P(db)$. However, we will see in the next subsection that the *maximum entropy* principle will enable a unique probability distribution to be defined. While this can be fruitfully exploited to define again a fixed-point semantics for a probabilistic knowledge model analogously to the relational case in Sect. 2, this can only be done for a non-recursive knowledge model. The reason is that if we fix the probability distribution at some stage of the fixed-point iteration, then in case of recursion additional equations may still be produced, which may invalidate the distribution. We will look at the recursive case separately.

3.2   Semantics for a Recursion-Free Set of Patterns

In Definition 8 we defined the semantics of a knowledge pattern P on $\mathscr{S}$ with respect to a probabilistic database db over $\mathscr{S}$ merely by a set of constraining equations for a probability distribution p rather than defining the distribution p itself. While in general there may be many solutions to such sets of constraints, the *maximum entropy* principle permits the selection of a distinguished solution, which comes with no additional probabilistic assumptions. For instance, if nothing is known about a probability distribution, the equal distribution gives rise to the maximum entropy.

Definition 9 The *entropy* of a probability distribution p over events Ω is defined as

$$\mathbf{H}(p) = -\sum_{\omega \in \Omega} p(\omega) \times \log p(\omega) \tag{2}$$

Using the maximum entropy principle we proceed with the definition of a fixed-point semantics for a probabilistic knowledge model $\mathbf{P}$ over $\mathscr{S} = \mathscr{R} \cup \mathscr{Q}$ with respect to a probabilistic database db over the extensional subschema $\mathscr{R}$. However, we will restrict the construction to recursion-free $\mathbf{P}$.

Definition 10 A probabilistic knowledge model $\mathbf{P}$ over $\mathscr{S} = \mathscr{R} \cup \mathscr{Q}$ is *recursion-free* if there exists a partition $\mathscr{S} = \bigcup_{i=0}^{k} \mathscr{S}_i$ satisfying the following two conditions:

- $\mathscr{S}_0 = \mathscr{R}$ and
- whenever $R \in \mathscr{S}_i$ (for $i > 0$) occurs as head predicate of a pattern rule π of $P \in \mathbf{P}$, then all predicates R_j in the body of π satisfy $R_j \in \mathscr{S}_{i(j)}$ with $i(j) < i$.

We call $\mathscr{S}_i$ the i'th *stratum* of the schema $\mathscr{S}$.

Now let $\mathbf{P}$ be a recursion-free probabilistic knowledge model over $\mathscr{S}$, let $\{\mathscr{S}_i\}_{i=0}^{k}$ denote the strata, and take a probabilistic database db over the extensional subschema $\mathscr{R}$. For any $j = 0, \ldots, k$ let $\mathbf{P}_j$ denote the set of probabilistic knowledge patterns from $\mathbf{P}$, the rules π of which contain a predicate $R \in \bigcup_{i=0}^{j} \mathscr{S}_i$. With this we define a sequence of probabilistic databases db_j over $\bigcup_{i=0}^{j} \mathscr{S}_i$ as follows:

- For $j = 0$ we have $db_0 = db$.
- For $j > 0$ let $E(\mathbf{P}_j)$ be the union of all the sets of constraints defined by the interpretations of probabilistic knowledge patterns $P \in \mathbf{P}_j$ with respect to db_{j-1}, i.e. $E(\mathbf{P}_j) = \bigcup_{P \in \mathbf{P}_j} P(db_{j-1})$, and define p_j as the probability distribution with maximum entropy satisfying the constraints in $E(\mathbf{P}_j)$, which determines the probabilistic database db_j.

Then the probabilistic database db_k over $\mathscr{S}$ defines the semantics of the probabilistic knowledge model $\mathscr{P}$ with respect to the probabilistic database db.

Note that the construction of the sequence of databases db_j is monotonic in the sense that the step from db_{j-1} to db_j only adds constraints for predicates in $\mathscr{S}_j$ and thus the probability distribution p_{j-1} is only extended to the distribution p_j, or equivalently $db_{j-1} \subseteq db_j$ holds for all $j = 1, \ldots, k$. In this sense it is justified to continue calling the semantics of $\mathscr{P}$ with respect to db an *inflationary fixed-point semantics*.

3.3 Semantics for a Recursive Set of Patterns

The last remark in the previous section remains only partly true in the general recursive case. If we start again with $db_0 = db$ and build the union of all the sets of constraints defined by the interpretations of probabilistic knowledge patterns $P \in \mathbf{P}$ with respect to db_0, we obtain again constraints for the probability distribution on ground atoms with intensional predicates R that appear in the head of pattern rules π, where the body contains only extensional predicates. However, different from the non-recursice case these may not be all pattern rules with R in the head. That is, if we fix a probabilistic database db_1 according to a probability distribution that satisfies these constraints, in particular using the maximum entropy principle, the continuation of the fixed-point iteration may lead to additional constraints that are not satisfied by this distribution. This holds analogously for each iteration step in the fixed-point construction.

However, building the interpretations of probabilistic knowledge patterns $P \in \mathbf{P}$ is still monotonic, as in each step constraints are only added. Therefore, it is sufficient to consider the set of all equations (1) defined by all instantiations Φ_t with $t \in \rho$ for all probabilistic knowledge patterns in $\mathbf{P}$.

Definition 11 The *interpretation* $\mathbf{P}(db)$ of a set $\mathbf{P}$ of probabilistic knowledge patterns $P = (\pi, \rho)$ on $\mathscr{S}$ with respect to a probabilistic database db over $\mathscr{Q}$ is the union of all sets $E(P)$ of all Eq. (1) defined by all instantiations Φ_t with $t \in \rho$ using all variable assignments with constants that appear in at least one tuple in a relation in db with positive probability.

Definition 12 The *fixed-point semantics* of a probabilistic knowledge model $\mathbf{P}$ on $\mathscr{S}$ with respect to a probabilistic database db over $\mathscr{Q}$ is the probabilistic database db_∞ that corresponds to the probability distribution p with maximum entropy satisfying all constraints in the interpretation $\mathbf{P}(db)$.

Due to the fact that we can still build the interpretation $\mathbf{P}(db)$ using an inflationary fixed-point construction it is justified to call the semantics a fixed-point semantics, though the interpretation can also be built in one step.

4 Discussion and Conclusion

We introduced probabilistic knowledge patterns extending the relational knowledge patterns from [28, 29] in two ways: (1) removing the restriction that intensional predicates in knowledge patterns must refer to equivalence relations, and (2) replacing the Horn clause in a relational knowledge pattern by a relational probabilistic conditional [17]. Then using a probabilistic ground database rather than a relational one we defined first the semantics of rule in a probabilistic knowledge pattern using the aggregate semantics for probabilistic implications. This gives rise to linear equations involving the probability distribution on instances of intensional predicates as unknowns. These linear constraints give rise to a unique distribution with maximum entropy [4, 22], by means of which we generalise the infationary fixed-point semantics for a set of relational knowledge patterns to a fixed-point semantics for the probabilistics extension. In doing so, we make a first step to integrate probabilities into knowledge patterns.

In doing so we achieve several results.

1. The integration can be seen as an exemplification of the theme of the NII Shonan workshop on the integration of explicit and implicit semantics. In the field of entity resolution the knowledge pattern approach exploits explicit knowledge about rules that define identities among entities, whereas common similarity-based methods exploit the existence of a distribution over identities that is implicitly manifested in the given data.

2. For applications to entity resolution the integration removes the restriction that entities are inferred to be either equal or not. Instead identities will come with probabilities as in similarity-based approaches, but the computation is based on the much stricter semantics of probabilistic logic with maximum entropy semantics. On one hand, clustering techniques will become applicable to the result of applying knowledge patterns, and on the other hands, the approach strengthens the foundations of statistical approaches in this area.

3. By removing the restriction on the intentional predicates to correspond to equivalence relations only we open up general data analysis applications beyond entity resolution. For instance, many methods in data analysis are grounded in statistical machine learning exploiting neural networks. Again, these methods exploit implicit semantics, knowing that an unknown distribution that is known to exist, can be obtained (up to some small error) from a well-chosen empirical distribution in the training data. If the network topology, i.e. the number of intermediate layers and the feedback connections, reflect appropriately the implicit distribution, the trained network will provably produce correct results.[2] On the other hand, each network realises a fixed-point computation, which is explicitly modelled in knowledge patterns. It is not yet known, if all machine learning algorithms can be represented by probabilistic knowledge patterns, but the prospective to further investigate this seems promising. In doing so, the whole area of data analysis would become an application area for the theory of knowledge patterns, which opens up almost unlimited opportunities for research and applications.

We expect to be able to generalise the results concerning the optimisation of knowledge patterns [29] to the probabilistic case and to preserve the ability of provenance-based explanation and minimal revision of inferences [35]. In particular, this would enable to learn exceptions to the default reasoning approach provided by knowledge patterns also under conditions of uncertainty. We further intend to generalise probabilistic knowledge patterns even more capturing different probabilities for different groundings of the same clause. We also believe that the contextual constraint theory that was sketched in [30] for the case of monoids, can be generalised capturing further constraint theories and coupling them with the probabilistic extension investigated in this paper.

---

[2]Note that this exploit deep results in probability theory and statistics that are not known to all researchers applying machine learning algorithms.

However, our research aims to go beyond a theory of probabilistic knowledge patterns for probabilistic default reasoning with contexts and exceptions. As claimed in the introduction we want to show that common probabilistic methods in machine learning can be captured by probabilistic knowledge patterns including the various similarity-based methods for entity resolution [9] and methods based on neural networks. This would enable known data-driven learning methods to be adapted to knowledge patterns, and provide opportunities for a true integration of logical and probabilistic methods in data analysis. It will further give rise to the interesting research problem, if the fixed-point construction on grounds of probabilistic logic with maximum entropy semantics can be exploited also directly in the common machine learning methods. All these problems define a rather large spectrum of open problems for future research.

References

1. S. Abiteboul, R. Hull, V. Vianu, *Foundations of Databases* (Addison-Wesley, Boston, 1995)
2. A. Arasu, M. Götz, R. Kaushik: On active learning of record matching packages, in SIGMOD, pp. 783–794 (2010)
3. C. Baral, M. Gelfond, J.N. Rushton, Probabilistic reasoning with answer sets. Theory Pract Logic Program **9**(1), 57–144 (2009)
4. C. Beierle, M. Finthammer, G. Kern-Isberner, Relational probabilistic conditionals and their instantiations under maximum entropy semantics for first-order knowledge bases. Entropy **17**(2), 852–865 (2015)
5. C. Beierle, G. Kern-Isberner: The relationship of the logic of big-stepped probabilities to standard probabilistic logics, in *Foundations of Information and Knowledge Systems, 6th International Symposium, FoIKS 2010* eds. by S. Link, H. Prade. Lecture Notes in Computer Science, vol. 5956 (Springer, Berlin, 2010), pp. 191–210
6. C. Beierle, G. Kern-Isberner, Semantical investigations into nonmonotonic and probabilistic logics. Ann. Math. Artif. Intell. **65**(2–3), 123–158 (2012)
7. J.S. Breese, Construction of belief and decision networks. Comput. Intell. **8**, 624–647 (1992)
8. M. Chavira, A. Darwiche, M. Jaeger, Compiling relational bayesian networks for exact inference. Int. J. Approx. Reason. **42**(1–2), 4–20 (2006)
9. P. Christen, *Data Matching. Data-Centric Systems and Applications* (Springer, Berlin, 2012)
10. W.W. Cohen, Data integration using similarity joins and a word-based information representation language. ACM Trans. Inf. Syst. **18**(3), 288–321 (2000)
11. T.M. Cover, J.A. Thomas, *Elements of Information Theory*, 2nd edn. (Wiley, Hoboken, 2006)
12. P. Domingos, D. Lowd, *Markov Logic: An Interface Layer for Artificial Intelligence*. Synthesis Lectures on Artificial Intelligence and Machine Learning (Morgan & Claypool Publishers, San Rafael, CA, 2009)
13. R. Fagin, J.Y. Halpern, Reasoning about knowledge and probability. J. ACM **41**(2), 340–367 (1994)
14. I.P. Fellegi, A.B. Sunter, A theory for record linkage. J. Am. Stat. Assoc. **64**(328), 1183–1210 (1969)
15. F.V. Jensen, T.D. Nielsen, *Bayesian Networks and Decision Graphs* (Springer, Berlin, 2007)
16. G. Kern-Isberner, *Conditionals in Nonmonotonic Reasoning and Belief Revision—Considering Conditionals as Agents*. Lecture Notes in Computer Science, vol. 2087 (Springer, Berlin, 2001)
17. G. Kern-Isberner, C. Beierle, M. Finthammer, M. Thimm, Comparing and evaluating approaches to probabilistic reasoning: Theory, implementation and applications, in *Transactions on Large-Scale Data- and Knowledge-Centered Systems VI*, pp. 31–75 (2012)

18. G. Kern-Isberner, M. Thimm, Novel semantical approaches to relational probabilistic conditionals, in *Principles of Knowledge Representation and Reasoning: Proceedings of the Twelfth International Conference, KR 2010*, eds. by F. Lin, U. Sattler, M. Truszczynski (AAAI Press, New York, 2010)
19. K. Kersting, L. De Raedt, Bayesian logic programming: Theory and tool, in *An Introduction to Statistical Relational Learning*, eds. by, L. Getoor, B. Taskar (MIT Press, Cambridge, 2007)
20. N.J. Nilsson, Probabilistic logic. Artif. Intell. **28**(1), 71–87 (1986)
21. D. Nute, C. Cross, Conditional logic, in *Handbook of Philosophical Logic*, eds. by D. Gabbay, F. Guenther, vol. 4 (Kluwer Academic Publishers, Dordrecht,2002), pp. 1–98
22. J.B. Paris, Common sense and maximum entropy. Synthese **117**(1), 75–93 (1998). https://doi.org/10.1023/A:1005081609010
23. J.B. Paris, A. Vencovská, In defense of the maximum entropy inference process. Int. J. Approx. Reason. **17**(1), 77–103 (1997). https://doi.org/10.1016/S0888-613X(97)00014-5
24. J. Pearl, *Probabilistic reasoning in intelligent systems—Networks of plausible inference* (Morgan Kaufmann, Burlington, 1989)
25. M. Richardson, P. Domingos, Markov logic networks. Mach. Learn. **62**(1–2), 107–136 (2006)
26. S. Sarawagi, A. Bhamidipaty, Interactive deduplication using active learning, in Knowledge discovery and data mining, pp. 269–278 (2002)
27. K.D. Schewe, Q. Wang, On the decidability and complexity of identity knowledge representation, in *Database Systems for Advanced Applications - 17th International Conference (DASFAA 2012)*, eds. by S. Lee, Z. Peng, X. Zhou, Y.S. Moon, R. Unland, J. Yoo. Lecture Notes in Computer Science, vol. 7238 (Springer, Berlin, 2012), pp. 288–302
28. K.D. Schewe, Q. Wang, Knowledge-aware identity services. Knowl. Inf. Syst. **36**(2), 335–357 (2013)
29. K.D. Schewe, Q. Wang, A theoretical framework for knowledge-based entity resolution. Theor. Comput. Sci. **549**, 101–126 (2014)
30. K.D. Schewe, Q. Wang, M. Rady, Knowledge-based entity resolution with contextual information defined over a monoid, in Model and Data Engineering - 5th International Conference (MEDI 2015), eds. by L. Bellatreche, Y. Manolopoulos. Lecture Notes in Computer Science, vol. 9344 (Springer, Berlin, 2015), pp. 128–135
31. P. Singla, P. Domingos, Object identification with attribute-mediated dependences, in *Knowledge Discovery in Databases (PKDD)* (Springer, Berlin, 2005), pp. 297–308
32. M. Thimm, G. Kern-Isberner, On probabilistic inference in relational conditional logics. Logic J. IGPL **20**(5), 872–908 (2012)
33. V.S. Verykios, G.V. Moustakides, M.G. Elfeky, A Bayesian decision model for cost optimal record matching. VLDB J. **12**(1), 28–40 (2003)
34. K. Wagstaff, C. Cardie, Clustering with instance-level constraints, in *AAAI*, p. 1097 (2000)
35. Q. Wang, K.D. Schewe, W. Wang, Provenance-aware entity resolution: Leveraging provenance to improve quality, in *Database Systems for Advanced Applications—20th International Conference (DASFAA 2015)*, eds. by M. Renz, C. Shahabi, X. Zhou, M.A. Cheema. Lecture Notes in Computer Science, vol. 9049 (Springer, Berlin, 2015), pp. 474–490

Proof Based Modelling

Proof-Based Modeling

An Explicit Semantics for Event-B Refinements

Pierre Castéran

Abstract We present a semi-shallow embedding in *Coq* of *Event-B*'s notions of abstract machine and refinement. The abstract structure of *Event-B* developments, including machines and refinement annotations, can be represented within *Coq*'s type system, using inductive and dependent types. This formalization allows us to reason at the meta-level on machines and their behaviors, considered as first-class citizens. We show how this formalization of *Event-B* structures into *Coq* allows us to model *Rodin*'s proof obligations, and prove how these obligations entail correctness properties of a given *Event-B* project. Moreover, the correctness of a given refinement is now an explicit theorem, instead of being an implicit consequence of a set of proof obligations.

1 Introduction

Within the paradigm of *correctness by construction*, the notion of *refinement* is a way to obtain a piece of software that is consistent with a given specification. For instance, the *Event-B* method [1], considers the whole development of a reactive system as a sequence of transition systems called *abstract machines*. The first element of this sequence is the given specification, and the last one should be ready to be effectively implemented through an automatic translation to a "classical" programming language. This method consists in proving that each machine of the sequence is a refinement of the preceding one, i.e., every execution of the refinement corresponds to some execution of the abstract machine.

The *Rodin* tool [2] helps the user to apply the *Event-B* method. It is mainly composed of a *proof obligation generator* and several automatic provers. Proof obligations (*PO*s) are theorem statements generated from the *Event-B* sources. The user of *Rodin* is happy when all the POs generated from the components of his/her project are proved. The meaning of each PO and why its validity entails the correctness

---

P. Castéran (✉)
LaBRI, University of Bordeaux, CNRS (UMR 5800) INP-Bordeaux, Bordeaux, France
e-mail: pierre.casteran@labri.fr

© Springer Nature Singapore Pte Ltd. 2021   155
Y. Ait-Ameur et al. (eds.), *Implicit and Explicit Semantics Integration
in Proof-Based Developments of Discrete Systems*,
https://doi.org/10.1007/978-981-15-5054-6_8

of the whole development is described either semi-formally in tutorials, or in the framework of simple set theory in [1].

We propose to go further, by the development of a formal, computer verified, description of the notions of abstract machines and refinements. Our plan is to express within *Coq*'s type system the abstract syntax of *Event-B* projects. Then we associate to any abstract machine a labeled transition system [3]. Finally, we study the relationship between *Event-B* constructs, the proof obligations generated from the machines and refinements, and the behavior of the associated transition systems.

For simplicity's sake, we do not take into account some parts of the *Event-B* formalism that are not directly related to machine refinement, mainly the WD proof obligations and the set inclusion $\mathbb{N} \subset \mathbb{Z}$. In *Coq*, all functions are total and $\mathbb{N}$ is not a subtype of $\mathbb{Z}$. We adopt this point of view in our examples: partial functions are expressed as total functions with possibly dependent types, and we use only natural numbers. A thorough treatment of all POs generated from an *Event-B* component remains to be done.

Convention: We shall use the `tt` font family for citing parts of *Coq* or *Event-B* code. For readability's sake, compound expressions will be surrounded by light parentheses, *which are not part of Coq's syntax*, like in $(x + y)$. Otherwise, we use *italics* for designing the current value of some variable like in "$x \geq 33$" or a meta-variable, like in "let *P* be some invariant".

2 A Small Example

For illustrating our definitions, we start by presenting a very simple and artificial example, consisting in an *Event-B* context, an abstract machine and a refinement that model a *counter*, a device that maintains a value bounded by some constant, with two possible actions: jumping—if possible—to a greater value, and resetting the counter value to 0.

2.1 Formalization in Event-B

The *Event-B context* C0 below declares the maximum value of the counter, as a stricly positive natural number N. Thus, all the further development is parameterized by this arbitrary value.

```
CONTEXT C0
CONSTANTS N
AXIOMS
  axm1 : N ∈ ℕ₁
END
```

```
MACHINE M0 SEES C0
VARIABLES
 x
INVARIANTS
 inv1 : x ∈ 0 .. N

EVENTS
 INITIALISATION≙
 x := 0 END

 jump≙
 ANY i  WHERE  0 < i ∧ x + i ≤ N
 THEN   x := x + i END

 reset≙
 WHEN x > 0  THEN x := 0  END
END
```

Fig. 1 The machine M0

The abstract machine M0 (see Fig. 1) models the counter through a variable x, the range of which is the set of integers between 0 and N. Its two main events are as follows:

jump: If the current value of x is less than N, then x is incremented by any value compatible with the constraint $x \leq N$.

reset: If $x > 0$, then x is set to 0.

We consider now a *refinement* M1 of M0, presented in Fig. 2, with three variables and four events.

The three events start, jump, and incr are controlled by a Boolean variable mode. More precisely, the *one-shot* event jump of machine M0 is refined in M1 by the following sequence:

1. The event start sets the Boolean mode to TRUE and the variable y to the value x + 1.
2. The event incr, which adds 1 to the value of y, is repeated any number of times, provided that the value of y does not exceed N.
3. Finally, jump copies the value of y into x.

Please notice that, only the event jump of M1 is declared to refine the event jump of machine M0. The events start and incr are *new events,* which correspond to no transition in M0. This is a usual pattern in *Event-B*, due to its *forward simulation semantics*: when some abstract event is refined by a sequence of events, only the last event of that sequence refines the abstract event.

```
MACHINE M1 Refines M0
VARIABLES
 x  y  mode
INVARIANTS
inv1 : mode = TRUE ⇒ x < y
inv2 : y ≤  N

EVENTS:
INITIALISATION≙
y, x, mode := 0, 0, FALSE

start≙
WHEN mode = FALSE ∧  x <  N
THEN  y, mode := x + 1, TRUE END

incr≙
WHEN mode = TRUE ∧  y <  N
THEN y := y + 1
END

jump≙
REFINES jump
WHEN mode = TRUE
WITH i: i = y - x
THEN x, mode := y, FALSE  END

reset≙
REFINES reset
WHEN mode = FALSE ∧ x > 0
THEN x, y := 0, 0  END

END
```

Fig. 2 The refinement M1 of M0

2.2 Interpretation of Machines as Transition Systems

Transition systems are a powerful and well-studied tool for describing reactive systems and their behavior. The reader may find in G. Tel's book [3] a mathematical treatment of transition systems and the notions of execution, invariant, fairness, etc. Associating a transition system to any *Event-B* machine gives us access to all these concepts and their properties.

Traditionally, labeled transition systems are represented as structures of four components:

- A set C of global states, also called *configurations*,
- A set L of event labels,
- A function tr that maps every event label to a binary "before/after" relation on C,
- A subset $I \subseteq C$ of *initial configurations*.

For instance the transition system associated with the machine M0 can be described as follows[1]:

- Since M0 contains only one variable, each of its configurations can be represented by a natural number: $C = \mathbb{N}$.
- The set of event labels is $L = \{\text{incr}, \text{reset}\}$.
- The transition relations associated with the events reset and jump of M0 are, respectively, described by the following expressions:

$$\text{reset}: \lambda x\, x'. (x > 0 \wedge x' = 0)$$
$$\text{jump}: \lambda x\, x'. \exists i \in \mathbb{N}. (0 < i \wedge x + i \leq \mathbb{N} \wedge x' = x + i).$$

- The set of initial configurations is reduced to the singleton $I = \{0\}$, in other terms the predicate $\lambda x. x = 0$.

2.3 On Event Parameters

In the description of the transition system associated with M0, the event parameter i is introduced by an existential quantifier in the relation associated with the event jump. But the scope of a quantifier is *local* to the sub-formula it occurs in, thus the variable i is unknown outside this quantification. This feature prevents us from keeping the value of the parameter in *traces* of execution, and from giving a logical meaning to the WITH clause of M1's event jump, which contains a reference to i.

In the next section, we will show how *dependent types* allow us to formalize properly transition systems with event parameters, giving a type to traces and WITH clauses in the refined events.

3   A Very Short Presentation of the *Coq* Proof Assistant

Coq [4] is an interactive *proof assistant*, i.e., a tool that makes it possible to develop proven programs and prove mathematical theorems, with a very high degree of confidence. For instance, it has been used for building a *certified* C-compiler [5] and for giving complete proofs of the four-color [6] and Feit–Thompson [7] theorems. *Coq* is based on a very expressive type system called the *Calculus of Inductive Constructions* (*CIC*). There is no room in this chapter for a presentation of the main features of *Coq*, thus we shall limit ourselves to present the constructions that are actually needed by our study.

---

[1]Following the *Coq* tradition, we represent sets [resp. binary relations] as unary [resp. binary] predicates on types, with λ as the binding symbol for abstraction. For instance, the set of even natural numbers is described by the predicate $\lambda i : \mathbb{N}, \exists j : \mathbb{N}, i = 2 \times j$.

Besides its powerful type system, *Coq* provides the user with a large collection of tools called *tactics* for building potentially large and complex proofs. The user is also able to define her/his own tactics for semi-automating proof construction in specific domains. Once built, every proof is checked to verify there is no missing part nor bad application of a logical rule.

The interested reader may consult tutorials on *Coq*'s page [4] and Software Foundations [8] or books like [9, 10].

3.1 Terms and Types

Unlike *Event-B*, which is based on classical set theory, *Coq* is based on type theory, and implements higher order intuitionistic logic. Nevertheless, its syntax is very close to mathematical logic, thus the reader will have no problem to understand the following definitions.

3.1.1 Typing Judgements

In *Coq*, every well-formed term has a type. The *judgement* "t has type A" is written "$t : A$". The validity of such a judgement may depend on a *context* formed with declarations and definitions. For instance, the following declaration introduces a natural number i, an arbitrary type A and some object of type A.

```
Variables (i:nat) (A: Type) (a: A).
```

3.1.2 Basic Types

The following types are already defined in *Coq*'s standard library.

- `nat` is the type of natural numbers
- `bool` is the type of Boolean values
- `unit` is a *singleton* type, with a unique value called `tt`; the type `unit` plays a role similar to *C*'s type `void`, and `tt` corresponds to `()`. In Sect. 4.2.2, `unit` is used as a "default type" for parameterless events.

3.1.3 Function Types

If A and B are types, then $A{\rightarrow}B$ is the type of total functions from A to B. Note that the operator is right associative, thus the type $A{\rightarrow}B{\rightarrow}C$ is an abbreviation of $A{\rightarrow}(B{\rightarrow}C)$. The application of a function f to x is simply noted ($f\ x$). Function application is left-associative, so the expression ($f\ x\ y$) is an abbreviation of

$((f \ x) \ y)$. The λ-abstraction is written in *Coq* with the keyword `fun`. For instance, the function that computes the arithmetic mean of two natural numbers is written (`fun x y : nat => (x + y)/2`).

3.1.4 Product Types and Records

If $A_1, \ldots, A_n$ are types, the product $A_1 \times \cdots \times A_n$ is the type of tuples $(a_1, \ldots, a_n)$, where $a_i : A_i$ for every i such that $1 \leq i \leq n$. In our formalization, we use also *records* with named fields, quite similar to structures in C. The basic use of records is commented in Sect. 4.1.1.

3.1.5 Propositions and Predicates

The *sort* Prop is the type of logical *propositions*. Predicates on type A are functions of type $A \rightarrow Prop$. Relations from A to B have type $A \rightarrow B \rightarrow Prop$.

Coq's standard library introduces the abbreviations (`Ensemble A`) for $A \rightarrow Prop$ and (`relation A`) for $A \rightarrow A \rightarrow Prop$.

For instance, the following expression describes the binary relation associated with integer division by 2:

```
fun x y:nat => x = 2 * y \/ x = 2 * y + 1.
```

According to the Curry–Howard correspondence, the same symbol $\rightarrow$ is used both for writing functional types and intuitionistic implication.

3.1.6 Dependent Types

Coq's type system includes dependent types, i.e., types that may depend on data, programs, or proofs. For instance "vector of length n", "prime number", and "divisor of p" are such types. Dependent types allow us to write formal specifications of programs. In this chapter, dependent types are used for giving a correct type to guarded events and execution traces in generic definitions. The book by Chlipala [10] presents a state of the art on the use of dependent types for writing "certified" programs.

3.1.7 Heterogeneous Lists

Heterogeneous lists [10] are sequences where the type of every item can be computed by calling some function. More precisely, let A be any type, and $f : A \rightarrow Type$ be some function that associates a type to any of its argument. Let now $s = a_0, \ldots, a_n$ be a list of values of type A. Then (`hlist f s`) is the type of the lists the ith element of which has type $f(a_i)$.

3.1.8 Module Notation

Coq's module system allows us to share several identifiers, like event labels, machine variables, etc., between various machines. Every machine description is considered as a module, so, in case of ambiguity, we will use the module notation like `MO.jump` or `MO.x` for designing the event `jump` or the variable `x` of machine `MO`.

4 Formalizing Abstract Machines in *Coq*

In *Event-B*, machines generally have several variables and invariants. In order to simplify our meta-theory, we will consider that machines have a single global state, called *configuration*, and only one invariant. This is not a loss of generality, since the type of configurations may be a record type with several fields, and an invariant may be a conjunction of more elementary invariants.

4.1 Configurations and Event Labels

Our definitions are parameterized with two type variables: the variable `C` for any type of configurations, and `L` for event labels. In *Coq*, one has to declare both variables of sort `Type`. This declaration is local to a *Section* named `Definitions`. At the end of this section, every definition and theorem that depend on local variables will be automatically generalized, making this part of the development fully *generic*.

```
Section Definitions.
Variable C : Type. (* Configurations *)
Variable L : Type. (* Event labels *)
```

An *instantiation* mechanism will allow us to replace these variables by more concrete values in our examples. For sake of clarity, we will specify, for each piece of *Coq* code, whether it is a generic definition or theorem or an application to our example.

4.1.1 Examples

At the object level, the variables `C` and `L` can be instantiated to data types which represent the configurations and event names of specific machines. For instance the machine `MO` contains a unique variable `x` and two event names: `jump` and `reset`. Thus, the variable `C` can be instantiated by a record type `config` with one field name `x` and `L` by an enumerated type `evt` with two values: `jump` and `reset`.

```
(* "variables"  *)
Record config := mk_conf {x : nat}.

(* Event labels *)
Inductive evt :=  jump | reset.
```

In machine M1, the types for configurations and event names are slightly more complex. Configurations are records with three components which correspond, respectively, to M1's variables.

```
(* Configurations *)
Record config := mk_conf {x : nat; y : nat ; mode : bool}.

(* Event names  *)
Inductive evt := incr | jump | reset | start.
```

The function mk_conf is a *constructor* that is used for building a new configuration when applied to two natural numbers and a Boolean. Reciprocally, if the variable γ has type config, then the term $(x\ \gamma)$ returns the value of the field x of the configuration γ.

4.2 Dependently Typed Events

In *Event-B*, events are composed of *guards* and *actions*. A guard is a condition that the current configuration must satisfy for the event to be triggered. The action part describes the "before/after" relation that holds between the old and new configurations.

The guard of the event M0.jump, depends on the parameter i of type nat. This parameter is also used in the action part of jump. On the contrary, the other events of machines M0 and M1 do not have any event parameter. For having a uniform treatment of events and machines, we have decided to state that every event has a unique parameter. *Rodin*'s parameterless events will be considered as parameterized with *Coq*'s singleton type unit. Events with several parameters will be considered as parameterized with Cartesian products or record types.

4.2.1 Guards and Actions

In *Event-B*, a guard is a proposition that depends on the machine's variables and the potential parameter of the event. In *Coq*, we can define a type of guards, indexed by an arbitrary type Pa of event parameter.

```
Definition guard_t (Pa:Type) := C -> Pa -> Prop.
```

For instance, the guard of event M0.jump is the predicate $\lambda \gamma\ i.\ (0 < i \wedge x \gamma + i \leq N)$ of type M0.config$\rightarrow$nat$\rightarrow$*Prop*, while the guard of M0.reset has type M0.config$\rightarrow$unit$\rightarrow$*Prop*.

The action part of an event describes the relation that links the configurations before and after the execution of the event (called the *before-after* relation in *Rodin*'s documentation). This relation may depend on the event parameter. We consider both *deterministic* (i.e., functional) or *non-deterministic* assignments.

```
Inductive assignment (Pa: Type) :=
  |  det (f :  C ->  Pa -> C)
  | ndet (r : C -> Pa -> Ensemble C).
```

The parameterized type `guarded_event` combines a guard and an assignment.

```
Record guarded_event (Pa:Type):  Type :=
    mk_evt {grd :  guard_t Pa;
            action :  assignment Pa}.
```

For instance `M0`'s events are built through the constructor `mk_evt`. Please note the argument of `guarded_event`, respectively, set to `nat` and `unit`.

```
Definition jump_evt : guarded_event nat :=
  mk_evt (fun gamma i  => 0 < i /\ x gamma + i <= N)
         (det (fun gamma i  => mk_conf (x gamma + i))).

Definition reset_evt  : guarded_event unit :=
  mk_evt (fun gamma  _ => 0 < x gamma)
         (det (fun gamma _   => mk_conf 0)).
```

4.2.2 Types of Abstract Machines

We have now all the ingredients for defining at the abstract level the structure of an *Event-B*-machine. Recall that the variable `C` denotes any type of configurations, and `L` any type of event labels. A machine is described by the three following components:

- A function `param_type` of type $L \rightarrow Type$ that maps any event name *e* to the type of its parameter.
- A function `evts` that maps any event label *e* to a guarded event. The type of `evts` is a *dependent product*, where `param_type` is applied to determine the type of the event's parameter:

  ```
  evts : forall e:L, guarded_event  (param_type e)
  ```

- Finally, a set of initial configurations

  ```
  Init : Ensemble C
  ```

4.2.3 Examples

The abstract machine `M0` can be described by instantiating the variables `param_type`, `evts` and `Init` or our model.

```
Definition param_type (e:evt) : Type :=
  match e with
    | jump => nat
    | reset => unit
  end.
```

The function that maps every label to a guarded event of the right type is defined as follows:

```
Definition evts (e:evt) : guarded_event (param_type e):=
  match e with
    | jump => jump_evt
    | reset => reset_evt
  end.
```

Finally, the set of initial configurations is described by its characteristic predicate.

```
Definition Init (gamma : config) := x gamma = 0.
```

Likewise, the abstract machine M1 is defined in *Coq* in Fig. 3.

Remark 1 Please note that, unlike *Event-B*, our description of M0 and M1 do not contain any refinement nor invariant description. In our formalization, transition

```
Definition param_type (_ : evt) : Type := unit.

Definition evts (e:evt) : guarded_event config (param_type e)
 := match e with
| start => mk_evt
  (fun gamma _ => mode gamma = false /\ x gamma < N)
  (det (fun gamma _ => mk_conf (x gamma)
                               (1 + x gamma)
                               true))
| incr => mk_evt
  (fun gamma _ => mode gamma = true /\ y gamma < N)
  (det (fun gamma _ => mk_conf (x gamma)
                               (1 + y gamma)
                               (mode gamma)))
| jump => mk_evt
  (fun gamma _ => mode gamma = true)
  (det (fun gamma _ => mk_conf (y gamma)
                               (y gamma)
                               false))
| reset => mk_evt
  (fun gamma _ => mode gamma = false /\ x gamma > 0)
  (det (fun gamma _ => mk_conf 0 0 (mode gamma)))
end.

Definition Init (gamma:config) :=
  y gamma = 0 /\ x gamma = 0 /\ mode gamma = false.
```

Fig. 3 Description of Machine M1 in Coq

systems are just defined by guards, events, and initial configurations. Invariants and refinement annotations *refer* to transition systems but are not parts of them as in *Event-B* syntax.

5   Operational Semantics of Abstract Machines

The behavior of a given machine can be expressed through the set of its executions. A (finite) execution in a machine M is a (finite) sequence of configurations $\gamma_0, \gamma_1, \ldots, \gamma_n$, where γ_0 is an initial configuration, and each γ_{i+1} is obtained from γ_i by the execution of some event of the machine. It is also possible in *Coq* to define infinite executions, but we do not need this extension here.

For defining the set of executions, we need to associate a relation to the events of the machine, i.e., to consider machines as transition systems. First, we associate a binary "before/after" relation on configurations to any assignment (deterministic or non-deterministic).

```
Definition a2rel {Pa: Type} (a:assignment Pa) gamma p  :=
    fun gamma' =>
      match a with
        | det f => gamma' = f gamma p
        | ndet r => r gamma p gamma'
      end.
```

Then, we restrict our relation to the configurations and parameters that satisfy the guard.

```
Definition event_ba_dep {Pa : Type}(e: guarded_event Pa)(p:Pa)
  := fun gamma gamma' =>  grd e gamma p  /\
                          a2rel  (action e)  gamma p  gamma'.
```

Finally, we abstract the parameter p, then the label event e, using existential quantifications:

```
Definition event_ba {Pa : Type} (e: guarded_event Pa) :=
    fun gamma gamma' => exists p, event_ba_dep e p gamma gamma'.

Definition tr e gamma gamma' : Prop :=
    event_ba  (Pa:= ev_params e) (evts e) gamma gamma'.

Definition transition: relation C :=
    fun gamma gamma' => exists e: L, tr e gamma gamma'.
```

The predicate `transition` is a description of the labeled transition system associated with a machine. The reflexive and transitive closure of `transition` allows us to define the following predicates:

- The configuration γ' is reachable from γ:

```
Definition reachable_from  : relation C :=
    clos_refl_trans  transition.
```

- The configuration γ is reachable (from some initial configuration):

```
Definition reachable  : Ensemble C :=
    fun gamma => exists gamma_i, Init gamma_i /\
                    reachable_from  gamma_i gamma.
```

5.1   On Invariants

The notion of invariant we use is the same as G. Tel's: an invariant is just a property on configurations that holds in any initial state and is preserved by the transition relation associated with the considered machine.

For instance, the following predicate is an invariant of machine M0:

```
Definition inv (gamma:config)  := x gamma <= N.
```

For proving that `inv` is an invariant of M0, we have just to prove that any initial configuration satisfies `inv`, and that the relations associated with the events `jump` and `reset` maintain this invariant. It is easy to write this proof using *Coq* tactics, but it presents no great interest, since it corresponds to *Rodin*'s INV proof obligations, which are solved automatically by this tool.

Nevertheless, our formalization of machines contains *at the meta-level* the theorem that states that, if P is an invariant of any machine M, then this predicate is "always true", i.e., holds in any reachable configuration.

```
Lemma inv_true (P :  C -> Prop)  :
    invariant P -> forall gamma, reachable gamma -> P gamma.
```

5.2   Deadlock Freeness

The theorem that ensures that a given machine M is deadlock-free can be defined at the abstract level. First we define a predicate on configurations that state there exists some event e whose guard is provable. Note that the following definition is generic and applies to any machine, thanks to the function `param_type` and the dependently typed function `grd`.

```
Definition DLF (gamma:C)  :=
    exists e, exists p:param_type e, (grd e) gamma p .
```

For proving that a given machine is alive, it suffices to prove that, for any reachable configuration γ, ($\text{DLF}\ \gamma$) holds, and that for every event e, the action associated with e is feasible, i.e., leads to some configuration γ', provided the guard of e is true.

5.3 Traces

The *trace* of an execution is a sequence of transitions leading from some configuration γ_0 to another configuration γ_n. Each step of this sequence can be described as a triple (e_i, p_i, γ_i), where e_i is the event label, p_i the value of the event parameter of e_i, and γ_i the configuration after the event execution.

Recall that the type of the parameter p_i depends on the event label, thus a trace must be represented as a sequence of values that may have different types.

The following definition implements traces as heterogeneous lists indexed by sequences of event labels; if e is such a label, the corresponding element of a trace is a pair composed of a parameter of type (`param_type e`) and a configuration.

```
Definition f_trace_t := hlist (fun e => ((param_type e) * C)).
```

If s is a sequence of event labels, then (`f_trace_t s`) is the type of traces associated with s. We proved formally in *Coq* that, if a configuration γ' is reachable from γ, then there exists some trace leading from γ to γ'.

6 Formalizing Refinements

In *Rodin* refinements are declared as parts of the "concrete" machine M_C that refines an "abstract" machine M_A, under the form of REFINES annotations at both the global level of the machine and local level of events. The invariants of M_C can refer to variables of M_A. The *Rodin* tool implicitly states that if a variable v is shared by M_A and M_C, then it must have the same value in both machines during "similar" executions.

As for the invariants in Sect. 5.1, we consider properties of refinements as predicates on machines, instead of being part of the machines.

First, let us define refinements at the level of events, then we will extend this notion to abstract machines.

6.1 Parameterization of the Definition

Our formal definition of refinement is parameterized by a concrete machine MC and an abstract machine MA. Let us introduce the description of both machines, through a set of variable declarations.

```
Variables CC  (*configurations of the concrete machine *)
           CA  (* configurations of the abstract machine *)
             : Type.

(* types of event labels *)
Variables LC LA : Type.

(* Types of event parameters *)
Variable param_typeC  : LC -> Type.
Variable param_typeA  : LA -> Type.

(* guarded events *)
Variable evtsC : forall e, guarded_event CC (param_typeC e).
Variable evtsA : forall e, guarded_event CA (param_typeA e).

(* initial configurations *)
Variable InitC : Ensemble CC.
Variable InitA : Ensemble CA.
```

6.2   Event Refinement

An event of the concrete machine M_C may refine or not several events of M_A. Event refinement is thus described through a binary relation between concrete and abstract event labels. Please note that in the current state of our formalization, we do not take into account *Event-B merge refinement* yet.

Let us call *new* any event that does not refine any event of M_A.

```
Variable evt_refines : LC -> LA -> Prop.
```

For instance, the REFINES annotation of M1 can be described as follows:

```
Definition evt_refines (eC: C_evt) (eA  : A_evt) : Prop :=
  match eC, eA  with
    jumpC, jumpA | resetC, resetA => True
  | _, _ => False
  end.
```

Possible constraints relating the parameters of a concrete and an abstract event are expressed by a dependently typed predicate. Please note that the hypothesis H restricts this constraints to event refinements. These constraints express the WITH annotations and the equality between parameters of the same name (for instance when a parametrized event is *extended*).

```
parameter_constraints: forall (eC:evt_C) (eA : evt_A)
    (H:evt_refines eC eA),
    CC -> ev_paramsC eC ->
    ev_paramsA eA -> Prop.
```

For instance, the relationship between the configurations and parameters of the pair of events M1.jumpC and M0.incrA is described by the following predicate:

```
fun (gC: CC) (_ :  unit)
    (i : nat) => (i = y gC - x gC))
```

6.3 Gluing Invariants

In *Event-B*, the invariants of the machine M_C may refer to M_A's variables. Thus, we propose to consider that, by default, an invariant is a gluing invariant, i.e., a predicate that relates configurations of machines M_C and M_A. Let us define a type of "gluing properties":

```
Definition GlueP := CC -> CA -> Prop.
```

Note that machine M_C's "proper invariants" can be easily transformed into gluing invariants by the following "coercion".

```
Definition Glue_lift (P : CC -> Prop) : GlueP :=
  fun (gammaC:CC) (gammaA: CA) => P gammaC.
```

6.3.1 Example

The two following predicates are the direct translation of the invariants inv1 and inv2 of Fig. 2.

```
Definition inv1 (gC: M1.config) (gA: M0.config)  :=
   mode gC = true -> x gC < y gC.

Definition inv2 (gC: M1.config) (gA: M0.config)  :=  y gC <= N.
```

6.3.2 Variable Sharing

The implicit convention on variable sharing can be made explicit by adding an invariant stating that the common variables have always the same value in the concrete and the abstract machine. In our example, the variable x is shared by the machines M1 and M0. This is expressed by the following predicate:

```
Definition x_share (gC: M1.config) (gA: M0.config)  :=
   x gA = x gC.
```

Thus, the gluing invariant associated to our refinement is the conjunction of inv1, inv2 and x_share.

Fig. 4 Relation associated
to event refinement

$$
\begin{array}{ccc}
\gamma_A & \xrightarrow{\;e_A/p_A\;} & \gamma'_A \\
\uparrow & & \uparrow \\
\gamma_C & \xrightarrow{\;e_C/p_C\;} & \gamma'_C
\end{array}
\qquad\qquad
\begin{array}{ccc}
\gamma_A & \xrightarrow{\;=\;} & \gamma'_A \\
\uparrow & & \uparrow \\
\gamma_C & \xrightarrow{\;e_C/p_C\;} & \gamma'_C
\end{array}
$$

$$(1) \qquad\qquad\qquad (2)$$

6.4 Proving Gluing Invariants

In order to give an operational definition of refinement, we have to relate executions
and traces of the concrete and the abstract machines.

Figure 4 describes in two parts the relation associated with the execution of an
event e_C with parameter value p_C.

- In diagram (1), e_C refines some event e_A. We assume that γ_C, p_C, and p_A satisfy
  the constraint described in Sect. 6.2.
- Diagram (2) corresponds to a "new" event e_C.

Both diagrams characterize a "before/after" relation between configurations of
the two machines. Let us denote this relation by $\gamma_C, \gamma_A \xrightarrow{e_C/p_C} \gamma'_C, \gamma'_A$.

Let $P : CC \rightarrow CA \rightarrow Prop$ be some predicate. We define the property "P is a gluing
invariant" if the two following propositions hold:

- For any $\gamma_C \in \text{InitC}$, there exists $\gamma_A \in \text{InitA}$ such that $P\ \gamma_C\ \gamma_A$ holds,
- For any reachable configurations γ_C and γ_A such that $P\ \gamma_C\ \gamma_A$ holds, and any transition $\gamma_C \xrightarrow{e_C/p_C} \gamma'_C$, there exists some configuration γ'_A such that $\gamma_C, \gamma_A \xrightarrow{e_C/p_C}$
  γ'_C, γ'_A and $P\ \gamma'_C\ \gamma'_A$ hold.

Proving that some predicate P is a gluing invariant corresponds tightly to the INV
proof obligations in *Rodin*.

6.4.1 Traces and Refinements

Figure 5 shows how to make correspond an execution in M_C with some execution
in M_A. It is an iteration of the relation described in Fig. 4. Considering the traces t_C
and t_A of the respective executions, we define a relation "$\gamma_C, \gamma_A \xrightarrow{t_C, t_A} \gamma'_C, \gamma'_A$,

Fig. 5 Trace
correspondance with respect
to a refinement

$$
\begin{array}{ccccccc}
\gamma_A & \xrightarrow{\;e_A/p_A\;} & \gamma_{A,1} & \xrightarrow{\;=\;} & \gamma_{A,2} & \dashrightarrow & \gamma'_A \\
\uparrow^P & & \uparrow^P & & \uparrow^P & & \uparrow^P \\
\gamma_C & \xrightarrow{\;e_C/p_C\;} & \gamma_{C,1} & \xrightarrow{\;e_{C,1}/p_{C,1}\;} & \gamma_{C,2} & \dashrightarrow & \gamma'_C
\end{array}
$$

By induction on traces, we prove at the meta-level the following statement, which describes formally the notion of machine refinement.

Theorem 1 *Let P be some gluing invariant for machines M_C and M_A. Assume $P \gamma_C \gamma_A$. Then for any execution in M_C leading to some configuration γ'_C with trace t_C, there exists some execution in M_A leading from γ_A to some configuration γ'_A with trace t_A, where the propositions γ_C, $\gamma_A \xrightarrow{t_C, t_A} \gamma'_C$, γ'_A and $P \gamma'_C \gamma'_A$ hold.*

Once proved within the context described in Sect. 6.1, this theorem is automatically completed by universal quantifications over configuration types, machines and invariants, and can be applied to specific *Event-B* components.

7 Conclusion and Future Work

The richness of *Coq*'s type system, and particularly dependent types, allows us to give a precise definition of the operational semantics of *Event-B* components. Thus, we are able to prove *explicit* theorems such as "every execution of M1 simulates some execution in M0", and all the corollaries that can be further deduced, with the help of the proof assistant.

Our objective is not to use *Coq* for proving *Rodin*'s proof obligations, but to explore formally the notions of machines and refinements considered as first-class citizens. This work can be extended in the following directions:

- Design tactics for proving that the new events in a refinement cannot run forever. In our example, we would say that the relation associated with the events start and incr of machine M1 is well founded.
- Take into account the refinement of several abstract events by a given abstract event ("merge refinement").
- Automatize the translation into *Coq* of real *Event-B* projects. The formalization of our examples have been typed by hand, but they should be obtained automatically from the abstract syntax of *Event-B* components.
- Prove formally the correctness of *Rodin*'s proof obligation generator, and make explicit the correspondance with *Rodin*'s POs and the hypotheses of theorems like Theorem 1. It should be interesting to apply such theorems to statements expressed and proved with *Rodin*.
- Prove meta-theorems like "being a refinement is a transitive relation on machines", and apply this result to a full sequence of refinements.
- Study at the generic level how to compose several refinements of the same machine, and certify some machine transformations [11].

Acknowledgements This work has been supported by the project Impex of the French "Agence Nationale de la Recherche". We thank the referees for their many helpful comments on this chapter.

References

1. J-R. Abrial, *Modeling in Event-B - System and Software Engineering* (Cambridge University Press, Cambridge, 2010)
2. The Rodin development team. The Rodin Tool.https://sourceforge.net/projects/rodin-b-sharp/
3. G. Tel, *Introduction to Distributed Algorithms* (Cambridge University Press, Cambridge, 2000)
4. The Coq development team. The Coq proof assistant. coq.inria.fr
5. Xavier Leroy, Formal verification of a realistic compiler. Commun. ACM **52**(7), 107–115 (2009)
6. G. Gonthier, Formal proof – the four-color theorem. Not. Am. Math. Soc. **55**(11) (2008)
7. Inria Marelle Team. Proof of the Feit-Thompson theorem. https://gforge.inria.fr/projects/coqfinitgroup/
8. B.C. Pierce et al., Software foundations. https://softwarefoundations.cis.upenn.edu/current/index.html
9. Y. Bertot, P. Castéran, *Interactive Theorem Proving and Program Development - Coq'Art: The Calculus of Inductive Constructions* (Springer, Berlin, 2004)
10. A. Chlipala, *Certified Programming with Dependent Types.* (MIT Press, Cambridge, 2013)
11. G. Babin, Y.A. Ameur, M. Pantel, Correct instantiation of a system reconfiguration pattern: A proof and refinement-based approach, in *17th IEEE International Symposium on High Assurance Systems Engineering, HASE 2016, Orlando, FL, USA, January 7-9, 2016* (2016), pp. 31–38

Contextual Dependency in State-Based Modelling

Souad Kherroubi and Dominique Méry

Abstract In conceptual modelling, context-awareness should be precisely high-lighted. In this chapter, we recall and detail preliminary results on contextualization and dependency in state-based modelling using the Event-B modelling language. The contextualization of Event-B models is based on knowledge provided from domains classified into constraints, hypotheses, and dependencies according to truthfulness in proofs. The dependency mechanism between two models makes it possible to structure the development of system models, by organizing phases identified in the analyzed process. We illustrate via two simple case studies and on a voting protocol.

1 Introduction

Various informal, semi-formal and formal methods, and techniques based on abstraction facilitate system design; they provide formal views of software-based systems of varying levels of preciseness. They are supported by tools which include simple tools such as editors (UML,SysML) as well as very sophisticated tools such as model-checkers [21], theorem provers or proof assistants [33, 36]. Tools are integrated into formal IDEs [9, 31, 34, 35], which provide platforms for assisting developers of a system following the correct-by-construction paradigm [26]. A very important feature in system design is the relevant acquisition of domain knowledge, allowing a better understanding of the problem to solve as well as facilitating better communication between designers, specifiers, and domain experts. In fact, a conceptual model integrates the *intention* of the designer and should provide a clear, correct,

This work was supported by grant ANR- 13- INSE- 0001 (The IMPEX Project http://impex.loria.fr) from the Agence Nationale de la Recherche (ANR).

S. Kherroubi · D. Méry (✉)
Université de Lorraine, LORIA UMR CNRS 7503, Campus scientifique, BP 239 - 54506, Vandœuvre-lès-Nancy, France
e-mail: Dominique.Mery@loria.fr

S. Kherroubi
e-mail: souad.kherroubi@inria.fr

and complete view of the system using a non-ambiguous semantics. Contextualism focuses precisely on details of a specific application, as well as on the modelling process itself. These details define a context and constitute the unique identity of each modelling task.

Contextualization is an abstraction mechanism [37] that provides an organization and structure to the collected data. The problem to solve is related to the differences in perception between actors in the system, thus facilitating the organization and rationalization of the perspectives of the same reality. The context is an important element for verifying systems: whereas the use of the context in the development phases makes it possible to situate the developed systems in space and time [32], its use during the development process improves the validation of the produced systems [14]. The validation is obtained by relating the real system to a model with respect to the intended use cases (configurations and scenarios) of the model. This chapter includes the notion of context and the introduction of a dependency relation among formal state-based models; it completes and extends previous works [19, 22, 23].

The *context* has been widely recognized as important according to the various research topics [7, 13, 27, 28]. For instance, the security domain, integrating knowledge from attack patterns with limited knowledge related to the target system vulnerabilities and potential threats strongly depends on the context of their application [6, 18]. The context is used to assess the impact and the *plausibility* of these attacks. The reliability of a system is based on a systematic approach to recover requirements and strategies [25], by identifying, detecting, and mitigating risks and threats, and developing secure mechanisms while helping designers to choose the appropriate countermeasures to reduce attackers' abilities. Thus, validation of assumptions made by designers is performed on the modelling of threats associated with contextual informations to protect the system against unauthorized modification of data or disclosure of information. The context is therefore a key element in the choice of the security patterns to be applied. Dines Bjørner [7] develops a domain engineering from his triptych domain: system, requirements, and domains define the context.

Logics and ontologies are two main scientific fields, where the context has been identified as an important notion. Recently, Barlatier [13] shows that this notion is not absolute and is always defined with respect to a focus [8], an activity [12, 16, 24], an intentional concept, i.e., an action [15]. Contexts are formal objects incrementally constructed from an existing one corresponding to the *context lifting* (see for instance, McCarthy [27]). The situation appears as a new parameter in predicates and thus, predicates depend on a situation. At the proof-theoretical level, the context establishes and validates trust relationships that provide valid interpretations in a specific domain for certification purposes. In this paper, we are interested in the contextualization and the context of proofs for the Event-B modelling language [1]. Event-B is a structuring framework that ensures, through its language, a formal model for reactive systems. The behavioral specifications in Event-B follow an assertional approach described by means of components called contexts and machines. Since we are using the term *context* in two different ways, we will use the term *Event-B context*, when we are referring to a component context of Event-B. An Event-B model is a collection of Event-B contexts and Event-B machines which are organized in Event-B projects.

Rather than considering contexts as formal objects like McCarthy, we consider the context as the minimal knowledge being part of a system focused on proof. The context is thus an effective part of proofs and constrains their semantics. We propose to decompose the context in Event-B into: *constraints*, *hypotheses*, and *dependencies* depending on whether the knowledge is acquired, supposed, or derived in a proof system. We define this new mechanism of dependency between Event-B models, where the property of stable termination is a necessary condition for its establishment. The constraints correspond to static properties defined in Event-B contexts. The dependency mechanism is then defined by a combination of states and static properties of Event-B contexts. The dependency is a relationship taking values from existing facts in a situation that requires the definition of a new proof obligation.

The chapter is organized as follows. We briefly present the Event-B modelling language in Sect. 2 and what we understand by context in this formalism. Section 3 defines dependency between Event-B models by providing a new proof obligation for establishing the dependency relationship. Section 4 illustrates the methodology for developing models of a system using the dependency relationship. Section 5 summarizes comments on the methodology and the role of dependency in the development of systems. Finally, Sect. 6 concludes our work and gives perspectives.

2 Modelling in Event-B

An Event-B model is built with two main ingredients: a specification language based on set theory and predicate logic, and a refinement mechanism that allows correct-by-construction development. An Event-B model $\mathcal{M}$ is composed of contexts and machines. A context $\mathcal{C}$ specifies the static part of a model and includes carrier sets s, constants c, axioms, and theorems that establish constraints and properties of static elements. A machine describes the dynamic part of the system by means of a finite list of variables x which describe the state of the system, possibly modified by a list of events $\{e_0, ..., e_n\}$. The change of state must preserve assertions called *invariants* $\mathcal{I}nv$ which must be maintained whenever events in the system are observed. An event is defined by a condition (guards) under which the event can be observed and actions that define the evolution of the variable values in the model for the next state. In Event-B, a situation $\mathcal{S}$ is called a state and a state includes variable values, as well as values of sets and constants. Values of sets and constants are members of a domain $\mathcal{D}$ which is used for interpreting terms as deferred sets or constants in Event-B contexts.

In an Event-B model $\mathcal{M}$, we have to express informations over the state as static informations in Event-B contexts and as dynamic informations in Event-B machines. The context $Context(\mathcal{M})$ of an Event-B model $\mathcal{M}$ is the minimal knowledge related to the behavior of any action in the system satisfying the safety properties in a given (state) situation. The context $Context(\mathcal{M})$ includes *axioms* and *theorems*. *Theorems* are derived from knowledges which are expressed as axioms or as others theorems.

The minimality is linked to a given starting point in the reasoning using the context. For instance, when considering the IEEE leader election protocol [2], the context contains axioms for defining what are the properties of acyclic graphs and a theorem expressing that, for any acyclic connected graph g and any node of the graph n, there is a spanning tree for the graph g rooted with n. The minimal knowledge is divided into three categories, explained in the following subsections.

2.1 Context as Constraints

The meaning of a model and its properties depends on the context of its conceptualization. The semantic aspect of a conceptualization is the point of view based on the model, its properties, and its context, which addresses the foundations of the system according to the existing form, structure, and facts of a *specific domain*. These features are called *constraints*. They correspond to the different concepts, roles, and values of the attributes of the current domain. Constraints can be physical (i.e., they exist in nature) such as the structure of the human body, the temperature in the atmosphere, or they are fully conceived by human or artificial artifacts. For instance, types of currency or types of vote (majoritarian voting, preferential voting, cumulative voting,…) which make it possible to compute and give meaning to the result of an election are constraints that define the context in which the proof is carried out. Constraints are expressed in Event-B by the structures defined in the context, namely, sets, constants, axioms, and theorems that establish their typing and their static characteristics or properties. We denote these constraints by the form C_AX and C_TM to illustrate axioms and theorems in a context of an Event-B model.

2.2 Context as Hypotheses

Assumptions or hypotheses on environment help to validate the system to build. They are established by designers: assumptions are not always verified but only assumed, accepted, or supposed. In the example of voting protocols, the objectives of each voting protocol as well as the type of attacks described determine whether proofs are successful or not. The use of cryptographic primitives to ensure the construction of secure voting guaranteeing the required properties is based on assumptions about difficulties in performing computations or solving some problems [17]. In the context of the Belenios system [10] based on the version of Helios [11] with credentials and zero-knowledge proofs, the proof scheme is constructed on the assumptions that the registration authority and the bulletin board are not simultaneously dishonest, and thus providing the correctness of the voting scheme is made under these trust assumptions. Other works impose stronger hypotheses than these, such as in [38], which assumes that bulletin boards are honest. We [19] have shown that these kinds of assumptions can be expressed in Event-B models, using refinement, by adding new variables, events and properties in the machines. Hypotheses are restrictions on static elements defined in Event-B contexts and are added to the Event-B context. Voting protocols have been verified by making assumptions about the format of the ballots. Thus, proofs are only possible using binary values of ballots (i.e., 1/0, yes/no, ...). These restrictions are expressed by adding new constants and axioms in the Event-B contexts. We note the hypotheses in the Event-B contexts by H_AX and H_TM to illustrate the additional axioms and theorems in the contexts of an Event-B model.

2.3 Context as Dependencies

McCarthy has shown that predicates defining context, are parametrized by situations, thus defining a *context lifting* in his work. While the ontological point of view considers that a context is a *moment universals*: A moment (from the German Moment) [15] is an individual that existentially depends on other individuals. Event-B is based on set theory and we have to consider a relation between Event-B and ontologies. For instance, Barlatier [5] establishes a correspondence between set theory and ontologies, namely, ontological concepts (i.e., types) are interpreted as sets of elements and roles as relations between the elements of different concepts. Starting from this observation, and by comparing with the situation theory, where situations are interpreted as states in Event-B, and constraints as elements defined in the Event-B contexts, we can define the context by a dependency relation between the Event-B models. This relation results from the combination of situations (of the values of states in the Event-B formalism) and of constraints expressed in Event-B contexts by interpreting *moment universals* as stable termination properties. This type of context can be illustrated by voting systems.

The voting process [19] consists mainly of three phases: *(1) The preparation phase*: At the end of this phase, the lists of nominated candidates, as well as the registered voters entitled to vote, shall be drawn up. The establishment of these lists depends on the local laws of each country; *(2) The vote registration phase*: This phase allows all eligible voters to express their choice on the basis of the list of nominated representatives in the previous phase; Thus, by using the voters list, the elector must authenticate himself as an eligible voter and cast his vote individually; *(3) The tallying phase*: It covers the counting and results of the reports arising from the recording phase. At the end of these phases, the interpretation of the results depends on the method of voting adopted which constrains the final decision. Each phase of the vote depends on the previous one. This notion of dependency exemplifies the parties integrated into the entire voting system (*"Whole"*), ensuring them a separate existence independent of the *"Whole"*.

The notion of *existential dependency* [20] states a general principle as follow: *Let the predicate ϵ denote existence. We have that an individual x is existentially dependent on another individual y iff, as a matter of necessity, y must exist whenever x exists, or formally* $ed(x, y) \overset{def}{=} \Box(\epsilon(x) \Rightarrow \epsilon(y))$.

The dependent constraints are expressed by D_AX and D_TM to illustrate dependent axioms and theorems in the contexts of an Event-B model.

We have sketched the contextualization in Event-B, and we will describe the notion of dependent Event-B models in order to enrich the structuring techniques in this formalism. As refinement is the main mechanism used in the Event-B formalism, we will explain in the next section how this mechanism is used to contextualize the systems under study, allowing conceptual models to have a better semantic integration and thus improved reasoning.

3 Dependency of Models

The dependency between two Event-B models $\mathcal{M}_1$ and $\mathcal{M}_2$ is informally defined as follows: *(i)* the Event-B contexts of $\mathcal{M}_1$ and $\mathcal{M}_2$ may be related; *(ii)* some variables of the first model $\mathcal{M}_1$ are *transformed* into *constants* in the target model $\mathcal{M}_2$; *(iii)* the predicate characterizing the termination of the first model implies the constraints defined in the Event-B context of the second component. In this case, a component corresponds to an Event-B model and possibly a set of models that refine this abstract one, and it is requested that the refinement level for each phase (for instance, preparation, vote, result) must be sufficiently elaborate so that the set of constants in the Event-B context of the phase that follows the current one finds its correspondent (variable) in the last refinement of the phase on which it depends. In the following, we adopt notations [4, 29], which express relational models and extensions to the temporal aspects following TLA.

Given two Event-B models $\mathcal{M}_i$, with $i \in 1..2$ defined by: $\mathcal{M}_i \stackrel{\wedge}{=} (\mathcal{T}h_i(s_i, c_i), x_i, Val_i, \mathcal{I}nit_i(x_i), \{e_{i0}, ..., e_{in_i}\})$

- $\mathcal{T}h_i(s_i, c_i)$ is the collection of Event-B contexts that define all elements and static properties of the model $\mathcal{M}_i$;
- the state space Val_i is the set of all possible values of the variables;
- $\mathcal{DC}_i(s_i, c_i)$ express the dependent constraints in the Event-B as the dependent axioms $D\_AX(s_i, c_i)$ and theorems $D\_TM(s_i, c_i)$;
- the set of initial states $\mathcal{I}nit_i$;
- a set of events $\{e_{i0}, ..., e_{in_i}\}$ and, for each event e, $BA(e)(x_i, x_i')$ is a relation between x_i and x_i' values for variables x_i;
- $Spec(\mathcal{M}_i) \stackrel{\wedge}{=} \mathcal{I}nit_i(x_i) \wedge \Box[NEXT_i]_{x_i} \wedge L_i$ with $NEXT_i \stackrel{\wedge}{=} \exists e.e \in \{e_{i0}, ..., e_{in_i}\} \wedge BA(e)(x_i, x_i')$; x_i' is the value of variables after the observation of the event e; $\Box[NEXT_i]_{x_i}$ means that all state pair satisfies the relation $NEXT_i$ and the values of variables remain the same or change; and L_i is a conjunction of weak and strong fairness constraints on the combinations of events e_i of the model $\mathcal{M}_i$.

A set of traces generated over Val_i, denoted by $Trace(\mathcal{M}_i)$, is attached to each relational model $Spec(\mathcal{M}_i)$ and are defined as follows:

$$Trace(\mathcal{M}_i) \triangleq \{\sigma \in \mathbb{N} \longrightarrow Val_i \mid \sigma_0 \in Init_i \wedge \forall j \in \mathbb{N}. \left(\begin{array}{c} (\sigma_j, \sigma_{j+1}) \in NEXT_i \\ \vee (\sigma_j = \sigma_{j+1}) \end{array} \right) \}$$

A trace $\sigma \in Trace(\mathcal{M}_i)$ is fair with respect to L_i, if σ satisfies L_i and $tfair(\mathcal{M}_i, L_i)$ is the set of traces fair with respect to L_i. To define the stable termination property, we modify the definition of the *Leads to*[1] operator [4, 29]. It expresses that we inevitably reach a future state satisfying Q and Q will remain true for any next states.

Definition 1 (*Stable termination*) The stable termination property $P \rightsquigarrow_s Q$ is valid for a trace $\sigma \in Trace(\mathcal{M}_i)$, if $\forall i.(i \geq 0 \wedge P(\sigma_i) \Rightarrow \exists j.(j \geq i \wedge Q(\sigma_j) \wedge \forall k.(k \geq j \wedge Q(\sigma_k))))$.

The notation $P \rightsquigarrow_s Q$ means that P leads to Q and when Q is true, it remains true. We are using this property for a specification $Spec(\mathcal{M}_i)$ of a model $\mathcal{M}_i$ under the fairness assumptions L_i as follows: $Init_i \rightsquigarrow_s T_i$, where T_i is a set of states considered as terminal states. $Init_i \rightsquigarrow_s T_i$ should be true for any trace $\sigma \in tfair(\mathcal{M}_i, L_i)$. The stability implies that the predicate that characterizes the final states will remain true for all reachable states afterwards. A proof of the progression by defining a variant on a well-founded order is necessary to show convergence in the system [4, 29].

Definition 2 (*Dependency between two Event-B models*) The two models $\mathcal{M}_i$, with $i \in 1..2$ (i.e., $\mathcal{M}_1$ and $\mathcal{M}_2$) are dependent, and we say that the model $\mathcal{M}_2$ contextually depends on the model $\mathcal{M}_1$ with respect to the property of termination T_1 and we denote $Dep(\mathcal{M}_1, T_1, \mathcal{M}_2, v_1, c_{21})$, if

1. $Spec(\mathcal{M}_1) \vDash Init_1 \rightsquigarrow T_1$ and T_1 is the predicate characterizing the set of final states at the stability of the model $\mathcal{M}_1$;
2. $Spec(\mathcal{M}_1) \vDash (\forall x_1, x_1'.(T_1(x_1) \wedge NEXT_1(x_1, x_1') \Rightarrow T_1(x_1')))$;
3. the dependent context extends the source context: $Th_2(s_2, c_2)$ *extends* $Th_1(s_1, c_1)$;
4. there is a non-empty subset v_1 of all set of variables x_1 ($v_1 \subseteq x_1$) in the first model $\mathcal{M}_1$ such that the following property is verified:

$$Th_2(s_2, c_2) \vDash (T_1(x_1) \wedge Inv_1(s_1, c_1, x_1) \wedge v_1 = c_{21} \wedge c_2 = c_{21}c_1c_{22}) \Rightarrow DC_2(s_2, c_2),$$

with c_2 is the set of constants defined in the context of the first model $\mathcal{M}_1$ and which is obtained by extension of the context Th_1 according to the item *3*, union the variables v_1 issued from the machine of the first model $\mathcal{M}_1$, to which the constants c_{22} are added newly introduced in the Event-B context of the second

---

[1]Leads to: Under the fairness assumptions L of the model $\mathcal{M}$, the specification of the model $Spec(\mathcal{M})$ satisfies the property $P \rightsquigarrow Q$, if for all traces $\sigma \in tfair(\mathcal{M}, L)$, the following property holds: $\forall i.(i \geq 0 \wedge P(\sigma_i) \Rightarrow \exists j.(j \geq i \wedge Q(\sigma_j)))$.

model $\mathcal{M}_2$. The values of c_{21} can be defined according to the values of the variables v_1 in the first model.

Concretely, a model $\mathcal{M}_2$ depends on another model $\mathcal{M}_1$, if the predicate characterizing the set of terminal states of the first model $\mathcal{M}_1$ satisfies the *initial configuration* of the $\mathcal{M}_2$ defined by the content of sets, and values of constants of the context $Th_2(s_2, c_2)$. $Th_2(s_2, c_2)$ is the structure defining the Event-B context of the second model, i.e., all static properties (axioms and theorems) in conjunction that establish the typing and the constraints of dependent and independent of the situations, while $\mathcal{DC}_2$ are the constraints that must be satisfied by all final states of the first model, i.e., which are dependent on situations. These situations are added to the context of the second Event-B model, in conjunction with the properties that are independent of the situations linked to the first model. This approach reflects the fact that at the stabilization of the first phase, no modification can be made on these elements as variables, since the latter in the first component maintain their values at termination. We then apply the new values of the variables defined in the first model to the operator defining the predicate of the static constraints, which takes the list of sets and constants defined in the context depending on. The correctness in Event-B is defined by the set of proof obligations. The proof obligation to prove the dependency mechanism between two models $\mathcal{M}_1$ and $\mathcal{M}_2$ which must be discharged is defined as follows.

Definition 3 (*Proof obligation of dependencies between $\mathcal{M}_1$ and $\mathcal{M}_2$*)

$$Th_2(s_2, c_2), Inv_1(s_1, c_1, x_1), v_1 \subseteq x_1, c_{21} \subseteq c_2, \mathcal{T}_1(x_1) \vdash \mathcal{C}_2(s_2, c_2)(v_1/c_{21})$$

The contexts[2] form a partonomic relation where a context $\mathcal{C}xt_1$ is a part of another context $\mathcal{C}xt_2$, if $\mathcal{C}xt_1$ contains at least all the constraints verified in $\mathcal{C}xt_2$. The second model becomes dependent on the first one.

4   Case Studies

The dependency relation provides a structuring mechanism, when developing models. We are considering three case studies illustrating how the dependency relation is used for. The first case study in Sect. 4.1 considers an experiment requiring the collection of data prior to the processing of those data. The process is modelled into two phases COLLECTING and INTERPRETING related by the dependency relation and the data to process are first collected in the phase COLLECTING and then the resulting set of data is assigned to a constant for the second phase INTERPRETING. The second case study in Sect. 4.2 describes operations of a ERP management system. It concerns the management of inventories by purchase and sale of articles, and a deferred accounting of revenues and expenditures. Finally, the last case study in Sect. 4.3 is addressing

---

[2]We will talk, indifferently, about Event-B contexts or models in partonomic relation.

the voting process which is modelled by three phases PREPARING, RECORDING, and TALLYING. The three phases are related as follow: the model RECORDING depends on the model PREPARING, and the model TALLYING depends on the model TALLYING.

4.1 Example of an Experiment

We illustrate required mechanisms using an experiment process. An experiment requires the collection of data and the processing of collected data to produce an indicator. We do not care what the experiment is and we consider the example as illustrating the phasing of two processes, namely, COLLECTING and INTERPRETING. The process COLLECTING is simply collecting data from sensors and storing them in a file; there is the possibility of invalid data, which is not stored in the file. One assumes that there is a filter which states when a datum is valid or not. The process INTERPRETING applies an analysis of the results which are stored in store at the end of the process COLLECTING. The problem is to evaluate the value of $sum(store)$ when $store$ is computed. The problem is clearly decomposed into two phases and is first expressed by a machine stating the two different computations: (1) The process COLLECTING produces the final value of $store$; (2) The process INTERPRETING returns the value of the summation of all the values of $store$. First, we define the initial context that helps to define the specification of the problem to solve. We define the function sum which returns the value of the summation of data of a given set. We do not care what is exactly sum and what we are really computing. The small example illustrates our methodology for analyzing this class of problems. The machine $exp0$ simply describes the two phases for producing the required result, namely, $s = sum(store)$, when the process is completed.

```
CONTEXT data0
SETS   data, PHASES
CONSTANTS   valid, tmax, val, sum, collecting, interpreting, final
AXIOMS
    axm1 : valid ⊆ data
    axm2 : tmax ∈ ℕ
    axm3 : tmax ≠ 0
    axm4 : val ∈ ℕ × data → ℕ
    axm5 : sum ∈ ℙ(ℕ × data) → ℕ
    axm6 : sum(∅) = 0
    axm7 : ∀i, e ↦ i ↦ e ∈ ℕ × data ⇒ sum({i ↦ e}) = val(i ↦ e)
    axm8 : ∀p, i, e ↦ p ⊆ ℕ × data ∧ e ∈ data ∧ i ∈ ℕ
              ⇒ sum(p ∪ {i ↦ e}) = sum(p) + val(i ↦ e)
    axm7 : partition(PHASES, {collecting}, {interpreting}, {final})
```

```
MACHINE exp0
SEES data0
VARIABLES   store, phase, s
INVARIANTS
  inv1 : store ⊆ ℕ × data
  inv2 : phase ∈ PHASES
  inv3 : s ∈ ℕ
  inv4 : phase = final ⇒ s = sum(store)
EVENTS
EVENT INITIALISATION
  BEGIN
    act1 : store := ∅
    act2 : phase := collecting
    act3 : s := 0
  END
EVENT collecting
  WHEN
    grd01 phase = collecting
  THEN
    act1 : phase := interpreting
    act2 : store : |(store' ⊆ ℕ × data ∧ ran(store') ⊆ valid)
  END
EVENT interpreting
  WHEN
    grd1 : phase = interpreting
  THEN
    act1 : phase := final
    act2 : s := sum(store)
  END
```

Each event (**collecting** and **interpreting**) describes a pre/post specification that is supposed to be developed in another session. Now, we develop two separate models that are modelling the two processes **collecting** and **interpreting**. Each event corresponds to a phase and defines a required *liveness* property which is derived from the definition of the process **experimenting**.

The main liveness property can be simply stating that the experiment starts and ends after a time which is explicitly stated in the experiment requirements.

- at(**experimenting**) $\land store = \emptyset \land s = 0 \land t = t0 \rightsquigarrow after$(**experimenting**) $\land$ $s = sum(store) \land t = tf$ and the process **experimenting** is decomposed into the two processes corresponding to models:
- phase *collecting*: at(**collecting**) $\land store = \emptyset \land s = 0 \rightsquigarrow after$(**collecting**) $\land$ $(store \subseteq \mathbb{N} \times data \land ran(store) \subseteq valid)$
- phase *interpreting*: at(**interpreting**) $\land (store \subseteq \mathbb{N} \times data \land ran(store) \subseteq$ $valid) \rightsquigarrow after$(**interpreting**) $\land s = sum(store)$.

The depends operation expresses that the variable *store* has a final value which is used as an constant *cstore* in the model interpreting. The validity of the depends operation is based on checking that the final value of *store* is satisfying the properties of *cstore*, which are defined as *axioms* in the development of interpreting. The depends operation is based on the definition and the proof of *liveness*barl properties and we are using the approach in [29] for combining and integrating the two phases. The two phases *collecting* and *interpreting* are defined in the same machine; in a next step, each event will be refined in separate machines which are dependent. The dependency is used for structuring the refinement-based development and for decomposing machines.

4.2 ERP Management System

The example presented here concerns the management of inventories by purchase and sale of articles, and a deferred accounting of revenues and expenditures. This modelling is a simplified representation and it does not describe all the management details in an ERP system. The first model denoted by M_sp is described by a context *sales_purchases_cxt* and a machine *sales_purchases_machine*. The system manages purchases (*buy* event) and the sale (*sale* event) of articles. Sales are performed according to the prices fixed in the context, while purchases are fixed by the market. The imputation of the deferred accounting entries allows transactions to be counted only at the time of the execution of the transfer operations.

CONTEXT *sales_purchases_cxt*
SETS
   ARTICLES
CONSTANTS
 deferred_period, *prices_art*
AXIOMS
 $axm1 : deferred\_period \in \mathbb{N}1$
 $axm2 : prices\_art \in ARTICLES \rightarrow \mathbb{N}1$

The journal of accounting entries for sales and purchases is configured on a deferred basis according to a closing period market *"deferred period"* noted *deferred_period*. Purchases and sales can be performed during this period (*grd3*).

This period is decremented by the convergent event *"forward_time"* which decrements the expression of the variant defined in this first machine, and is under a weak fairness assumption. The printing of the various sales and purchases operations is done through the two variables *incomings* and *expenses*, respectively.

```
MACHINE sales_purchases_machine
SEES sales_purchases_cxt
INVARIANTS
  inv1 : period ∈ 0 .. deferred_period
  inv2 : sold_art ⊆ ARTICLES
  inv3 : incomings ∈ sold_art → ℕ
  inv4 : purchased_art ⊆ ARTICLES
  inv5 : expenses ∈ purchased_art → ℕ
  inv6 : purchased_art ∩ sold_art = ∅
  inv7 : ∀art, p ↦ (art ↦ p ∈ incomings
        ⇒ art ↦ p ∈ prices_art)
VARIANT
  deferred_period − period
EVENTS
EVENT forward_time   convergent
  WHEN
    grd1 : period ∈ 0 .. (deferred_period − 1)
  THEN
    act1 : period := period + 1
  END
```

```
EVENT sale
  ANY art
  WHERE
    grd1 : art ∈ ARTICLES
    grd2 : art ∉ sold_art ∧ art ∉ purchased_art
    grd3 : period ∈ 0 .. (deferred_period − 1)
  THEN
    act1 : sold_art := sold_art ∪ {art}
    act2 : incomings(art) := prices_art(art)
  END
EVENT buy
  ANY art, p
  WHERE
    grd1 : art ∈ ARTICLES ∧ p ∈ ℕ₁
    grd2 : art ∉ purchased_art ∧ art ∉ sold_art
    grd3 : period ∈ 0 .. (deferred_period − 1)
  THEN
    act1 : purchased_art := purchased_art ∪ {art}
    act2 : expenses(art) := p
  END ...
```

The *closure* of an operation period implies that the accounting of the income and expenditure can *begin* when the value of the *period* variable will be equal to *deferred_period*, i.e., at the stability of the first component. This is expressed by the dependency relation between the two models. Accounting is described by the model $\mathcal{M}\_accounting$ defined by the machine *accounting_machine* which sees the context *accounting_cxt* depending on the machine *sales_purchases_machine*. This context extends the first one *sales_purchases_cxt* and contains all the constants defined as variables in this latter (*sales_purchases_machine*) with constant *period* having as its value *deferred_period* in the dependent context. We note the dependency between the two models by

$Dep(\mathcal{M}\_sp, period = deferred\_period, \mathcal{M}\_accounting, v_1, c_{21})$, with: v_1 all variables defined by the invariants $inv1,...,inv7$ in the first machine; and c_{21} corresponds to the same elements, but defined as constants in the dependent context *accounting_cxt*, to which we add the following constraint: $axm11 : period = deferred\_period \Rightarrow (purchased\_art \neq \varnothing \lor sold\_art \neq \varnothing)$. The dependent constraint in this example is $period = deferred\_period \land axm11$. The accounting consists of the computation using the variable *balance*, which is initialized to 0, of the difference between the *incomings* and *expenses* by means of the constant *total* defined in the dependent context as a function that allows to sum the values of arguments that it takes.

CONTEXT *cxt_compt* $EXTENDS$ *cxt_achat_vente*
CONSTANTS
  period, sold_art,
AXIOMS
  $axm1 : period = deferred\_period$
  $axm2 : sold\_art \subseteq ARTICLES$
  $axm3 : incomings \in sold\_art \rightarrow \mathbb{N}$
  $axm4 : purchased\_art \subseteq ARTICLES$
  $axm5 : expenses \in purchased\_art \rightarrow \mathbb{N}$
  $axm6 : purchased\_art \cap sold\_art = \varnothing$
  $axm7 : \forall art, p \mapsto (art \mapsto p \in incomings \Rightarrow art \mapsto p \in prices\_art)...$

MACHINE *accounting_machine* / ∗ *dependent machine* ∗ /
SEES *accounting_cxt* / ∗ *dependent context* ∗ /
INVARIANTS
  $inv1 : balance \in \mathbb{Z} \land accounted \in BOOL$
  $inv2 : period = deferred\_period \land accounted = FALSE$
    $\Rightarrow balance = 0$
EVENT accounting
  WHEN
    $grd1 : accounted = FALSE$
  THEN
    $act1 : balance := total(incomings) - total(expenses)$
    $act2 : accounted := TRUE$
  END

This relation can be generalized to any number of Event-B models, as shown in the diagram of Fig. 1. If we take for instance the example of a management system, it is then possible to define a dependency between *Human Ressources Service* and *Payroll Service*, and between this last service and the *Accounting Service*. Thus, the dependency relation is irreflexive and transitive, since the extensions between Event-B contexts have a meaning only in one way, i.e., if c_2 *extends* c_1, then c_1 does not extend c_2.

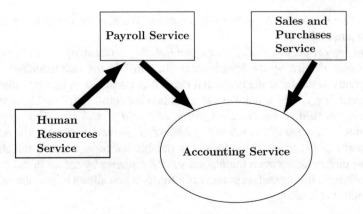

Fig. 1 ERP management system

4.3 Applying the Dependency Mechanism for Voting Protocols

We have developed a voting system [19] as an Event-B composition via the dependency mechanism. The models in our development are described by means of Event-B contexts and machines linked by refinement describing the constants and the dynamics of the system (see Fig. 2). The advantages of such modelling is that proofs are more easily realized, and verification properties can be expressed separately. We have, for instance, expressed *Eligibility*, *No double voting*, *Confidentiality*, in the recording phase; and the *Verifiability* property in the tallying phase.

```
VARIABLES rec_votes, timer
INVARIANTS
   inv1 : rec_votes ∈ Sig ↔ Choices
   inv2 : timer ∈ start_time .. end_time
VARIANT end_time − timer
INITIALISATION
   act1 :rec_votes := ∅
   act2 :timer := start_time
EVENT register_votes
   WHEN
      grd1 :timer ≥ start_time ∧ timer < end_time
      grd2 :∀i, j ↦ i ↦ j ∈ interrupt_sequences ⇒ timer ∉ i .. j
   THEN
      act1 :rec_votes : |rec_votes′ ∈ (Sig ↔ Choices)
   END
```

```
EVENT forwarding_time
  STATUS convergent
  WHEN
    grd :timer < end_time
  THEN
    act :timer := timer + 1
  END
EVENT finish
  WHEN
    grd1 :timer = end_time
  THEN
  act :skip
  END
```

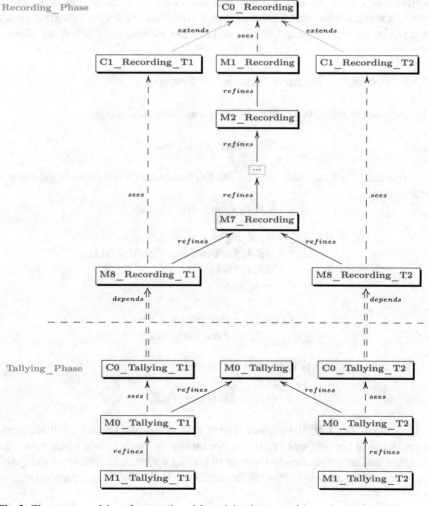

Fig. 2 The structure of the refinement-based formal development of the voting system

In the first abstract model *M1_Recording* the state of the system is characterized by two variables that represent the registered votes and the elapsed time in
the system. The votes are modelled as a relationship between all signatures (*Sig*)
and the electors' choices (*Choices*). The invariant in this machine simply provides
a means for typing these variables. The precondition for this phase, as expressed by
the initialization event, is that the time is equal to the opening time of the offices
fixed in the context *C0_Recording* and that no vote has been recorded. A vote
modifies the variable *rec_votes* which is performed by the event *register_votes*.
In this model, we distinguish only the values of variables *rec_votes* which take
their values in *Sig* $\leftrightarrow$ *Choices* without making precise the undertaken actions. The
event *forwarding_time* changes the value of the variable *timer* introduced in this
machine to express the progression of time in the system. The variable value is incremented by the action of the event *forwarding_time* until the closing time of the
offices *end_time* is reached. We note that this event has a convergent status under
which a weak fairness assumption is made. Thus, this event (*forwarding_time*)
will not be observable when the value of the *timer* variable has reached *end_time*.

4.3.1 Proving the Required Termination Property

The required termination property is defined as follows. Consider

$$T_1 \mathrel{\widehat{=}} timer = end\_time$$

We have $x_1 = \{rec\_votes, timer\}$. The fairness assumptions made in the current
machine are

$$L_{\mathcal{M}_1} \mathrel{\widehat{=}} WF_{x_1}(forwarding\_time)$$

$$
\begin{aligned}
NEXT_1 \mathrel{\widehat{=}} &\vee BA(register\_votes)(x_1, x_1') \\
&\vee BA(forwarding\_time)(x_1, x_1') \\
&\vee BA(finish)(x_1, x_1') \\
&\vee (x_1 = x_1')
\end{aligned}
$$

$$
\begin{aligned}
\mathcal{I}nit_1(x_1) \mathrel{\widehat{=}} &timer = start\_time \\
&\wedge rec\_votes = \varnothing
\end{aligned}
$$

The stability property to prove is the following:

$$\Phi_{\mathcal{M}_1} \mathrel{\widehat{=}} \mathcal{I}nit_1(x_1) \rightsquigarrow_s T_1$$

which means that, from the opening time of polling stations, i.e., at the initialization,
when no vote has yet been registered, we inevitably reach a state where time will
progress until reaching the closing hour of polling station, with possibly a change of
the values of the recorded votes. We define: $\mathcal{I}nit_{x_1} \mathrel{\widehat{=}} P_0$, $V(t) \mathrel{\widehat{=}} end\_time - timer = t$.

The property of termination $\Phi_{\mathcal{M}_1}$ can be expressed as follows:

$$\Phi_{\mathcal{M}_1} \hat{=} \{V(t) \rightsquigarrow V(t-1) \qquad \Phi_1$$
$$, (\exists t.V(t)) \rightsquigarrow V(0) \qquad \Phi_2$$
$$\}$$

- Φ_1 indicates that inevitably that we will reach a state where the counter value decrease by 1;
- Φ_2 indicates that we will inevitably reach a state where the difference will converge to the closing time of the polling station, the value for which the variant will be equal to zero;

By letting $T_1 \hat{=} V(t)$ and $T_2 \hat{=} V(t-1)$, we have

$$P_0 \wedge [NEXT_1]_{x_1} \Rightarrow (T_1' \vee T_2')$$

We split the proof of this implication into the following proofs, according to the definition of the $NEXT_1$:

- $T_1 \wedge BA(forwarding\_time)(x_1, x_1') \Rightarrow (T_1' \vee T_2')$
- $T_1 \wedge BA(register\_votes)(x_1, x_1') \Rightarrow (T_1' \vee T_2')$
- $T_1 \wedge BA(finish)(x_1, x_1') \Rightarrow (T_1' \vee T_2')$
- $T_1 \wedge x_1 = x_1' \Rightarrow (T_1' \vee T_2')$

The only event that ensures progress of time values is $forwarding\_time$, thus, the value of the counter is modified only by the actions in this event. This event is observable until the value of $timer$ will reach end_time. The weak fairness assumption made on the event $forwarding\_time$ lets say that from (a): $T_1(x_1) \wedge BA(forwarding\_time)(x_1, x_1') \Rightarrow T_2(x_1')$, and from the feasibility condition of the event $forwarding\_time$ we can deduce that (b): $T_1(x_1) \Rightarrow (\exists x_1'. BA(forwarding\_time)(x_1, x_1'))$ are satisfied by the event $forwarding\_time$. Thus, we can deduce from (a), $T_1 \wedge \langle NEXT \wedge forwarding\_time \rangle_{x_1} \Rightarrow T_2$ and from (b) we can deduce that $T_1 \Rightarrow ENABLED\langle forwarding\_time \rangle_{x_1}$, with $ENABLED\langle forwarding\_time \rangle_{x_1} \hat{=} (\exists y_1.BA(forwarding\_time)(x_1, y_1))$. The WF1 rule allows the deduction that time will progress to decrease the variant, and since no event disturbs the property $V(t)$, except of the event $forwarding\_time$. By applying the LATTICE rule over the set of well founded of natural integers, we can deduce that the system will converge toward the value $V(1)$. We can then deduce that: $Spec(\mathcal{M}_1) \vdash \Phi_1$.

Then we have: $T_1 = 1 = V(1)$. The observation of the event $forwarding\_time$ reaching $V(1)$, and the weakest fairness assumption made on the same event, allow us to conclude that: $Spec(\mathcal{M}_1) \vdash \Phi_2$, from where the termination with stability is reached, because once the variant reaches its minimum value, only the event $finish$ will be activated in the system, and no changes can be made on the variables in the system, since none of the events $register\_votes$ and $forwarding\_time$ will be activated.

In all refinements that follow this abstract model, termination proofs are the same. Since all events that will be introduced will also be guarded by the guards $grd1$ and $grd2$ of the event $register\_votes$, no event changes the time and $forwarding\_time$ and $finish$ remain unchanged. Termination with stability is also defined and proved in the tallying phase of the patterns defined for voting protocols.

4.3.2 Condition for Dependencies

A vote is validated only when all the constraints defined in the dependent Event-B context of the tallying phase are valid. The validation of these constraints is based on facts or data generated during the recording phase. This implies the existence of states in the model $M8\_Recording\_T1$ (Fig. 2) satisfying these constraints that we call **context deduced or combination of situations and constraints**. The satisfaction of axioms thus defined, particularly the axioms **dep_axm23** and **dep_axm24**, expresses the *"initial configuration"* of this phase of the vote:

$$C0\_Tallying\_T1(s2, c2) \wedge Init_2$$

where s_2 and c_2 are, respectively, sets and constants of the B context $C0\_Tallying\_T1$. This relationship expresses a dependency between these two components. In particular, the two axioms **dep_axm23** and **dep_axm24** should be validated by properties over values of state variables of the previous phase.

The states that validate these constraints are the states which, in addition to satisfying the axioms **axm1** ...**axm22**, must also fulfill the conditions defined in the axiom **dep_axm24** which expresses constraints, such as

- the closing time of polls has arrived: $timer = end\_time$;
- no corrupt signature has been recorded: $alone\_corrupt\_signatures = \varnothing$;
- no corrupt choices assigned to an envelope have been recorded: $corrupt\_choices\_envelopes = \varnothing$;
- choices and signatures are registered in the polls provided the voters who made these choices have signed at offices where they were registered to vote:
  $\forall\ s, v, h \mapsto (s \mapsto v \in rec\_votes \wedge h \mapsto (s \mapsto v) \in registred\_votes\_offices \Rightarrow \exists\ elec \mapsto ((elec \mapsto s) \mapsto h \in voters\_hosting))$;
- the number of correct votes is the same as the number of recorded envelopes: $correct\_choices\_envelopes \in valid\_choices \twoheadrightarrow valid\_envelopes$;
- a recorded vote (with valid choices and signatures) cannot belong to two different offices: $\forall\ v1, b1, b2 \mapsto (b1 \mapsto v1 \in registred\_votes\_offices \wedge b2 \mapsto v1 \in registred\_votes\_offices \Rightarrow b1 = b2)$;
- the number of correct choices is the same as the number of recorded bulletins: $choice\_bulletins \in valid\_choices \twoheadrightarrow recorded\_bulletins$;

- at the closure of polling stations, we have the guarantee that all voters who expressed their choice and have signed the electoral list will have a bulletin that will be counted, and vice versa, that all the bulletins that will be counted, were expressed by voters who expressed their choice and have correctly signed.

5   Contextualizing Systems Versus Refinement of Event-B Models

The case study [19] illustrates how refinement mechanism in its various forms (horizontal and vertical) contextualizes the target systems and provides a good compromise between expressiveness and rigorous reasoning. Refinement can be seen as an a posteriori approach, where implicit knowledge can be explicit by integration into abstract models, thus avoiding some conflicts and ambiguities. Implicit semantics play an important role in the identification and evaluation of the functionalities of the studies systems.

Our investigations show that as long as there are semantically equivalent relations between constraints, refinement mechanism remains a good integration approach for the verification of systems that reconciles the different views of different actors and parties involved in the system. The examples [3, 30] illustrate this point of view. The work [3] deals with the design of an avionics system, where the part that produces information such as altitude and flight speed communicates them to the party responsible for their display via a unidirectional channel. These data are exchanged by converting values expressed in inches, meter, and kilometer using constants, axioms defined in Event-B contexts. While the component responsible for calculating this information expresses the altitude in inches and the speed in meters per hour, their display is carried out in meters and kilometers per hour, respectively. The system [30] estimates wheel speed in kilometers per hour, while the calculation for determining ground speed is estimated in miles per hour. This example also introduces conversion of constants between the various units used to express the requirements.

We have also modelled a toll system that considers a case of currency. In this case study, the ticket price is a number that takes its values in the set of integers. However, this value is interpreted for the system designer and for the users of the services offered on the highway, it corresponds to the currency used to pay for the access service to the highway. We have shown by refinement that this semantics can be explained because there is a relation which establishes an equivalence between the different values of the different currencies that can be used to settle the sum to be paid. This is the *exchange rate* between two currencies which defines the rate at which one currency will be exchanged for another. This is achieved by: adding new sets, constants, and axioms by extending the B contexts; by modifying the events concerned and by introducing the appropriate gluing invariants to better express the semantics of this context. The context treated in the various cases cited corresponds to the *context of constraints*.

Accordingly, the carrier sets and logic quantifiers on which the Event-B is based also allow a good parameterization of the models for the verification by an automatic construction of invariants on systems covering enough behaviors to conceive design patterns. The instantiation of the obtained patterns consists of configuring the system to specify the values of the sets in Event-B contexts, this case has a link with the validation issues and does not give rise to additional proof obligations; or to introduce other refinements for the specific needs or requirements of designers. The refinement in the Event-B formalism is defined by the addition of machines that refine other machines to better define behaviors in the systems or to introduce other behaviors that do not exist in the abstract models. This approach requires the use of Event-B contexts that defined the static aspects in the models. Often, it is done by extension of these Event-B contexts that one can integrate new concepts, useful mechanisms for system functionalities. This method has a great advantage that consists in factoring the efforts of proofs to be realized for possible reuse of the proofs and therefore of the developed models.

Contrary to what we have claimed in the above, when no interpretation exists between the constraints to elaborate the semantic of the context, refinement via different extensions between the Event-B contexts is possible. The example of the voting systems that we developed in [19] illustrates this case. Different elections have different modes/types of voting and voting theory analyzes the advantages and disadvantages of each. For example, a *majoritarian voting* where a presidential candidate must be elected is represented by paper ballot where every candidate has the option to vote and each paper corresponds to one candidate (vote or poll). This constraint is shown in Event-B by the following constant: $axmt1$: *bulletins_representatives* $\in$ *Representatives* $\rightarrowtail$ *Bulletins*, where *Representatives* corresponds to the set of all representatives needed for a specific election including designations that may be chosen by a voter. For instance, this set can contain: $candidat_1$, $candidat_2$,..., $candidat_n$, *None*_of_the_*above*, in the case of a presidential election. It may also contain *favorable*, *unfavorable*, if the choice in a referendum is an adherence to any law.

In the case of a *preferential voting* or *cumulative voting*, voters should make their choice on paper ballot, where all candidates are listed on all these papers. This choice corresponds to a preference order mentioned next to each candidate on the same paper ballot. This constraint corresponds to a Cartesian product presented as follows in the Event-B method: $axmt2$: *bulletins_representatives* $=$ *Representatives* $\times$ *Bulletins*. These constraints situate our development and thus contextualize the proofs. We have shown that constraints rely on the static part in the system, and we qualify them as a **context of constraints**. Each type of voting is defined in a different Event-B context as shown in Fig. 2.

6 Conclusion

The *context lifting* of McCarthy involves situations or times. Furthermore, the parto-nomic relations between contexts in [13] express a change of structures which corre-sponds to a change of models in our case. The context as knowledge is a notion which depends on space and time, it can therefore be defined as a dependency between mod-els in Event-B, where the termination as defined must be established. The principle of dependency is a dual principle to the principle of invariance in Event-B machines, claiming that states are constrained by invariants in order to establish safety in a proof system. As well as improving productivity, this decomposition can also improve soft-ware/system quality by providing guarantees with respect to avoidance of security risks and attacks in the case of voting protocols. This modelling approach also makes it possible to exploit the dependent Event-B contexts for validating ontologies.

The proof context in [14] refers to the environment in which the target system is designed and interacts with this system. This environment is constructed as a component with a set of finite but exhaustive behaviors. The context is the perimeter, i.e., constraints and conditions, which characterize a set of interactions of the model with its environment. Such a system is a model represented by context automata. The modelling of the context is carried out in a context description language called *CDL*. While in this formalism the analysis of the safety and liveness properties is carried out by a composition of the model to be verified with the one containing the environmental specification, in the Event-B formalism, this notion is part of the developed models, since the considered systems are closed systems where the environment is included in their modelling.

Future work will explore the use of the dependency relationship, when developing technical systems, and questions related to the mechanization depends on relation over Event-B components.

References

1. J-R. Abrial, *Modeling in Event-B: System and Software Engineering* (Cambridge University Press, Cambridge, 2010)
2. Jean-Raymond Abrial, Dominique Cansell, Dominique Méry, A mechanically proved and incremental development of IEEE 1394 tree identify protocol. Formal Asp. Comput. **14**(3), 215–227 (2003)
3. Y. Ait Ameur, D. Méry, Making explicit domain knowledge in formal system development. Sci. Comput. Program. **121**(100–127) (2016)
4. M.B. Andriamiarina, *Développement d'algorithmes répartis corrects par construction* (Uni-versité de Lorraine; Loria & Inria Grand Est, Thése, 2015)
5. Patrick Barlatier, *Conception et implantation d'un modèle de raisonnement sur les contextes basée sur une théorie des types et utilisant une ontologie de domaine* (Université de Savoie, Thése, 2009)
6. N. Benaïssa, D. Mér,. Cryptographic protocols analysis in event B, in *Perspectives of Systems Informatics, 7th International Andrei Ershov Memorial Conference, PSI 2009, Novosibirsk,*

Russia, June 15-19, 2009. Revised Papers, ed. by A. Pnueli, I. Virbitskaite, A. Voronkov, volume 5947 of *Lecture Notes in Computer Science* (Springer, 2009), pp. 282–293

7. Dines Bjørner, Manifest domains: analysis and description. Formal Asp. Comput. **29**(2), 175–225 (2017)

8. P. Brézillon, C. Tijus, Représentation contextualisée des pratiques des utilisateurs, in *Extraction des connaissances : Etat et perspectives (Ateliers de la conférence EGC'2005)*, ed. by J-M. Petit, N. Vincent, F. Cloppet, vol. E-5 of *RNTI*, Cépaduès-Éditions (2005), pp. 81–88

9. ClearSy. Atelier B. http://www.atelierb.eu/

10. Véronique Cortier, Georg Fuchsbauer, David Galindo, BeleniosRF: a strongly receipt-free electronic voting scheme. IACR Cryptol. **2015**, 629 (2015)

11. Véronique Cortier, David Galindo, Stéphane Glondu, Malika Izabachène, A generic construction for voting correctness at minimum cost - application to helios. IACR Cryptol. **2013**, 177 (2013)

12. J.L. Crowley, J. Coutaz, G. Rey, P. Reignier, *Perceptual Components for Context Aware Computing* (Springer, Berlin, Heidelberg, 2002), pp. 117–134

13. Richard Dapoigny, Patrick Barlatier, Modeling contexts with dependent types. Fundam. Inform. **104**(4), 293–327 (2010)

14. Philippe Dhaussy, Frédéric Boniol, Mise en œuvre de composants MDA pour la validation formelle de modèles de systèmes d'information embarqués. Ingénierie des Systèmes d'Information **12**(5), 133–157 (2007)

15. P. Dockhorn Costa, J.P. Andrade Almeida, L. Ferreira Pires, G. Guizzardi, M.J. van Sinderen, Towards conceptual foundations for context-aware applications, in *AAAI Workshop on Modeling and Retrieval of Context 2006*, ed. by T.R. Roth-Berghofer, S. Schulz, D.B. Leake, *AAAI Technical Report*, vol. WS-06, Menlo Park, CA, USA (AAAI Press, 2006), pp. 54–58

16. Paul Dourish, Seeking a foundation for context-aware computing. Human-Comput. Interact. **16**(2–4), 229–241 (2001)

17. P-A. Fouque, *Le partage de clés cryptographiques: Théorie et Pratique*. Thèse de doctorat, Université Paris 7 (2001)

18. Igor Nai Fovino and Marcelo Masera. Through the description of attacks: A multidimensional view. In Janusz Górski, editor, *Computer Safety, Reliability, and Security, 25th International Conference, SAFECOMP 2006, Gdansk, Poland, September 27-29, 2006, Proceedings*, volume 4166 of *Lecture Notes in Computer Science*, pages 15–28. Springer, 2006

19. J. Paul Gibson, S. Kherroubi, D. Méry, Applying a dependency mechanism for voting protocol models using event-B, in *Formal Techniques for Distributed Objects, Components, and Systems - 37th IFIP WG 6.1 International Conference, FORTE 2017, Held as Part of the 12th International Federated Conference on Distributed Computing Techniques, DisCoTec 2017, Neuchâtel, Switzerland, June 19-22, 2017, Proceedings*, ed. by A. Bouajjani, A. Silva textitLecture Notes in Computer Science, vol. 10321 (Springer, 2017), pp. 124–138

20. G. Guizzardi, *Ontological Foundations for Structural Conceptual Models*. Ph.D. thesis, University of Twente, 2005. Published as the book "Ontological Foundations for Structural Conceptual Models", Telematica Instituut Fundamental Research Series No. 15, ISBN 90-75176-81-3 ISSN 1388-1795; No. 015; CTIT PhD-thesis, ISSN 1381-3617; No. 05-74

21. G. Holzmann, The spin model checker. IEEE Trans. Softw. Eng. **16**(5), 1512–1542 (1997)

22. S. Kherroubi, D. Méry, Contextualisation et dépendance en Event-B, in *Approches Formelles dans l'Assistance au Développement de Logiciels (AFADL)*, Montpellier, France (2017)

23. S. Kherroubi, D. Méry, Contextualization and dependency in state-based modelling - application to event-B, in *7th International Conference on Model and Data Engineering (MEDI 2017)*, Model and Data Engineering, Barcelona, Spain (2017)

24. A. Kofod-Petersen, J. Cassens, Using activity theory to model context awareness, in *Modeling and Retrieval of Context, Second International Workshop, MRC 2005, Edinburgh, UK, July 31 - August 1, 2005, Revised Selected Papers*, ed. by T. Roth-Berghofer, S. Schulz, D.B. Leake, *Lecture Notes in Computer Science*, vol. 3946 (Springer, 2005), pp 1–17

25. G. Kotonya, I. Sommerville, *Requirements Engineering: Processes and Techniques*, 1st edn. (Wiley Publishing, New York, 1998)

26. G.T. Leavens, J-R. Abrial, D.S. Batory, M.J. Butler, A. Coglio, K. Fisler, E. C.R. Hehner, C.B. Jones, D. Miller, S.L. Peyton Jones, M. Sitaraman, D.R. Smith, A. Stump, Roadmap for enhanced languages and methods to aid verification, in *GPCE*, ed. by S. Jarzabek, D.C. Schmidt, T.L. Veldhuizen (ACM, 2006), pp. 221–236
27. J. McCarthy, Notes on formalizing context, in *Proceedings of the 13th International Joint Conference on Artifical Intelligence - Volume 1*, IJCAI'93, San Francisco, CA, USA (Morgan Kaufmann Publishers Inc, 1993), pp. 555–560
28. J. McCarthy Notes on formalizing context, in *Proceedings of the 13th International Joint Conference on Artificial Intelligence. Chambéry, France, August 28 - September 3, 1993*, ed.by R. Bajcsy (Morgan Kaufmann, 1993), pp. 555–562
29. D. Méry, M. Poppleton, Towards an integrated formal method for verification of liveness properties in distributed systems. Softw. Syst. Model. (SoSyM) (2015)
30. D. Méry, S. Rushikesh, A. Tarasyuk, Integrating domain-based features into event-B: a nose gear velocity case study, in *Model and Data Engineering - 5th International Conference, MEDI 2015*, ed. by L. Bellatreche, Y. Manolopoulos, LNCS, vol. 9344 (Springer, Rhodes, Greece, 2015), pp. 89–102
31. project RODIN. Rigorous open development environment for complex systems. http://rodin-b-sharp.sourceforge.net/ (2004). 2004–2007
32. A.G. Sutcliffe, S. Fickas, M. Moore Sohlberg, PC-RE: a method for personal and contextual requirements engineering with some experience. Requir. Eng. **11**(3), 157–173 (2006)
33. The Coq Development Team. *The Coq Proof Assistant*. INRIA, http://coq.inria.fr (1999–2017)
34. The FoCaLiZe Development Team. *FoCaLiZe*. INRIA, http://focalize.inria.fr/
35. The Frama-C Development Team. *Frama-C* . CEA, https://frama-c.com/
36. The Isabelle Development Team. *Isabelle*. Cambridge University and TUM, http://www.cl.cam.ac.uk/research/hvg/Isabelle/index.html (1988–2017)
37. M. Theodorakis, A. Analyti, P. Constantopoulos, N. Spyratos, Contextualization as an abstraction mechanism for conceptual modeling. Technical Report TR255, University of Crete (1999)
38. Y. Tsiounis, M. Yung, On the security of elgamal based encryption, in *Public Key Cryptography, First International Workshop on Practice and Theory in Public Key Cryptography, PKC '98, Pacifico Yokohama, Japan, February 5-6, 1998, Proceedings*, ed. by H. Imai, Y. Zheng, *Lecture Notes in Computer Science*, vol. 1431 (Springer, 1998), pp. 117–134

Configuration of Complex Systems—Maintaining Consistency at Runtime

Azadeh Jahanbanifar, Ferhat Khendek, and Maria Toeroe

Abstract For management purposes, the sub-systems of a system are generally described through various configurations, each fragment focusing on a specific sub-system, e.g., platform, middleware, etc. To form a consistent system configuration, these independently developed configurations, also known as partial configurations or configuration fragments, need to be integrated together. This integration is a challenging task, mainly because of overlapping entities (different logical representations of the same system resource) in the configuration fragments and/or complex relationships among the entities of the different configuration fragments. At runtime, the system may be reconfigured to meet new requirements or in response to performance degradations for instance. These changes may lead to inconsistency as they may violate some constraints between entities. Maintaining the consistency, i.e., satisfying the defined constraints, and adjusting the system configuration at runtime is another challenging task. In this book chapter, we describe our overall model-based framework for tackling these important issues. We discuss briefly the modeling and other approaches that compose this framework and elaborate on the runtime configuration validation approach. With this approach, the runtime reconfiguration requests are checked, before applying them, against a reduced set of the consistency rules instead of the complete set of rules, and the reconfiguration requests are applied only if they are safe, i.e., they preserve the configuration consistency and satisfy the constraints. Using a reduced set of consistency rules reduces the validation time, which is important for dynamic/runtime reconfigurations.

A. Jahanbanifar · F. Khendek (✉)
Engineering and Computer Science, Concordia University, Montreal, Canada
e-mail: khendek@encs.concordia.ca

A. Jahanbanifar
e-mail: az_jahan@encs.concordia.ca

M. Toeroe
Ericsson Inc, Montreal, Canada
e-mail: maria.toeroe@ericsson.com

© Springer Nature Singapore Pte Ltd. 2021
Y. Ait-Ameur et al. (eds.), *Implicit and Explicit Semantics Integration in Proof-Based Developments of Discrete Systems*,
https://doi.org/10.1007/978-981-15-5054-6_10

1 Introduction

A system, e.g., new composite application or a system of systems such as in the cloud architectures, is built by assembling together independently developed Commercial Off-The-Shelf (COTS) components. Each of these components/sub-systems may have its own perspective of the system described as a configuration. This configuration specifies the organization and the characteristics of the resources the component/sub-system is aware of and potentially manages. A system can also be viewed from different perspectives or aspects (such as performance, security, or availability) and thus have multiple configurations. Therefore, a composite system may be described through various independently developed configuration fragments. One of the main challenges of such systems is the integration of these configuration fragments in a consistent manner that reflects the relations and constraints between the entities of the different configuration fragments and ensures that the resulting system meets the required properties such as availability, performance, and security. The complexity of the integration task stems from the overlapping entities of the different configuration fragments (i.e., different logical representations of the same physical entity) and from the complex relationships among the entities of the different configuration fragments. Manual and ad hoc integration of the fragments is difficult and error-prone.

Furthermore, at runtime a system actor (i.e., the administrator or a management application) may need to modify the configuration. These reconfigurations may be needed to meet new requirements, respond to performance degradations, for elasticity or upgrade purposes. Changes made to one configuration entity may have an impact on other entities of the configuration because of the relations and dependencies between the entities some of which are only known by the integration framework, but not by the actors requesting the changes. Nevertheless, the changes should be conducted in a safe way not to compromise the consistency of the system configuration and therefore jeopardize the system operations. Thus, the proposed changes should be checked and the modified configuration needs to be validated to guarantee its consistency and therefore to protect the system from malfunctioning. According to [1], the consistency of a configuration is defined as the correctness of the data which requires the satisfaction of the structural integrity requirements and the application/domain constraints. The system configuration, especially for large systems, or composite systems, may consist of thousands of entities each with several attributes and complex relations between the entities. In such systems, the management and control of the reconfiguration side-effects with an ad hoc or manual approach is a difficult and error-prone task as the actor must know and take care of all the relations and constraints. This problem is even worse for real-time and highly available systems as the validation and reconfiguration time should be minimal. Moreover, such systems should not be shut down or restarted because of the reconfiguration. Therefore, an automated and efficient approach is required to manage the reconfiguration and protect the system consistency from invalid modifications.

To address the aforementioned issues, we defined a model-based configuration management framework for the integration, runtime validation, and adjustment of system configurations. In this framework, Domain-Specific Modeling Languages (DSML) capture the concepts, their relations, and consistency rules (constraints) of configuration fragments. The Unified Modeling Language (UML) and its profiling mechanism [2, 3] is our choice for defining the DSMLs. The constraints coming from the application domain restrict the configuration entities and their relations by governing both their structure and behavior. The constraints are expressed using the Object Constraint Language (OCL) [4].

In this book chapter, we provide an overview of the framework and elaborate on one of the approaches, namely, on the runtime configuration validation. Such runtime validation is a prerequisite for systems with dynamic reconfiguration capabilities as it detects the potential inconsistencies that can be caused by the reconfigurations (changes to a single system entity or a bundle of changes to a number of entities). Runtime reconfigurations often target only parts of the system and an exhaustive validation, i.e., checking against all the constraints, can be time and resource consuming. We propose a partial validation of the configurations, which can be used at runtime to reduce the validation time and overhead. A configuration model is validated against the configuration profile including the OCL constraints. However, in our partial validation, only the constraints that are affected and need to be checked are selected— hence it is referred partial—as the other ones remain valid. The approach consists of filtering the set of constraints to be rechecked, categorizing them, and then validating them. We perform an evaluation of the approach and provide a semi-formal proof of its correctness.

The rest of the chapter is organized into five sections. In the next section, we describe the modeling framework. In Sect. 3, we provide an overview of the configuration management framework. In Sect. 4, we discuss the partial validation approach. We review related work in Sect. 5 before concluding in Sect. 6.

A semi-formal proof of our partial validation approach is given in the appendix.

2   Modeling Framework

A system configuration is a set of configuration entities and their relations. These configuration entities are logical representations of the system resources. The configuration defines the arrangement and the rules that the system should obey with respect to the represented resources. The granularity and the definition of the configuration entities depend on the application domain represented and its requirements. Components, groups of components, sub-systems, virtual machines, and hardware elements are examples of the resources the configuration entities represent in the context of this chapter. For the representation and manipulation of configurations, a configuration schema is required to specify the structure of the configuration entities and their relations. We use UML and its profiling mechanism to define the configuration schema. To capture the domain semantics and additional restrictions, we added the consistency rules to the configuration profile as OCL constraints. In this section, we explain our extension to OCL and its need.

Although the standard OCL is suitable for many applications, it is not always sufficient. By extending OCL we can add more information to the constraints. An example of extension is the addition of severity and a description to the constraints as introduced in [5]. The configuration constraints are restrictions on the attribute values and relations of the configuration entities. They are defined when the configuration schema is designed to reflect the requirements of the system/application domain. In the case of configurations, a constraint puts some restrictions on some entities but it does not characterize the role of these entities in the constraint, i.e., if some entities in the constraint influence the others. These dominant and dominated roles of entities cannot be expressed by standard OCL. Knowing these roles, however, will help us in identifying the constraints that need to be checked and the order in which they must be checked whenever needed.

To illustrate our idea we use an example from the Open Virtualization Format (OVF) domain [6] defined by the Distributed Management Task Force (DMTF) [7]. OVF is a packaging standard, which describes an extensible format for the packaging and distribution of software products for virtual systems. It enables the cross-platform portability by allowing software vendors to create pre-packaged appliances for which the customers can have different choices of virtualization platforms [6]. The example is shown in Fig. 1 where the upper part shows the structure of a simple two-tier Petstore appliance. It consists of a Web Tier and a Database Tier. The Database Tier itself consists of two Virtual Systems for fault tolerance. So, three Virtual Systems (Web Server, DB1, and DB2) and three Virtual System Collections (Petstore, Web Tier, and DB Tier) are included in the Petstore OVF package.

The OVF package definition allows for specifying the deployment of Virtual Systems with specific proximity needs through the definition of placement group

Fig. 1 The structure of the Petstore OVF package

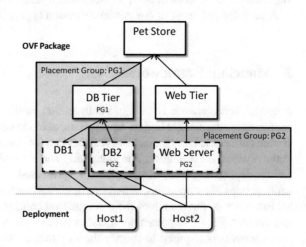

Placement Group PG1 Policy - Availability
Placement Group PG2 Policy - Affinity

policies for the Virtual Systems and Virtual System Collections. The policies are as follows[6]:

- **Affinity Policy**: It is used to specify that two or more Virtual Systems should be deployed closely together, for example, because they need fast communication.
- **Availability Policy**: It is used to specify that two or more Virtual Systems should be deployed separately because of HA or disaster recovery considerations.

In the illustrated Petstore appliance of Fig. 1, the DB Virtual Systems (DB1 and DB2) should be deployed on different hosts for fault tolerance. Thus, the PG1 placement group with the availability policy is specified for the Virtual System Collection of the DB Tier. PG1 is a property of the DB Tier. On the other hand, the DB2 and Web Server Virtual Systems should be deployed on the same host for fast communication, so a placement group, i.e., PG2, with the affinity policy is specified for these two Virtual Systems. PG2 is defined as a property for each the DB2 and the Web Server Virtual Systems.

Figure 2 illustrates the relation between these entities of a simplified OVF domain model. The restrictions that the policies impose on the deployment of Virtual Systems are expressed with the OCL constraints included in the figure.

At deployment time, the Virtual Systems with their placement groups dictate how they should be deployed on the Hosts that are shown at the bottom of Fig. 1. Note that the placement group may be defined for the Virtual System (e.g., in the Web Server Virtual System), implied by the parent Virtual System collection (e.g., DB1 Virtual System), or a combination of these two cases (e.g., DB2 Virtual System). The DB Tier has a placement group PG1, which in turn has the "Availability" policy, thus, all the Virtual Systems of the DB Tier (DB1, DB2) should be hosted on different Hosts as shown on Host1 and Host2. On the other hand, the placement group PG2 defined for DB2 and for the Web Server has the "Affinity" policy, thus, they should be placed on the same Host, i.e., Host2. An OCL constraint captures the restrictions on the relation between the Virtual Systems and their Host(s) imposed by the placement group policies. However, an OCL constraint cannot capture the role of the Virtual Systems in the constraint as determining the Host entity selection. In the relation

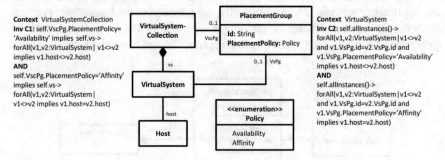

Fig. 2 Partial model of the Virtual Systems, their collection and placement policy in an OVF package

between the Virtual System and the Host entities, the Virtual System entity has a leader role and drives the Host entity, that is, the follower. This means that if the Virtual System entity (including its Placement Group) changes and the constraint becomes violated, the Host of the Virtual System should change too to follow the Virtual System change and satisfy the constraint. On the other hand, if the Host of the Virtual System changes and this change violates the constraint, the Virtual System and its Placement Group cannot be changed as the role semantics does not allow the leader entity to adjust to the changes of the follower entities. For instance, in the Petstore example, the DB Tier (i.e., DB1, DB2) and the Web Server Tier are the leader entities while Host1 and Host2 are the followers. Now let us assume that Host1 fails. Since PG1 does not allow the collocation of DB1 and DB2, DB1 cannot be re-deployed on the remaining Host2. On the other hand, if we want to change the placement group of the DB Tier from PG1 to PG2, this change of the leader results in changing the Host entity (follower) which means that now DB1 and DB2 should be placed on the same Host (Host2).

As the standard OCL cannot express these roles for entities, we extended the OCL by defining roles for the constrained entities to show the influence of some entities over others in the constraint. Considering the semantics of the relations between the entities we can identify a leadership flow between them. In other words, in a constraint with multiple entities involved, changes in some entities (*Leader*) may impact the others (*Follower*), but not the other way around. In other relations where the entities have equal influence over each other, we call them *Peer* entities.

Figure 3 shows the extension of the constraints with the leadership information. We extended the OCL to enrich the constraints without changing its grammar and metamodel so that parsers and validators designed for the standard OCL remain usable. We consider OCL together with our extension as our constraint profile.

The roles of the entities in the constraints of OVF example are shown in Fig. 4. In this figure, constraints are shown as ovals. The participation of each entity in a constraint is represented by an edge between the constraint and the constrained entity. The role of the entity in the constraint is shown as a label on this edge (e.g., label "L" represents the Leader role). This representation focuses on the role of entities in the constraints and depicts how the constrained entities can affect each other.

It is worth mentioning that the roles of the constrained entities may change with the application scenario. More specifically, we may define the leader/follower/peer roles for the entities differently for design time and runtime. At design time we generate the configuration according to an optimal design method. Once the system

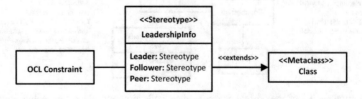

Fig. 3 OCL constraint enriched by the leadershipInfo

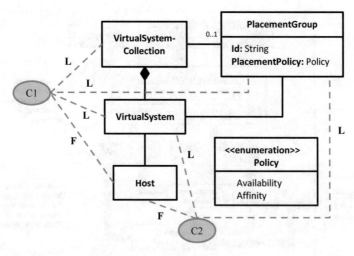

Fig. 4 Representation of entity roles in constraints

is deployed we may be limited in the changes allowed. For example, due to budget reasons, we may not add new hosts to the system and as a result we want the Virtual Systems (including the software products) adopt and follow the Host restrictions in this respect. This means that now the Host entity becomes a leader and the Virtual System and Virtual System collection are followers. Note that the standard portion of the OCL constraint between them remains unchanged. Defining the roles for the entities through the leadership information has the advantage that we can define and change the roles whenever it is needed without affecting the constraints themselves.

3   Overview of the Configuration Management Framework

As shown in Fig. 5 our model-based framework for configuration management includes a module for configuration generation at system design time and a module for system runtime change management. In this section, we briefly discuss the different parts of our configuration management framework.

3.1   System Configuration Design—Integration of Configuration Fragments

As mentioned earlier, sub-systems of a system are described through several configuration fragments. These configuration fragments need to be integrated together to form a consistent system configuration to ensure correct operation of the system.

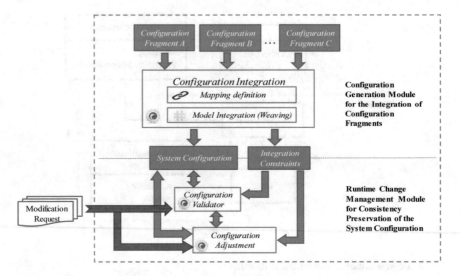

Fig. 5 Overview of configuration management framework

The system configuration should ensure that the resulting system meets the required properties.

We extend the model weaving technique [8, 9] and use model transformation techniques to devise an approach for the integration of configuration fragments targeting specific system properties. We define the semantics of the relations between the entities of the configuration fragments as links at a higher level of abstraction which has several advantages:

- It increases the reusability of the defined links (relations) to map other entities between configurations.
- It allows adding/modifying the interpretation of the links and embedding them into the integration process without modifying the links.
- It is easily extendible as various configuration profiles can be added to the integration process using predefined or new links.
- The integration process is automated. The system configuration can be generated automatically with the same rules for different input configurations.

We define the integration semantics as integration constraints to enrich the system configuration profile. The integration constraints (describing the semantics of the relation between the fragments) in addition to the union of the constraints of the fragments form the system configuration constraints which guard the consistency of system configuration models against unsafe runtime modifications. In other words, the consistency is formally defined by the set of OCL constraints generated automatically from the weaving and the constraints of all the fragments, which are by construction non-contradictory. More details on the integration technique can be found in [10, 11].

3.2 System Runtime—Consistency Preservation with Change Management

A reconfiguration may be performed for many reasons, e.g., in response to environmental changes or for fine-tuning. These changes may compromise the consistency of the configuration. To manage configuration changes, we propose the architecture shown in Fig. 6 which includes a configuration validator to check the change requests and an adjustment agent, which attempts to add complementary modifications to resolve the potential inconsistencies detected during the validation phase.

Runtime Configuration Validation
At runtime the administrator or the management applications may need to change the system configuration to control/manage the resources they are responsible for. These changes may compromise the consistency of the system configuration and jeopardize the system's operation. Thus, the requested changes should be checked and the modified configuration has to be validated to guarantee its consistency. As shown in Fig. 6 the configuration validator is responsible for performing the validation with respect to the system configuration profile and its constraints. The validation may result in one of the three following cases:

(a)   The requested changes do not violate the configuration constraints and respect the profile. Therefore, the changes are safe and can be applied.
(b)   The requested changes violate one or more constraints of the profile and these violations cannot be resolved as there is no chance to propagate the changes to other entities of the violated constraints to resolve the violations. Thus, the requested changes are rejected.

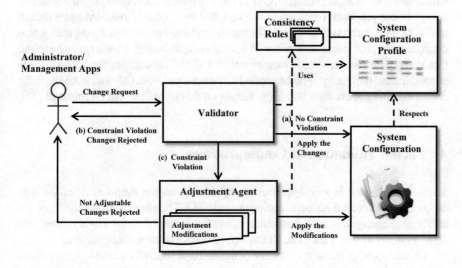

Fig. 6 Runtime configuration validation and adjustment

(c) The requested changes violate one or more constraints of the profile; however, the changed entities/attributes can impact other entities/attributes of the violated constraints. Therefore, there may exist a chance to resolve the constraint violation by changing other constrained entities. Thus, the result of the validation will be passed to an adjustment agent.

The decision of rejecting the requested changes (i.e., case b) or trying to adjust the configuration (i.e., case c) is made based on the ability of the changed entities to impact other entities of the violated constraints. The possibility of the impact is determined based on the leadership information of the constraint. In the next section, we elaborate on our technique for runtime configuration validation. We aim at an efficient technique that can be applied at runtime.

Configuration Adjustment

Although a configuration validator can detect the constraint violations caused by unsafe/incomplete change requests, it cannot tell if such violations or conflicts might be resolved by adding complementary modifications that complete the initial set of changes. The potential inconsistencies detected by the validation technique can be due to the incompleteness of the set of changes as performed by the administrator who is not aware of all the relations between all involved entities/attributes. In order to resolve such inconsistencies, we devise a technique for complementing an incomplete set of changes and therefore adjust automatically the configuration at runtime. The adjustment consists of modifications of other entities/attributes that re-establish the configuration consistency. We achieve this by propagating the changes in the configuration according to the system constraints following the possible impacts of the configuration entities on each other. We minimize the complementary modifications to control the side-effects of the changes. The problem is formulated as a Constraint Satisfaction Problem (CSP) [16, 17] which we solve using a constraint solver. In our proposed framework, this task is done by the adjustment agent shown in Fig. 6. It takes the validation result from the validator as input and uses the system configuration profile and constraints. If a set of complementary changes can be found that along with the requested changes satisfies all the constraints, the adjustment is successful and the changes can be applied to the configuration. Otherwise, the initially requested changes are rejected. More details on this technique can be found in [12].

4 Partial Validation of Configurations

Runtime validation is a prerequisite for dynamic reconfiguration as detection and correction of potential inconsistencies are required. The capabilities of dynamic (or runtime) reconfiguration and runtime validation are needed for Highly Available (HA) systems as they cannot be shut down or restarted for reconfiguration.

A system configuration may consist of hundreds of entities, with complex relations and constraints. Runtime reconfigurations (changes to a single system entity or a bundle of changes to a number of entities) often target only parts of the configuration. An exhaustive validation that checks all the consistency rules is not always required and can be substituted by a partial validation, in which the number of consistency rules (constraints) to be checked are reduced and this results in reduction of the validation time and overhead. In a partial validation, only the constraints that are impacted by the changes are selected and checked. The other constraints remain valid and do not need to be rechecked. Thus, by checking only a subset of the constraints we can claim if the configuration is valid or not.

4.1 Partial Validation Technique

To validate a configuration model, we check its conformance to the configuration profile. The profile defines the stereotypes, their relations (the structure of the model) along with a set of constraints over these stereotypes and relations to assure well-formedness. When a request for changing some entities of the configuration model is received, the modified model needs to be checked for conformance against its profile. The full validation can be time and resource consuming. Such overhead is not desirable in live systems, especially in real-time and HA systems. A solution to reduce the overhead and improve performance is to reduce the number of constraints to be checked, i.e., check only what needs to be checked against. We refer to this as partial validation.

In our approach, we minimize the number of constraints to be checked based on the requested changes. This typically also leads to the reduction of the number of configuration entities to be checked. We check only the entities whose stereotypes are involved in the selected constraints. This new set of configuration entities includes at least the changed entities and the ones related to them through their constraints. In the appendix, we provide a semi-formal proof to establish that the results of our partial validation approach are equivalent to the results of the full validation where all the constraints are checked for every change request.

Figure 7 shows an example, a model and its profile, in which changes affect only some of the constraints. The model entities conform to their respective stereotypes in the profile, e.g., A1, A2 entities of the model conform to stereotype A, and B2, B3 entities conform to stereotype B and so on. The constraints of the profile are shown as blue ovals (i.e., C1, C2, etc.). Assuming that the change set includes model entities B2 and D1, to validate the model instead of checking all the constraints of the profile (C1–C4), it is enough to select only the ones that are affected by these changes, i.e., constraints C2 and C3.

To reduce the validation overhead the time to select this set should be negligible compared to the time saving we achieve by the partial validation. Note that in cases where the modification request includes entities of many different stereotypes, the number of constraints that need to be selected is considerable and the selection may not be worthwhile anymore. The case is similar when although only a few constraints

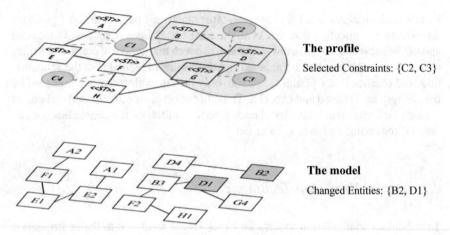

Fig. 7 Model changes and affected constraints in the profile

are selected but they apply and should be checked for a large number of configuration entities.

4.1.1 Filtering and Categorizing the Constraints

To identify the reduced set of constraints we filter the constraints based on the modification request. A request may consist of many changes each of which applies to one or more entities of the configuration model—we call them the change set. We represent the configuration entities of the change set as a model, which conforms to a change profile. Figure 8 shows the change profile and an example change set. The change profile has a stereotype called CEntity which extends the NamedElement metaclass of UML. The CEntity stereotype represents the configuration model entities to be changed—referred to as changed entity. In the Petstore, for example, we may need to reconfigure the Web Server, i.e., have it as the changed entity. The operation requested on the model entities is represented by the Operation stereotype in the change profile. It is specialized as the Add, Update, and Delete stereotypes. Regardless of the requested operation, the constraints in which the entity is involved should be checked. A change model is an input to the constraint filtering process.

Having the change model and the change profile as well as the configuration profile, the stereotypes applied on each changed entity can be identified. These are the stereotypes of the configuration profile and also of the CEntity stereotype. For the validation, the stereotypes of the configuration profile are considered. The constraints of the configuration profile are defined over these stereotypes and we also captured their roles in the constraints through the leadership concept. By looking up the stereotype of each changed entity we can select from the constraints of the configuration profile those that have the same stereotype as the stereotype applied to the entity of

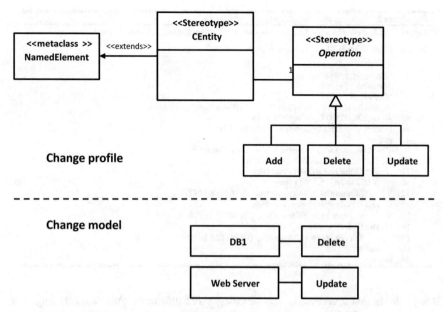

Fig. 8 Change profile and a simple change model

the change model. The role/leadership information determines the relevance of the constraint.

We also categorize each selected constraint based on the roles of the changed entities in the constraint. For this purpose, we use three sets: FConstraint, LConstraint, and PConstraint. Algorithm 1 describes the filtering process and the categorization of the filtered constraints. It starts with an empty set of stereotypes, set A, and three empty sets, LConstraint, FConstraint, and PConstraint. The constraints of the configuration profile are all in the set Constraint. In set A, we collect the stereotypes of all changed entities (lines 6 to 8). For each constraint in the set Constraint, we consider its LeadershipInfo and compare the Leader/Follower/Peer stereotypes with the stereotypes in set A. If a common stereotype is found, then the constraint is added to one of the sets, LConstraint, FConstraint, or PConstraint, while making sure that each constraint will be added only to one of these output sets (line 10 to 19). If the stereotype applied on a changed entity has a leader role in the constraint, we add the constraint to LConstraint. Similarly, if the stereotype of a changed entity plays the follower role, the constraint is added to FConstraint. The set PConstraint is for the constraints whose entities have a peer role in the constraint and also appear in the change model. The filtering and categorizing process of Algorithm 1 are implemented using transformations.

Algorithm 1. Filtering and Categorizing the constraints

Input: ConfigurationProfile, ChangeProfile, ConstraintProfile, Constraint, ChangeModel

Output: LConstraint, FConstraint, PConstraint

```
 1:   A := {}
 2:   LConstraint := {}
 3:   FConstraint := {}
 4:   PConstraint := {}
 5:   // Find all the stereotypes applied to the entities of the ChangeModel
 6:   for each ENTITYj in ChangeModel do
 7:      A:= A ∪ {ENTITYj.getAppliedStereotypes()}
 8:   end for
 9:   // Filtering and categorizing constraints of the Constraint
10:   for each CONSTRAINTi in Constraint do
11:      K:= CONSTRAINTi->LeadershipInfo
12:      if {K.Leader} ∩ A ≠ {} then
13:         LConstraint := LConstraint ∪ {CONSTRAINTi}
14:      else if {K.Follower} ∩ A ≠ {} then
15:         FConstraint := FConstraint ∪ {CONSTRAINTi}
16:      else if {K.Peer} ∩ A ≠ {} then
17:         PConstraint := PConstraint ∪ {CONSTRAINTi}
18:      end if
19:   end for
```

It is possible that a constraint can be categorized into more than one category. For example, when both the leader and follower entities of a constraint are changed within the same change set, the constraint can be categorized as FConstraint or as LConstraint and added to either category. Considering the Petstore again a request may, for example, change DB1, an instance of the DB Tier (leader) and Host1, an instance of the Host (follower) in the same change set. In such a case, we have to make a choice and add the constraint to the least restrictive set, i.e., LConstraint, to allow for potential adjustments. Since at least one leader entity is involved in case of constraint violation its follower(s), except the ones that are in the change request, can be adjusted to satisfy the constraint. Similarly, the PConstraint category is preferred over the FConstraint category.

4.1.2 Validation of the Constraints

After filtering and categorizing the constraints, the validation process considers first the least flexible constraints before the more flexible ones. The constraints in the FConstraint category are the least flexible ones because if they are violated, no adjustment can be made within the context of the change set to resolve the inconsistency as the follower entities cannot affect the leader entities. Thus, in case of detecting a violation of a constraint in FConstraint, the requested change is rejected and the validation process stops. Next, the LConstraint and the PConstraint sets are checked. If the validation fails in these cases, we consider this as a potential violation of configuration consistency because we may be able to resolve it with additional changes as leader entities can affect their followers and similarly peer entities can affect their peers. The adjustment module will try to resolve the inconsistency through additional modifications relying further on the leadership concept. Initial results on the partial validation technique have been reported in [18].

4.2   Evaluation

In this section, we present an evaluation of our partial validation approach using a prototype implementation.

4.2.1   Setup and Scenarios

We used the UML profile of the Entity Type File (ETF) [13] as a configuration profile and applied our partial validation approach to its instances. This ETF UML profile had 28 stereotypes and 24 OCL constraints defined over these stereotypes.

We implemented the partial validation method in the Eclipse Modeling Framework (EMF) [14], using the Atlas Transformation Language (ATL) [15] for constraint selection and also the EMF OCL APIs for constraint validation in a standalone java application. The experiments were performed on a machine with an Intel® Core™ i7 with 2.7 GHz and 8 Gigabytes RAM and a Windows 7 operating system.

We created an initial ETF model that conforms to the ETF UML profile. This model had 50 entities. We considered three change sets to be applied to this model. In each case, a certain number of model entities were randomly selected and changed. This number was 10, 20, and 30 for the three cases, respectively. The selections were made independently from each other. For the cases, we compared the number of constraints selected in the constraints model and the total number of constraint checks performed during the partial validation. We also measured the execution time of the full validation and each of the cases of the partial validation. Each validation test was executed five times and the average was considered as the validation time.

4.2.2   Results and Analysis

Table 1 presents the results for the different cases of partial validation in comparison with the full validation. The first row of the table represents the results of the full validation of the model in which all entities are checked for all applicable constraints, i.e., as if all entities have been changed. The second, third, and fourth rows present the results for the partial validations for 10, 20, and 30 changed entities. As the number of changed entities increases, more constraints are selected, more times they are checked and the validation time increases.

As it was expected the validation time is proportional to the number of selected constraints which in turn depends on the number of changed entities. However, the number of selected constraints is not proportional to the number of changed entities, which is due to the characteristics of the ETF profile. In the ETF profile, some entity

Table 1 Partial validation performance results

	Number of changed entities	Number of selected constraints	Total number of constraint checks performed	Partial validation time (ms)
Initial model	50	24	70	6933
Test CASE 1	10	8	41	4432
Test CASE 2	20	15	55	5413
Test CASE 3	30	18	57	5845

types have only a few tagged values and constraints while others have a relatively large number of each. Also, the frequency of use of different entity types in an ETF model is different. In a given ETF model the number of component types is typically higher than the number of other entity types. This means that in a random selection of changes the probability is higher to select a change in a component type and with that more constraints are selected. This is the reason that with only 10 changed entities the number of selected constraints is already 8 and these constraints are checked 41 times. The high ratio of checks is further explicable by the fact that the component type is specialized into several specific component types (using the UML generalization in the profile). Thus, each child component type inherits the tagged values and the constraints of its parent component types. That is, if the constraint of a parent stereotype is in the selected constraint set, then that constraint should be checked over all the child entities of that parent.

The experiments show that the stereotype of the changed entity has a determining role whether using partial validation results in the expected time gain. The characterization of the configuration profile is necessary to determine whether partial validation is beneficial and for which kind of change set.

In this evaluation, we did not include the constraint categorization as in these preliminary measurements we focused on the time gain resulting from the partial validation. In this respect, we have shown that there is a time gain,;however, the results are also showing that further analysis is needed to determine the circumstances. This is important as the constraint selection process itself takes some time and it becomes an additional overhead.

5 Related Work

The work reported in this book chapter touches upon several aspects, including system configuration generation, configuration adjustment, but elaborated mainly on configuration validation as prerequisite to dynamic reconfiguration. A lot of work has been done in each of these aspects. For a full review and discussion of related work on configuration design, we refer the reader to [11], for the configuration adjustment aspect we refer to [22]. Hereafter, we discuss the work related to configuration validation.

The validation of configurations against runtime changes has been thoroughly investigated; it is not only critical for preventing invalid changes that risk the integrity and consistency of the configuration but also because it is a necessity for self-managed systems. Some work in this context focus on the structural checking of the functional configuration parameters [19–21], e.g., type correctness, checking the validity of the values of the system entities' attributes with respect to the constraints of each entity and the relation between the entities. In the SmartFrog configuration management framework [20], the components consist of three parts: the configuration data, the life cycle manager, and the functionality of the component itself. Constraints of each component are considered within its configuration data by attaching the conditions as predicates on a description. For combining the components, the configuration data should be extended and the conditions are propagated and additional predicates may be added grouping the old and new predicates. The component developer is in charge of defining the conditions (restrictions) for the components and their combination in the configuration data templates. The authors indicate that the validation of the configuration data happens by checking these conditions; however, they do not mention how the conditions are checked. In addition, the constraints for combining the components can be expressed as simple conditions but it might not be possible to describe more sophisticated constraints (coming from special requirements of the domain) with the conditions in the configuration data templates.

In [22, 23], the authors propose a solution for dynamic reconfiguration. They consider the validation of the structural integrity and runtime changes. They use the predefined constraints for the validation of the requested modification on the structural and current operational conditions of the system. The solution consists of a model repository for storing the reference model and the constraints and an online validator for performing the dynamic constraint evaluation. The online validator takes the configuration modification request and the current system state as input and validates the request by checking the configuration instance against the reference model and the constraints. The solution uses an exhaustive validation, i.e., checks all the system constraints, which can degrade the validation performance especially for large configuration models. In our work, we check the structural integrity and also validate only the constraints affected by the changes.

Some existing approaches for re-validating models after changes also try to reduce the number of constraints and/or the number of model entities to check. In [24, 25], a list of events that can violate the OCL constraints is defined and added to their configuration schema to check the constraints only if changes are related to these constraints and only on relevant entities. This approach cannot handle complex constraints containing recursion, loops, or complex iterations. Bergmann et al. [26] use a query language (IncQuery) based on graph pattern formalism on EMF models. The queries are stored permanently in memory and they update the values of the partial matches used in queries after each model change. This approach has considerable memory consumption. In [27], an approach for incremental validation of OCL constraints has been proposed. The validation log of each constraint over the model entities is stored. A re-validation is triggered when the stored parts are changed. The work is extended in [28] to improve the performance by only checking parts of the constraints that are affected by changes and to avoid checking all the constraints. This improvement achieves a better performance but consumes more memory.

6 Conclusion

In this book chapter, we provided an overview of a model-based framework for configuration management to integrate configuration fragments in a consistent manner at design time and to preserve the consistency of the generated system configuration at runtime by validating and adjusting the configuration modification requests whenever necessary and possible. In this framework, configuration profiles are defined using the UML profiling mechanism to capture the concepts of each configuration domain, relation, and constraints between the concepts. We extended the OCL by defining roles for the constrained entities to represent the impact of the entities on each other in a constraint, i.e., the leadership concept. We extended the concept of model weaving for the purpose of semantic configuration fragments integration.

We discussed in detail our partial validation approach for the validation of the configurations at runtime. In our partial validation, only the subset of constraints affected by the modifications is selected and checked as the other constraints remain valid. Constraints are also categorized to specify the order in which they need to be checked. For evaluating the partial validation approach, we semi-formally proved its equivalence to the exhaustive validation which checks all the constraints. Also, a quantitative evaluation demonstrates reduction of the validation time compared to the exhaustive validation.

As future work, we will consider the validation of the complete framework with real case studies.

Acknowledgements This work has been partially supported by Natural Sciences and Engineering Research Council of Canada (NSERC) and Ericsson.

Appendix: A Semi-formal Proof for the Partial Validation Approach

In this appendix, we provide a semi-formal proof of our validation approach which shows that the reduced set of constraints contains all the required ones to guarantee the validity of the whole configuration model.

Note that we do not distinguish the leader/follower/peer roles of the constrained entities for the proof. This is because to prove the correctness of the validation we are only concerned about the sufficiency of the selected (filtered) constraints rather than their categorization.

Definitions

A Profile (P) is defined as a set of Stereotypes (ST_P), the set of Relations between them (R_P), and their set of Constraints ($Cons_P$).

$$P(ST_P, R_P, Cons_P)$$

For referring to the sets of stereotypes, relations, and constraints of a given Profile we use the Profile's name as the index of the set, e.g., ST_{P1} is the set of stereotypes of Profile P1.

Each relation Rl in the set R_P consists of a source stereotype (Rl.SrcST) and a destination stereotype (Rl.DstST). They specify the two ends of the relation Rl. A tuple of Lowerbound and Upperbound also specifies the minimum and maximum number of instances of the DstST in relation with a SrcST.

$$Rl(Rl.SrcST, Rl.DstST, (L, U))$$

For simplicity, the relations in this definition are considered to be associations. Other types of relations (generalization, dependency, etc.) can be added with appropriate modifications of the definition.

Each constraint ConsX in the set of $Cons_P$ consists of an *Invariant* (a Boolean expression) and a set of stereotypes based on which the invariant is defined, i.e., the constrained stereotypes. The set of constrained stereotypes is shown as ST_{ConsX}.

$$ConsX(Invariant, ST_{ConsX})$$

A Model M is defined as a set of entities (en_M) and a set of relations (r_M).

$$M(en_M, r_M)$$

Each relation rl ($rl \in r_M$) has a source entity represented as rl.SrcEn and a destination entity represented as rl.DstEn.

$$rl(rl.SrcEn, rl.DstEn)$$

In order to be valid, a model should conform to its profile. This means that each entity of the model should respect the stereotype(s) of the profile that is applied to and also all the constraints of the profile should be valid in the model. These functions are defined as follows:

Let us assume profile P is applied on model M. The function AST^1 for the input of a model entity (that belongs to the entity set of a model M) returns as the output the stereotype (that belongs to the stereotype set of the profile P), which is applied on the entity.

$$S = AST(e), \ e \in en_M, \ s \in ST_P$$

AtomicValid is defined over a constraint () and a subset of entities and relations () that belong to a model (M). The result of this function is a Boolean value which shows whether is satisfied with the values of the entities and relations in or not. Thus, is a subset of M and contains the entities and relations that are related to constraint . The entities in are defined as those entities of the model on which the stereotypes of the constraint are applied. If the stereotypes of constraint are applied on the member ends of a relation, the relation is included in . (en_M and r_M represent the set of entities and relations in model M and ST_x represents the set of stereotypes of the constraint).

$$AtomicValid(K, x)$$

$$K(en_K, r_k) = \{(e, r) | e \in en_M, AST(e) \in ST_x, r \in r_M, (AST(r.SrcEn) \in ST_x \ \text{AND} \ AST(r.DstEn) \in ST_x)\}$$

If the result of the AtomicValid is true, it means that the constraint is satisfied over a subset of entities and relations of the model M (entities on which the stereotypes of the constraint are applied). So we can conclude that the constraint is satisfied in model M or in other words, the validity of the constraint x over model M is true. Another function Valid is used to represent this statement.

$$AtomicValid(K, x) \leftrightarrow Valid(M, x)$$

Conformance of a model $M(en_M, r_M)$ to a profile $P(ST_P, R_P, Cons_P)$ is defined through the conform function which returns true if all the constraints of the P are valid over the model M and also all the entities and relations of the model respect the stereotypes and relations of the P. $_P$ and ST_P are the sets of constraints and stereotypes of the profile P, respectively. en_M and r_M are the set of entities and relations of model M. The Respect function is used to check if the entities and relations of the model respect the stereotypes and relations of the profile.

$$Conform(M, P) \leftrightarrow (\forall x \in Cons_P, Va(M, x)) \ \text{AND}$$

---

[1] Applied StereoType.

$$(\forall e \in en_M, AST(e) \in ST_P, Respect(e, AST(e))$$
$$AND(\forall t \in r_M, \exists z \in R_P, AST(t.SrcEn)$$
$$= z.SrcST, AST(t.DstEn) = z.DstST, Respect(t, z))$$

Modifying the Model

We assume that we have an initial model M1 which is valid according to the profile Pr, i.e., Conform(M1, Pr). The Change function takes the changeSet model and M1 as input and results in a new model M2 with the modified entities and relations, i.e., applies the changeSet to M1.

$$changeSet(en_{changeSet}, r_{changeSet})$$

$$M2 = Change(M1, changeSet)$$

To verify whether the changed model (M2) is also valid, we need to validate it by checking its conformance to the reference profile (Pr). To do so instead of performing a full validation and using Pr, we consider a second profile Pv which is created from the reference profile Pr with the same stereotypes and relations as Pr but with a reduced set of constraints. A filtering reduces the constraints of Pr based on the entities of the changeSet. As a result Pv is a subset of Pr.

$$Pv = Filter(Pr, changeSet), Pv \subseteq Pr$$

According to the filtering function:

$$\forall y \in en_{changeSet}, (if\ \exists\ x \in Cons_{Pr}, AST(y) \in ST_x) \rightarrow x \in Cons_{Pv}$$

$$AND\ \forall z \in r_{changeSet}, if\ \exists\ g \in Cons_{Pr},$$
$$(AST(z.SrcEn) \in ST_g AND\ AST(z.DstEn) \in ST_g) \rightarrow x \in Cons_{Pv}$$

$$AND\ \forall s \in ST_{Pr} \rightarrow s \in ST_{Pv}$$

$$AND\ \forall r \in R_{Pr} \rightarrow r \in R_{Pv}$$

The Proof of Partial Validation

We prove by contradiction that the partial validation has the same result as the full (or complete) validation. We make the assumption that the initial configuration model (to which the changes should be applied) is valid, i.e., it conforms to its profile. Using the mentioned definitions we prove that if a modified model (M2) conforms to the filtered profile (Pv) then it also conforms to Pr. This means:

$$\text{Conform}(M2, Pv) \to \text{Conform}(M2, Pr)$$

Prove by contradiction technique is used, which means that we assume that the above statement is not true and show considering the other assumptions a contradiction.
We add the negation of this statement to our assumptions:

$$\text{Conform}(M2, Pv) \text{ and Conform}(M2, Pr)$$

Based on the definition of the conform function we can say that there is at least one constraint of Pr that is not valid in M2 or at least one of the entities or relations of M2 does not Respect the profile Pr.

$$\text{Conform}(M2, Pr) \to (\exists\, e \in en_{M2}, \text{AST}(e) \in ST_{Pr}, \text{Respect}(e, \text{AST}(e))$$
$$\text{or } (\exists t \in r_{M2}, \nexists z \in R_{Pr}, \text{AST}(t.\text{SrcEn})$$
$$= z.\text{SrcST}, \text{AST}(t.\text{DstEn}) = z.\text{DstST} \text{ Respect}(t, z)))$$
$$\text{or } (\exists x \in \text{Cons}_{Pr}, \text{Valid}(M2, x))$$

At first we show that if the first part of the "or" statement would be true, we face a contradiction:

$$\exists\, e \in en_{M2}, \text{AST}(e) \in ST_{Pr}, \text{Respect}(e, \text{AST}(e))$$

From the definition of the Pv:

$$\forall s \in ST_{Pr} \to s \in ST_{Pv}$$

As the $ST_{Pr} = ST_{Pv}$, the ST_{Pr} in the first statement can be replaced with ST_{Pv} and thus:

$$\exists e \in en_{M2}, \text{AST}(e) \in ST_{Pv}, \text{Respect}(e, \text{AST}(e))$$

This is in contradiction with the assumption that Conform(M2, Pv) is true, because:

$$\text{Conform(M2, Pv)} \leftrightarrow (\forall x \in \text{Cons}_{\text{Pv}}, \text{Valid}(M2, x)) \text{ AND}$$
$$(\forall e \in \text{en}_{M2}, \text{AST(e)} \in \text{ST}_{\text{Pv}}, \text{Respect(e, AST(e))})$$

Similarly, it can be shown that if a relation of model M2 does not respect Pr, a contradiction is encountered.

In the next step, we show that if there is a constraint in Pr which is violated by M2, it would contradict to our initial assumptions. Three cases are possible:

First: The constraint x already belongs to Pv:

$$x \in \text{Cons}_{\text{Pv}}$$

which is in contradiction to the assumption that M2 conforms to Pv because:

$$\exists x \in \text{Cons}_{\text{Pv}}, \text{Valid }(M2, x) \leftrightarrow \text{Conform }(M2, \text{Pv})$$

Second: The constraint x does not belong to Pv (i.e., $\notin$ ‘Pv), and constraint involves the change set entities, which means the stereotype set of constraint x has at least one stereotype which is applied to at least one of the entities of the change set or constraint x has stereotypes that are applied to the member ends (entities) of a changed relation in the change set:

$$\exists y \in \text{en}_{\text{changeSet}}, \text{AST(y)} \in ST_x \text{ OR}$$
$$\exists z \in \text{r}_{\text{changeSet}}, (\text{AST(z.SrcEn)} \in \text{ST}_x \text{ AND AST(z.DstEn)} \in \text{ST}_x)$$

According to the Filter function:

$$\forall y \in \text{en}_{changeSet} (if \ \exists x \in Cons_{\text{Pr}}, \text{AST(y)} \in ST_x) \rightarrow x \in Cons_{\text{Pv}}$$
$$\forall z \in \text{r}_{changeSet}, if \ \exists x \in \text{Cons}_{\text{Pr}},$$
$$(\text{AST(z.SrcEn)} \in ST_x \text{ AND AST(z.DstEn)} \in ST_x) \rightarrow x \in Cons_{\text{Pv}}$$

And as $\in_{\text{Pr}}$, it can be concluded that x should also belong to $_{\text{Pv}}$, that is:

$$x \in Cons_{\text{Pr}}, \exists y \in \text{en}_{changeSet}, \text{AS(y)} \in ST_x \rightarrow x \in Cons_{Pv} \text{ OR}$$
$$x \in Cons_{\text{Pr}}, \exists z \in \text{r}_{changeSet},$$
$$(\text{AST(z.SrcEn)} \in ST_x \text{ AND AST(z.DstEn)} \in ST_x) \rightarrow x \in Cons_{\text{Pv}}$$

This means that if such constraint exists in Cons_{Pr}, it should have been already added in the Cons_{Pv} too because all the constraints that are relevant to the change set should be in the Cons_{Pv}.

Third: The constraint x does not belong to the constraint set of Pv (i.e., $\notin_{\text{Pv}}$), and constraint does not involve the change set entities, which means the stereotype set of constraint x does not have any stereotype which is applied to at the entities of the change set or the member ends (entities) of the changed relations in change set:

$$x \notin \text{Cons}_{\text{Pv}}, \nexists y \in \text{en}_{\text{changeSet}}, AS(y) \in STx$$
$$x \notin \text{Cons}_{\text{Pv}}, \nexists_z \in r_{\text{changeSet}}, (AST\,(z.\text{SrcEn}) \in \text{ST}_x \text{ AND } AST(z.\text{DstEn}) \in \text{ST}_x)$$

Based on our assumption:

$$\text{Valid}\,(M2, x) \leftrightarrow \text{AtomicValid}(K, x)$$

$$K(\text{en}_K, r_K) = \{(e, r)|e \in \text{en}_{M2}, AST(e) \in ST_x, r \in r_{M2}, (AST\,(r.\text{SrcEn}) \in ST_x \text{ AND } AST(r.\text{DstEn}) \in \text{ST}_x)\}$$

The constraint x does not have any stereotypes which is applied to the entities or member ends of relations in the change set, so the intersection of the two sets change set and K (set of entities and member ends of the relations of the model $M2$ on which stereotypes of constraint x is applied) is empty:

$$K \cap \text{changeSet} = \emptyset$$

When none of the entities of K belongs to the changeSet, it can be deducted that all the entities of K are in $M1$ model:

$$K \subseteq M1$$

Thus $M2$ model can be replaced with $M1$ in the previous assumption and state that:

$$K(\text{en}_K, r_K) = \{(e, r)|c \in \text{en}_{M1}, AST(e) \in ST_x, r \in r_{M1}, (AST\,(r.\text{SrcEn}) \in ST_x \text{AND} \\ AST(r.\text{DstEn}) \in \text{ST}_x)\}$$

And because K is in common between $M1, M2$, then:

$$\text{Valid}\,(M2, x) \leftrightarrow \text{AtomicVali}(K, x) \leftrightarrow \text{Valid}(M1, x)$$

And this is a contradiction to our first assumption because:

$$\text{Valid}\,(M1, x) \leftrightarrow \text{Conform}\,(M1, \text{Pr})$$

Thus, we can conclude that the filtered constraints are sufficient for validating the model.

References

1. K. Moazami Goudarzi, Consistency preserving dynamic reconfiguration of distributed systems. Ph.D. Thesis (Imperial College, London, UK, 1999)
2. Object Management Group: Unified Modeling Language—Superstructure Version 2.4.1, formal/2011–08-05. http://www.omg.org/spec/UML/2.4.1/
3. L. Fuentes-Fernández, A. Vallecillo-Moreno, An introduction to UML profiles. Upgrade Eur. J. Inf. Prof. **5**(2), 5–13 (2004)
4. Object Management Group, UML 2.0 OCL Specification, Version 2.4, formal/2014-02-03. (2014). http://www.omg.org/spec/OCL/2.4/
5. J. Bézivin, F. Jouault, Using ATL for checking models. In: *Proceedings of the 4th Int. Workshop on Graph and Model Transformation (GraMoT 2005), Electronic Notes in Theoretical Computer Science*, vol. 152 (2006) pp. 69–81
6. Open Virtualization Format Specification, DMTF Standard, Version 2.1.0, December 2013. http://www.dmtf.org/sites/default/files/standards/documents/DSP0243_2.1.0.pdf
7. Distributed Management Task Force. http://dmtf.org/
8. M. Didonet del Fabro, P. Valduriez, Towards the efficient development of model transformations using model weaving and matching transformations. J. Softw. Syst. Model. (SoSym) **8**(3), 305–324 (2009)
9. M. Didonet del Fabro, J. Bezivin, F. Jouault, P. Valduriez, Applying generic model management to data mapping. In: *Proceedings of the Journées Bases de Données Avancées (BDA05)* (Saint-Malo, France, 2005)
10. A. Jahanbanifar, F. Khendek, M. Toeroe, A model-based approach for the integration of configuration fragments. In: *Proceedings of 11th European Conference on Modelling Foundations and Applications (ECMFA)* (Italy, 2015), pp. 125–136
11. A. Jahanbanifar, F. Khendek, M. Toeroe, Semantic weaving of configuration fragments into a consistent system configuration. Inf. Syst. Front. Spring. **18**(5), 891–908 (2016)
12. A. Jahanbanifar, F. Khendek, M. Toeroe, Runtime adjustment of configuration models for consistency preservation. In: *Proceedings of 17th Int. Symposium on High Assurance Systems Engineering (HASE)* (Orlando, USA, 2016), pp. 102–109
13. Service Availability Forum, Application Interface Specification, Software Management Framework, SAI-AIS-SMF-A.01.02. http://www.saforum.org/HOA/assn16627/images/SAI-AIS-SMF-A.01.02.AL.pdf
14. Eclipse Modeling Framework, EMF. http://www.eclipse.org/emf
15. Atlas Transformation Language (ATL) website. http://www.eclipse.org/atl/
16. E. Tsang, Foundations of constraint satisfaction. (Books on Demand, 2014)
17. V. Kumar, Algorithms for constraint satisfaction problems: A survey. AI Magaz. **13**(1), 32–44 (1992)
18. A. Jahanbanifar, F. Khendek, M. Toeroe, Partial validation of configuration at runtime. In: *Proceedings of the 18th Int. Symposium on Real-Time Distributed Computing (ISORC)* (Auckland, New Zealand, 2015), pp. 288–291
19. I. Warren, J. Sun, S. Krishnamohan, T. Weerasinghe, An automated formal approach to managing dynamic reconfiguration. In: *Proceedings of the 21st IEEE/ACM Int. Conference on Automated Software Engineering* (Washington, USA, 2006), pp. 37–46
20. P. Goldsack, J. Guijarro, S. Loughran, A. Coles, A. Farrell, A. Lain, P. Murray, P. Toft, The smartfrog configuration management framework. ACM J. SIGOPS Oper. Syst. Rev. **439**(1), 16–25 (2009)
21. M. Burgess, A.L. Couch, Modeling next generation configuration management tools. In: *The Proceedings of the 20th Conference on Large Installation System Administration (LISA)* (Washington, USA, 2006), pp. 131–147
22. L. Akue, E. Lavinal, T. Desprats, M. Sibilla, Runtime configuration validation for self-configurable systems. In: *The Proceedings of the IFIP/IEEE Int. Symposium on Integrated Network Management (IM)* (Ghent, Belgium, 2013), pp. 712–715

23. L. Akue, E. Lavinal, T. Desprats, M. Sibilla, Integrating an online configuration checker with existing management systems: application to cim/wbem environments. In: *The Proceedings of the 9th Int. Conference on Network and Service Management (CNSM)* (Zurich, Switzerland, 2013), pp. 339–344

24. J. Cabot, E. Teniente, Incremental integrity checking of UML/OCL conceptual schemas. J. Syst. Softw. **82**(9), 1459–1478 (2009)

25. J. Cabot, E. Teniente, Computing the relevant instances that may violate an OCL constraint. In: *The Proceedings of the 17th Int. Conference on Advanced Information Systems Engineering (CAiSE'05)*, eds. by O. Pastor, J. Falcão e Cunha, vol. 3520 (LNCS, Springer, 2005), pp. 48–62

26. G. Bergmann, Á. Horváth, I. Ráth, D. Varró, A. Balogh, Z. Balogh, A. Ökrös, Incremental evaluation of model queries over EMF models. In: *The Proceedings of the 13th Int. Conference on Model Driven Engineering Languages and Systems (MODELS)*, Part I (Springer, 2010), pp. 76–90

27. I. Groher, A. Reder, A. Egyed, Incremental consistency checking of dynamic constraints. In: *The Proceedgins of Fundamental Approaches to Software Engineering (FASE)*, vol. 6013, eds. by D.S. Rosenblum, G. Taentzer (LNCS, Springer, 2010), pp. 203–217

28. A. Reder, A. Egyed, Incremental consistency checking for complex design rules and larger model changes. In: *The Proceedings of the 15th Int. Conference on Model Driven Engineering Languages and Systems (MODELS)*, vol. 7590 (Springer, 2012), pp. 202–218

Assurance Cases

Towards Making Safety Case Arguments Explicit, Precise, and Well Founded

Valentín Cassano, Thomas S. E. Maibaum, and Silviya Grigorova

Abstract The introduction of safety cases into the practice of safety assurance has revolutionized safety engineering. Via a 'safety argument', a safety case aims to explicate, and to provide some structure for, the kind of reasoning involved in demonstrating that a system is safe. To date, there are several notations for writing down safety arguments. These notations suffer from not having a well-founded semantics, making them deficient w.r.t. the requirements of a serious approach to engineering. We consider that a well-founded semantics for safety arguments ought to be based on logical principles in the form of a logical calculus. Logic is the basis for reasoning in mathematics, philosophy, and science, and the same should be true for safety reasoning. With this goal in mind, we take some steps towards constructing a logical calculus for safety arguments by exploring some of the features of this calculus. Moreover, we look into the essential role that evidence plays in safety arguments. Evidence sets apart safety arguments from their traditional logical counterpart, as assumptions in safety arguments must be grounded on (i.e., justified by) data from the empirical world. We present our thoughts on these matters, and illustrate them by means of examples. We consider that our work establishes a framework for discussing safety arguments in a more rigorous manner.

1 Introduction

The introduction of safety cases into the practice of safety assurance aims to make explicit and to organize the justification for a claim that some engineered artefact is safe. (What 'safe' means is, however, a totally different issue, which we choose to

V. Cassano · T. S. E. Maibaum (✉) · S. Grigorova
McMaster Centre for Software Certification, McMaster University, Hamilton, Canada
e-mail: tom@maibaum.org

V. Cassano
e-mail: cassanv@mcmaster.ca

S. Grigorova
e-mail: grigorsb@mcmaster.ca

© Springer Nature Singapore Pte Ltd. 2021   227
Y. Ait-Ameur et al. (eds.), *Implicit and Explicit Semantics Integration in Proof-Based Developments of Discrete Systems*,
https://doi.org/10.1007/978-981-15-5054-6_11

set aside here.) A safety case is defined as: 'A structured argument, supported by a body of evidence that provides a compelling, comprehensible, and valid case that a system is safe for a given application in a given operating environment' (see [35]). This widely accepted definition of a safety case as a structured argument is a big step towards a more precise definition. Moreover, it sets a standard against which competing definitions of a safety case can be assessed [16]. See also Sect. 2.

A significant amount of useful work has been accomplished in turning the idea that a safety case is a structured argument into practical notations to support their development (see [1, 33]). However, to date, existing notations for safety cases have no semantics. This entails that, when presented with a safety case in one of these notations, we have no means for deciding whether the structured argument it formulates is syntactically well formed, never mind whether the reasoning it purports to represent is sound. In other words, though safety cases have been developed and used with some success, they seem to be largely supported by intuition and experience. This is in stark contrast to other disciplines of engineering where mathematical rigour is the norm. The situation is worrisome. Can we reasonably expect to deal with the increasing complexity of systems such as cyber-physical systems or autonomous cars largely based on intuition? It is well known that, in the end, intuition always fails us when confronted by complexity. Would we have entrusted the lives of astronauts to outer space missions had space shuttles been engineered based on intuition and not science? Of course not.[1] So why do we not hold the development of safety cases to the same high standards? It does certainly seem to be appropriate. Moreover, how do we teach new safety engineers the necessary rigour required in their field without a proper scientific basis? Do we appeal to intuition and experience? Intuition only takes us so far, and certainly not far enough to justify the safety of complex systems. The moral of the story is that history clearly demonstrates that notations lacking a well-founded semantics are deficient w.r.t. the requirements of a serious approach to engineering. This state of affairs in safety assurance has persisted for too long. We consider that it is time to bring this issue to the fore.

We do not think that the use of the term 'structured argument' is incidental in the definition of a safety case.[2] For this reason, our view is that a well-founded semantics for safety cases should be based on logical principles in the form of a logical calculus. We view the development of this logical calculus in the light of Logic Engineering (see [36]). Logic Engineering addresses the development of logical frameworks for specific purposes. In our case, the specific purpose is safety reasoning. As logic engineers, we then need to identify among the available logical calculi if there is one that is adequate for safety reasoning, and in case there is none, to construct one (possibly by combining or borrowing elements from those that exist).

---

[1]This does not mean that we expect engineering to be perfect. Engineers do make mistakes. However, engineers learn by experience and codify that knowledge in mathematical analyses and engineering methods, on which they can rely to build systems that are reliable.

[2]Tim Kelly, the developer of one of the most commonly used notations for safety cases, the Goal Structured Notation, see [33], and his PhD. supervisor John McDermid in [25] directly linked the notation to the argument language developed by Toulmin in [34]. Moreover, the UK MoD standard definition of a safety case, see [35], also links safety cases to arguments.

As a disclaimer, we do not propose to reduce safety reasoning to an existing logical calculus, nor will we fully develop a logical calculus for safety reasoning; we do not yet know enough to do so. Instead, we take some steps towards a logical calculus for safety reasoning by presenting and discussing some of its main elements in the form of *working definitions*. More precisely, we provide a precise definition of the notion of a structured argument in a safety case and discuss some of the elements of safety reasoning in relation to it. An important part of this reasoning is the essential role that evidence plays in safety cases. Evidence sets apart the kind of arguments involved in safety cases from their traditional logical counterpart, as assumptions in safety cases must be grounded on (i.e., justified by) data from the empirical world. An important source for ideas in this regard is the work of epistemologists such as Carnap, Hempel, and Popper, amongst others, see [5, 18, 29]. In particular, we have taken inspiration from Carnap's Two Level Language of Science. This language is a logical formalism that has a limited logic for observational reasoning, i.e., about evidence, which is included in another language, the so-called theoretical language, that is used for reasoning about universal generalizations. The current observations about safety cases and the distinctions between reasoning about evidence versus inferential reasoning of a more general nature are directly rooted in the ideas outlined above.

Our work has two main outcomes. First, we set up a framework for discussing the kind of reasoning involved in safety cases. Second, we set up a standard against which progress can be measured by providing some working definitions. Working definitions are the basis of science and engineering and are an essential tool against which to measure scientific progress. Working definitions allow us to make further progress in transforming safety cases into a properly grounded engineering tool, enabling a systematic and rigorous construction and analysis. But, of course, our working definitions should not be seen as defining a dogmatic position; we will happily make changes as we learn more and are able to justify their necessity.

Structure: In Sect. 2, we begin with some preliminary observations on safety cases. In Sects. 3 and 4, we discuss some of the main elements of safety reasoning and offer some working definitions. In Sect. 5, we put these elements together in an attempt to present a coherent picture. In Sect. 6, we offer some conclusions and comment on next steps.

2 Preliminary Observations

The first observation regarding safety cases concerns notations for their presentation/development (e.g., [1, 33]). To save us from having to continually refer to all of them, we use the diagrams in the Goal Structured Notation (GSN) as our witness. In our view, GSN diagrams do not present arguments in the usual logical sense of the term. Instead, they present decomposition structures for safety goals, i.e., a strategy S related to goals $\{G_i\}_{i \in n}$, G expresses a decomposition of G into $\{G_i\}_{i \in n}$. This is reminiscent of problem solving by decomposition, a well-known technique for coping

Fig. 1 Goal decomposition

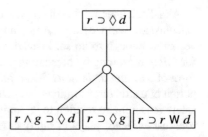

with the complexity of large problems (see [28]), where solutions to sub-problems are combined in a prescribed way to solve the original problem. More closely, this is reminiscent of goal structured requirements approaches such as KAOS (see [23]), which apply problem decomposition ideas to requirements definition. (In fact, GSN researchers often refer to the notation as supporting safety goal decomposition.) But decomposition structures for safety goals and structured arguments in safety cases are different: while a goal decomposition structure breaks down a complex goal G into more manageable goals $\{G_i\}_{i \in n}$, as stated in some underlying logical calculus, a structured argument substantiates that G follows from the set $\{G_i\}_{i \in n}$. This difference is immediate if we draw an analogy with goal decomposition structures in KAOS. The following example, adapted from one presented in [9], clarifies this point. Figure 1 illustrates the decomposition of a goal G stating that 'if the train is *r*eady to depart, then, it eventually *d*eparts' into goals G_1, G_2, and G_3 stating that 'if the train is *r*eady to depart and the go signal turns green (*g*o), then, it eventually *d*eparts'; 'if the train is *r*eady to depart, then, the go signal turns green (*g*o) eventually'; and 'if the train is *r*eady to depart, then, it remains in that state at least until it *d*eparts'; respectively. In Fig. 1, goals G and G_1 – G_3 are formulated in ($\mathcal{TK}$; see [24]) as $r \supset \Box \Diamond d, r \wedge g \supset \Diamond d, r \supset \Diamond g$, and $r \supset r \mathsf{W} d$, respectively. Figure 2 illustrates what we consider to be a structured argument in the context of this goal decomposition. In Fig. 2, single lines correspond to inference steps of $\mathcal{TK}$, and double lines to the use of lemmas in $\mathcal{TK}$, i.e., combinations of inference steps (see [24]). Though related, it can be readily seen that decomposition structures for goals and structured arguments are not the same. GSN diagrams fall short at presenting structured arguments. So-called strategies do not involve structured arguments akin to that presented in Fig. 2, and what they do better at representing, a goal decomposition structure, while very important, is hindered, from a scientific and engineering point of view, by not having a properly defined semantics (as exists in KAOS).

Another popular notation for safety cases is CAE developed by Adelard (see [1]). The ideas and motivations are similar to those of GSN in many ways. In CAE claim decompositions can be viewed as logical conjunctions of the sub-claims. This links directly intended logical meaning of claim decomposition to reasoning about the claims. However, such a 'semantics' of claim decomposition is very limiting and simplistic. As noted above, in problem solving by decomposition, the way sub-solutions are put together to obtain a solution for the original problem must be well defined, but the composition method may be more complicated than conjunction.

Fig. 2 Structured argument

$$\frac{\dfrac{r \supset (r \text{ W } d)}{r \supset ((r \text{ U } d) \vee \Box r)}}{\dfrac{r \supset \Diamond g \qquad r \supset (\Diamond d \vee \Box r)}{\dfrac{r \supset (\Diamond g \wedge (\Diamond d \vee \Box r))}{\dfrac{r \supset ((\Diamond g \wedge \Diamond d) \vee (\Diamond g \wedge \Box r))}{\dfrac{r \wedge g \supset \Diamond d \qquad r \supset ((\Diamond g \wedge \Diamond d) \vee \Diamond (g \wedge r))}{\dfrac{r \supset ((\Diamond g \wedge \Diamond d) \vee \Diamond \Diamond g)}{r \supset \Diamond d}}}}}$$

Similarly, when we build proof procedures, another example of the principle of problem solving by decomposition, putting proofs together may be much more complex than simply 'build the conjunction' of the proofs. Breaking claims down to ones that can be put together by conjunction enormously, and unnecessarily, complicates the decomposition problem.[3]

There is also something to be said about attempts at reducing safety reasoning directly to First-Order Logic (FOL) and using automated deduction support and proof calculi for expressing a safety argument (see [31, 32]). These attempts try to provide a strong, well defined, foundation for eliciting what is meant by a structured argument. But they must also face up to the fact that safety reasoning is not FOL reasoning (and, more generally, not that captured by classical deductive logical calculi). There are several reasons for this. We proceed to enumerate some:

(i) Safety reasoning contains textbook examples of fallacies in FOL (e.g., arguments from authority, such as expert opinions). Independently of how we express them, including a fallacy in a proof renders the proof a fallacy, and thus a no-proof.

(ii) Safety reasoning makes use of inductive generalizations (as in inductive reasoning, see [12]). An obvious example of this occurs when it is concluded from a test set extracted from an universe of data, where every test case is successful, that a corresponding property of the universe of data is the case. This kind of generalization requires a truly inductive reasoning step. FOL is not the logic for dealing with inductive generalizations.[4]

---

[3]When trying to define how components could be composed 'in parallel', researchers also proposed that the semantics was conjunction. This was found to be very limiting, failing to deal well with interaction and communication between components and was soon replaced by the use of categorical operations, such as co-limit, applied to diagrams of components and morphisms in an appropriate category.

[4]The position that inductive generalizations correspond to reasoning at the level of evidence, that, once this is sorted out, we can move to a more 'pure' form of reasoning, and that the non-evidential part of the reasoning in a safety case can be done in FOL is difficult to sustain. There is no clear distinction between when reasoning at the level of evidence stops and when we can move to 'pure' reasoning. In fact, the related literature categorically contradicts this position. When reasoning about statements that make assumptions about evidence, it seems implausible, at least, that FOL will do.

(iii) Safety reasoning includes elements of defeasible reasoning (as discussed in the field of non-monotonic logics; see [3]). Often safety reasoning makes inferences from incomplete information, i.e., neither are we certain that a property holds for an artefact, nor that it does not, yet we still conclude something about the artefact. Moreover, safety reasoning makes use of defeasible inferences. These inferences are defeasible because further investigation may invalidate the conclusions drawn from them, forcing their revision, or withdrawal. Defeasible reasoning falls outside of the scope of FOL.

(iv) FOL is inadequate for reasoning about actions, modalities, and agency. This part addresses the idea that modal reasoning can be better dealt with in FOL. This does not mean that actions, modalities, and agency cannot be reasoned about in FOL. It simply means that they are better dealt with by logics which were developed with that particular purpose in mind (see [10, 27]). From the perspective of logic engineering, these logics provide a more suitable formalism for the task at hand.

(v) Safety case reasoning sometimes also uses a form of reasoning called eliminative induction (see [15]). Eliminative induction, first developed by Francis Bacon, and taken up by philosophers such as John Stuart Mill, John Maynard Keynes, Karl Popper, Jonathan Cohen, et al., works like this: Suppose that we conclude property A and that, at the same time, we identify that A may not be true in the presence of one or more properties $B_1, \ldots, B_n$. The set of P_is associates some uncertainty to P. If none of the P_is can be concluded, then, the uncertainty associated with P is reduced. This form of reasoning is in fact an example of a form of probabilistic reasoning that departs from the frequentist based reasoning of probability and is more related to confidence (as in confidence in a scientific theory). Confidence underlies reasoning about scientific theories, legal cases, and other domains, and some valuable lessons can be learned from those domains. For example, confidence is the basis on which semantics for statements in law like 'beyond a reasonable doubt' or 'on the balance of probabilities' can be defined. (Toulmin includes 'qualifiers' as elements in the logical statements he uses in his arguments. He would recognize the two examples we just presented as examples of qualifiers. Safety case examples are replete with qualified statements such as 'sufficiently' safe or acceptably safe.) In safety reasoning, confidence is absolutely necessary for it manifests scientifically the conventional wisdom that safety cannot be absolutely guaranteed, and, therefore, the degree of confidence becomes an essential aspect of reasoning. Again, confidence falls outside of the scope of FOL.

(vi) Safety reasoning has a global rather than a compositional, inductive, nature. Defeasible and probabilistic reasoning exhibit this particularity. In these forms of reasoning it is not generally possible to put consequences together in a soundness-preserving way (see [2]). This has grave consequences for the possibility of devising incremental safety approaches that support the well tried and understood concept of incremental design improvement (see [37]). Lack of compositionality is not a feature of FOL.

(i)–(vi) lead to the observation that FOL is not a suitable framework for safety reasoning. There might be a need to look elsewhere for a logic for safety reasoning.

To summarize, it is unsurprising that safety reasoning presents a challenging topic for research. The practical implications of this are plainly evident. Taking on this challenge, we take some steps towards establishing a logical calculus for safety reasoning.

3 Structured Arguments in Safety Reasoning

The concept of a safety case is a cornerstone of safety reasoning. But what do we exactly mean by a safety case? A safety case is commonly defined as: 'A structured argument, supported by a body of evidence that provides a compelling, comprehensible, and valid case that a system is safe for a given application in a given operating environment' (see [35]). The introduction of safety cases into safety reasoning is a step in the right direction. Safety cases make a serious attempt to explicate, and to provide some structure for, the inference licenses, a.k.a. rules of inference, used in guaranteeing that a system is safe. Nonetheless, a striking feature of the definition of a safety case just given is its logical vagueness. It is unclear what is to be taken as constituting a structured argument, as in, what are its defining characteristics, and how is such a structured argument to be assessed in terms of the soundness of the reasoning it involves. In this section we discuss these issues from the perspective of a logical calculus. This presentation extends and clarifies an earlier work of ours (see [6]).

3.1 Background

We begin by introducing some basic definitions and comments on Gentzen's Calculus of Natural Deduction for Classical First-Order Logic ($\mathcal{NK}$ for short; see [8, 14]). With this, we aim to provide a well-defined context for discussion. In the process, we will fix the terminology that we will use in what follows, and we will make this terminology precise.

As explained in [14, p. 291], with his $\mathcal{NK}$, Gentzen intended to provide: 'A formalism that reflects as accurately as possible the actual logical reasoning involved in mathematical proofs'. Gentzen offers as an example of this kind of reasoning:

$(\exists x \forall y F x y) \supset (\forall y \exists x F x y)$. The argument runs as follows: Suppose there is an x such that for all y $F x y$ holds. Let a be such an x. Then for all y: $F a y$. Now let b be an arbitrary object. Then $F a b$ holds. Thus there is an x, viz., a, such that $F x b$ holds. Since b was arbitrary, our result therefore holds for all objects, i.e., for all y there is an x, such that $F x y$ holds. This yields our assertion. (See [14, p. 292].)

In essence, the program laid out by Gentzen in [14] consists of the integration of the kind of mathematical proofs carried out above in an exactly defined calculus, his

$\mathcal{NK}$. To this end, Gentzen provides precise definitions of so-called *symbols*, *expressions*, and *figures*. Symbols are the alphabet of Classical First-Order Logic (FOL for short). Expressions are the language of FOL, i.e., the set of all formulæ defined recursively over the alphabet of FOL. We will need to refer to arbitrary formulæ in the language of FOL. We indicate these arbitrary formulæ with uppercase boldface letters. Figures are *inference figures* or *proof figures*. Inference figures consist of a finite set of formulæ called upper formulæ and a single formula called a lower formula. Regarding inference figures Gentzen explains in [14, p. 291] that: 'We shall have inference figures and they will be stated for each calculus as they arise'. The permissible inference figures which make up the $\mathcal{NK}$ correspond to the well-known rules of introduction and elimination of the logical connectives of the alphabet of FOL and the law of the excluded middle (see [14, pp. 292–295]). Gentzen states these permissible inference figures via a set of inference figure schemata. An inference figure schema is to be understood as: The permissible inference figure obtains from the inference figure schema by instantiating the syntactical variables for formulæ by corresponding formulæ. Figure 3 illustrates an inference figure schema (corresponding to the introduction of material implication). Observe that in this inference figure schema **A**, **B**, and **A** $\supset$ **B** are not sentences, they are variables or templates for sentences. Figure 4 illustrates an instance of the inference figure schema in Fig. 3. In this inference figure, $\{(\exists x \forall y F x y), (\forall y \exists x F x y)\}$ is the set of upper formulæ, instantiating **A** and **B**, respectively, and $(\exists x \forall y F x y) \supset (\forall y \exists x F x y)$ is the lower formula, instantiating **A** $\supset$ **B**.

Proof figures, also called *derivations*, combine a number of formulæ to form inference figures such that: 'Each formula is a lower formula of at most one inference figure; each formula (with the exception of exactly one: the *endformula*) is an upper formula of at least one inference figure; and the system of inference figures is non-circular, i.e., there is in the derivation no cycle [...] of formulæ of which each upper formula of an inference figure has the lower formula as the next one in the series' (see [14, p. 291]). Figure 5 illustrates the result of incorporating the mathematical proof given above in Gentzen's $\mathcal{NK}$. (Numbering annotations in Fig. 5 identify where formulæ are *discharged* and they are solely used for bookkeeping purposes.)

Introducing some further terminology that we will use later on, Gentzen calls the formulæ of a derivation that are not lower formulæ of an inference figure *initial*; the formulæ of a derivation *D-formulæ*; the inference figures of a derivation *D-inferences*; and a series of D-formulæ in a derivation, whose first formula is an initial one and whose last formula is the endformula, and of which each formula but the last is an

Fig. 3 Inference figure
schema

$$\frac{\begin{array}{c}[\mathbf{A}]\\ \mathbf{B}\end{array}}{\mathbf{A} \supset \mathbf{B}} \supset\text{-}I$$

Fig. 4 Inference figure

$$\frac{\begin{array}{c}[\exists x \forall y F x y]\\ \forall y \exists x F x y\end{array}}{(\exists x \forall y F x y) \supset (\forall y \exists x F x y)} \supset\text{-}I$$

Fig. 5 Proof figure (a.k.a. derivation)

$$\frac{\dfrac{\dfrac{[\forall y F a y]^1}{F a b}\ {}^{\forall\text{-}E}}{\exists x F x b}\ {}^{\exists\text{-}I}}{\dfrac{[\exists x \forall y F x y]^2 \quad \forall y \exists x F x y}{\dfrac{\forall y \exists x F x y}{(\exists x \forall y F x y) \supset (\forall y \exists x F x y)}\ {}^{\supset\text{-}I_2}}}\ {}^{\exists\text{-}E_1}}\ {}^{\forall\text{-}I}$$

upper formula of a D-inference figure whose lower formula is next in the series, a *branch*. Note that, in Gentzen's formulation of the $\mathcal{NK}$, it is possible for some of the initial formulæ of a derivation not to be discharged. We call such initial formulæ *premisses*. At times, we need to refer to derivations without making their structure explicit. For this purpose, we use symbol $\vdash_{\mathcal{NK}}$. We understand this symbol as a relation between sets of formulæ and formulæ. The source of $\vdash_{\mathcal{NK}}$ is the set of undischarged formulæ in the derivation, the target of $\vdash_{\mathcal{NK}}$ is the endformula of the derivation. For example, we indicate the derivation in Fig. 5 as $\{\} \vdash_{\mathcal{NK}} (\exists x \forall y F x y) \supset (\forall y \exists x F x y)$.

3.2 Some Concepts

We make some observations about Gentzen's $\mathcal{NK}$ as a prelude to what follows. First, via the integration of mathematical proofs into $\mathcal{NK}$, Gentzen provides a precise definition of what is a mathematical proof, enabling an analysis of its scope and limits. For us, the importance of this cannot be underestimated, in particular, because, to a certain extent, the notion of a mathematical proof stands in analogy with that of a structured argument in a safety case, or a safety argument for short: while a mathematical proof aims at capturing the kind of reasoning involved in mathematics, a safety argument aims at capturing the kind of reasoning involved in safety reasoning. In that respect, we consider that safety arguments should be given a definition akin to the one that Gentzen provides for mathematical proofs. Without such a definition it is impossible to judge whether a proposed safety argument is indeed such. If logic, logical methods, and their history have taught us anything at all, it is that only through the provision of precise definitions and their analyses can we avoid fallacious reasoning steps. Two of the most important results about Gentzen's definition of a derivation are the Soundness and Completeness Theorems (see [8]); having, at the very least, a soundness theorem for a logical calculus for safety reasoning would greatly improve the state of the art in this domain of knowledge.

In light of the previous paragraph, we offer some clarifications to avoid any subsequent confusion. We are not saying that mathematical reasoning and safety reasoning are one and the same. There are most definitely some points of departure between the two, some of which we have already mentioned in Sect. 2, some of which we will make clear below. Neither are we saying that without a definition of a safety argument that stands on grounds analogous to Gentzen's definition of a derivation, safety reasoning is vacuous. Though with some reservations, even in the absence of such a definition of safety argument, we see no major reason preempting logical progress in safety reasoning. (After all, it is not as if mathematical reasoning could not be carried out before Gentzen's definition of a derivation.) Lastly, we are not saying that the aforementioned definition of a safety argument can or shall be given from the outset. This would be a clear impossibility given the current state of the art of safety reasoning. Instead, our remarks are oriented towards the formulation of a working definition of a safety argument that is (i) suitable for capturing as accurately as possible the actual logical reasoning involved in safety assurance, and (ii) amenable for the logical analyses that are needed to establish the well-formedness and the soundness of the inference licenses to be used in safety assurance. It is relative to (i) and (ii) that a Logical Engineering approach proves its worth. We hope that by discussing and refining such a working definition we can establish a strong logical foundation on which to improve safety reasoning and ultimately develop a logical calculus for safety reasoning.

One final discussion may be of importance in clarifying what we are trying to do. It has been widely recognized that safety reasoning includes at least two forms of reasoning: so-called evidential reasoning, to incorporate experimental observations that might be relevant to our conclusions about the safety of a system, and so-called inferential reasoning, to enable us to manipulate statements that are not directly about experimental data. There seems to be a consideration that inferential reasoning is logical, as in FOL, while evidential reasoning is not logical, but so-called epistemological, based on conventional probability notions (see [31, 32]). Now, though there are good reasons to distinguish the two kinds of reasoning, there seems to be no good reason to demote evidential reasoning from the realm of logic. There is a century-old history of trying to do exactly the opposite. Carnap's Two Level Language of Science was an attempt at characterizing the logic behind scientific reasoning (see [5]). As in safety cases, Carnap had to deal with the incorporation of observation in science with the more general forms of reasoning that any mathematician, scientist, or logician would recognize. He divided his logical language into two parts: one that has to do with observations, the other that has to do with general, universal reasoning, e.g., about universal laws. The observational language was of limited expressive power; it included observations as ground atomic formulae (e.g., 'this glass is blue', 'the output of this program run in this test harness, when the input is a, is b', etc); the observational logic had the usual connectives, but limited inferential power (e.g., universal generalization is not allowed), only so-called empirical generalizations (e.g., 'all the swans we have observed are white'). The so-called Theoretical Level of discourse, on the other hand, was more like FOL and allowed universal generalization. This latter logic incorporated the former. Thus, reasoning about evidence (observations) and inferential reasoning are integrated into a single, coherent whole. When we refer below to a logic of safety cases, we have in mind a logic analogous to Carnap's. It incorporates elements for evidential reasoning as well as general (inferential) reasoning. It seems to us that making evidential reasoning not logical just leaves us with the non-trivial problem of integrating the two parts.

Fig. 6 s-inference figure
schema

$$\frac{A_1 \dots A_n}{B} \; \langle R \rangle$$

Fig. 7 s-derivation

$$\frac{r \wedge g \supset \Diamond d}{r \supset (\Diamond g \supset \Diamond \Diamond d)} \quad \frac{\dfrac{\dfrac{\dfrac{(s \cup g) \wedge (g \cup s)}{(s \cup g)}}{\Diamond g}}{r \supset \Diamond g} \langle \neg v \rangle \quad \dfrac{r \supset (r \, W \, d)}{r \supset (\Box r \vee \Diamond d)}}{r \supset ((\Diamond g \wedge \Diamond d) \vee \Diamond (g \wedge r))}}{\dfrac{r \supset ((\Diamond g \supset \Diamond \Diamond d) \wedge ((\Diamond g \wedge \Diamond d) \vee \Diamond (g \wedge r)))}{r \supset \Diamond d}}$$

Following from these preliminary observations, similarly to Gentzen's aim of incorporating mathematical proofs into a logical calculus, Gentzen's $\mathcal{NK}$, what we have in mind is also the integration of safety arguments into a logical calculus, which we refer to as $\mathcal{SK}$. This integration provides the sought after definition of a safety argument. In working towards this end goal, we make precise first the concept of a s-derivation. Reminiscent of Gentzen's derivations, s-derivations consist of a number of s-formulæ which are combined to form s-inference figures. For each s-derivation, each s-formula is a lower s-formula of at most one s-inference figure; each s-formula (with the exception of exactly one, the s-endformula) is an upper s-formula of at least one s-inference figure; and the system of s-inference figures is non-circular. We consider a s-inference figure to be an instance of the s-inference figure schema in Fig. 6.[5] In the s-inference figure schema in Fig. 6, $A_1, \dots, A_n$, **B**, **R** are variables for s-formulæ; the part corresponding to $\langle R \rangle$ is optional. We will have to consider a particular s-inference figure schemata in the definition of our sought after $\mathcal{SK}$, and these we will have to state precisely, but we are not in a position to do so yet. Following Gentzen's terminology, for a s-inference figure schema such as the one given above, we call the instances of $A_1, \dots, A_n$ upper s-formulæ and the instances of **B** lower s-formulæ. We call the instances of **R** s-rebuttals. We will return to them shortly for they occupy a special place in s-derivations. We call the s-formulæ participating in a s-derivation s-formulæ and the s-inference figures participating in a s-derivation s-inference figures. Moreover, we call the s-formulæ of a s-derivation that are not lower formulæ of a s-inference figure initial s-formulæ. The initial s-formulæ of a s-derivation can be discharged (in the sense given by the introduction of appropriate conditionals, modalities, or quantifiers) or not. We call those initial assumptions of a s-derivation that are not discharged s-premisses. At times, we need to refer to s-derivations but not their structure. For this purpose, we use symbol $\vdash_{SK}$. We understand this symbol as a relation between sets of s-formulæ and s-formulæ. The source of $\vdash_{SK}$ is the set of undischarged s-formulæ in the s-derivation, the target of $\vdash_{SK}$ is the s-endformula of the s-derivation. We label $\vdash_{SK}$ with the s-rebuttals of the s-inference figures in the s-derivation.

---

[5]Technically, the s-inference figure schema in Fig. 6 is a s-inference figure schemata. We obtain a s-inference figure schema for each value of n.

Fig. 8 Goal decomposition

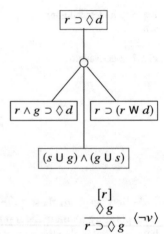

Fig. 9 s-inference Figure

$$\frac{[r]}{\Diamond g} \langle \neg v \rangle$$
$$\overline{r \supset \Diamond g}$$

We illustrate what a s-derivation may look like in Fig. 7.[6] In this s-derivation, $\{r \wedge g \supset \Diamond d, (s \cup g) \wedge (g \cup s), r \supset (r \text{ W } d)\}$ is its set of s-premisses, and $r \supset \Diamond d$ is its s-endformula. With the exception of the s-inference figure in Fig. 9, to which we will return, the s-inference figures used in this s-derivation are obvious; they are inference figures of Temporal Logic (see [24]). We indicate the s-derivation in Fig. 7 as $\{r \wedge g \supset \Diamond d, (s \cup g) \wedge (g \cup s), r \supset (r \text{ W } d)\} \vdash_{SK}^{\{\neg v\}} r \supset \Diamond d$.

We can understand the s-derivation in Fig. 7 in light of the goal decomposition in Fig. 8. In this goal decomposition, we broke the top goal $r \supset \Diamond d$ into goals $r \wedge g \supset \Diamond d$, $r \supset (r \text{ W } d)$, and $(s \cup g) \wedge (g \cup s)$. We borrowed $r \supset \Diamond d$, $r \wedge g \supset \Diamond d$, and $r \supset (r \text{ W } d)$ from the goal decomposition example in Sect. 2 and they have the same intuitive meaning. With $(s \cup g) \wedge (g \cup s)$ we capture the idea that 'the go signal turns from green (go) to red ($\underline{s}top$) and from red ($\underline{s}top$) to green (go)'. With $\neg v$ we capture the idea that 'the go signal is not $\underline{v}$isible to the operator of the train'. Relative to the goal decomposition in Fig. 8, the s-derivation in Fig. 7 identifies clearly and definitely a structured argument substantiating that the top goal follows from the (sub)goals it has been broken into.

The definition of $\mathcal{SK}$ concludes with the definition of the language of s-formulæ, and with the formulation of the permitted s-inference figures via s-inference figure schemata. We envision the language of s-formulæ as the $\mathcal{SK}$ counterpart of the claims involved in safety arguments, safety claims for short, and the permitted s-inference figures as the $\mathcal{SK}$ counterpart of the inference licenses used in the formulation of safety arguments. Their precise formulation is, however, an open research question and part of what makes the definition of a safety argument, via its integration into an exactly defined calculus, a working definition.

---

[6]In the s-derivation in Fig. 7, we assume that the language of Temporal Logic is part of the language of s-formulæ and that the inference figures of Temporal Logic are permissible (see [24]).

3.3 Some Comments on the Logic of Safety Arguments

A significant and non-trivial part of our SK needs to be completed. We need to: (i) provide a formal definition of s-formulæ; (ii) formulate the s-inference figure schemata for the permissible s-inference figures of SK; and, more importantly, (iii) integrate a basic stock of examples into SK. However, even at this early stage, the definition of a s-derivation allows us to discuss technically certain important issues regarding safety reasoning.

3.3.1 Regarding s-Formulæ

The following observation made by Gentzen in [14] provides some context for discussion: 'To the concept of "object", "function", "predicate", "theorem", "axiom", "proof", "inference", etc., in logic and mathematics there correspond, in the formalization of these disciplines, certain symbols or combinations of symbols'. What Gentzen implicitly assumes is the translation of some ordinary language of mathematics into the formal language FOL. Arguing about the faithfulness of the translation of statements in the ordinary language of mathematics into that of FOL is a moot point, first, because, to a large extent, the language of FOL has been designed having in mind the ordinary language of mathematics, and second, because statements in mathematics are rigorously precise and unambiguous. After all, no one will doubt that the ordinary statement of mathematics 'there is no natural number whose successor is zero' is expressed by the formula $\neg \exists n (S(n) = 0)$.

More generally, a faithful translation of an ordinary language, such as English, into a formal language, such as that of FOL, brings with it a number of non-trivial issues to address. In particular: Is there then a formal language in which to provide a precise definition of s-formulæ that caters for a faithful translation of safety claims formulated, say, in plain English? The answer to this question is, however, non-trivial.

It is not at all clear how to faithfully translate logical connectives in an ordinary language such as English into a formal language. For instance, we have chosen to translate the English claim 'if the train is $r$eady to depart, then, it eventually $d$eparts' into the formula $r \supset \Diamond d$. The problem with this is that, if we assume that the inference figure schemata ruling the introduction and elimination of $\supset$ in the SK are similar to those in Gentzen's NK for $\supset$, i.e., if $\supset$ is like the material conditional, then, we can establish $r \supset \Diamond d$ from $r \vee b \supset \Diamond d$. But there is something counter-intuitive in this situation; in particular, if we understand the formula $r \vee b \supset \Diamond d$ as a faithful translation of the English claim 'if the train is $r$eady to depart or it is $b$roken, then, it eventually $d$eparts'; for, clearly, we would not want a broken train to depart. The problems of conditional statements in ordinary English and material implication discussed in [7] offer some further food for thought on this issue.

To further complicate matters, a quick perusal of some safety claims reveals a heavy use of vaguely defined modal logical connectives, e.g., 'acceptably', 'sufficiently', 'adequately', in combination with quantifiers of a restricted nature, e.g.,

'All identified hazards'. It is well known in classical logical studies that these are not easily dealt with, and adding modal logical connectives intertwined with logical quantifiers to the mix does not simplify matters.

In addition to the above, there are also issues related to reasoning about actions, and about qualifiers on actions, that pose some challenges in their own right.

At this point, some may wonder: Why should we even bother in developing and proposing a formal language of s-formulæ if it is so devilishly complicated? First, because formal languages are unambiguous, easier to provide a clear semantics for, and, ultimately, more amenable to analyses and tool support. Second, because the unrestricted use of ordinary languages, e.g., English, is known to be prone to paradoxes, e.g., 'This sentence has five words', or the heinous 'This sentence is false'.

We are of the opinion that a version of a paradox of language is already present in safety reasoning. To explain this observation, we draw an analogy between reasoning about safety, and reasoning about correctness in Hoare's Calculus ($\mathcal{HK}$; see [20]).

Hoare's $\mathcal{HK}$ is a formalism enabling us to reason deductively about programs. Its lore involves claims such as: 'The program S is correct w.r.t. its precondition P and its postcondition Q'. But we note here an important point: There is no formula in the formal language of Hoare's $\mathcal{HK}$ capturing such claims about correctness. The formal language of Hoare's $\mathcal{HK}$ consists of triples $\{P\}$ S $\{Q\}$. These triples are the formal counterpart of claims of the form: 'If (the precondition) P is true before the initiation of (the program) S, then, (the postcondition) Q will be true upon the completion of S'. Claims such as 'The program S is correct w.r.t its precondition P and its postcondition Q' are formulated outside of Hoare's $\mathcal{HK}$ and refer, from outside the calculus, to the existence of a derivation, inside the calculus, which has the triple $\{P\}$ S $\{Q\}$ as an endformula. In other words, 'The program S is correct w.r.t its precondition P and its postcondition Q' is defined to be 'There is, in the $\mathcal{HK}$, a derivation which has the triple $\{P\}$ S $\{Q\}$ as an endformula'. Figure 10 illustrates what a derivation in Hoare's $\mathcal{HK}$ looks like. (In this figure, single lines correspond to inference steps of Hoare's $\mathcal{HK}$, and double lines to the use of lemmas in the calculus, i.e., combinations of inference steps).

The point is that the claim 'The program $x := x + y; y := x - y; x := x - y$ is correct w.r.t. its precondition $x = X \wedge y = Y$ and its postcondition $x = Y \wedge y = X$' refers to the derivation in Fig. 10. But the claim of correctness does not belong to the language of Hoare's $\mathcal{HK}$. Including a formula that can refer to correctness inside of Hoare's $\mathcal{HK}$ yields a calculus which can refer to its own notion of derivation, giving rise to all sorts of logical problems, not to mention fallacies.

Our observation is that, though programs and systems are distinct entities, and so is reasoning about correctness and safety, we consider that in the same way that the correctness of a program in Hoare's $\mathcal{HK}$ refers to the existence of a derivation in the calculus, a statement about the safety of a system, whether acceptably, sufficiently, adequately, etc., refers to the existence of a s-derivation and, as such, it is not part of the s-derivation itself. In other words, a goal such as 'The system is acceptably/sufficiently/adequately safe' can never be the top-level goal of any safety argument. Instead, this top-level goal should correspond to a property akin to precon-

$$\cfrac{\cfrac{\{x = X + Y \wedge y = Y\}}{\text{y} := \text{x} - \text{y}}{\{x = X + Y \wedge y = X\}} \quad \cfrac{\{x = X + Y \wedge y = X\}}{\text{x} := \text{x} - \text{y}}{\{x = Y \wedge y = X\}}}{}$$

$$\cfrac{\cfrac{\{x = X \wedge y = Y\}}{\text{x} := \text{x} + \text{y}}{\{x = X + Y \wedge y = Y\}} \qquad \cfrac{\{x = X + Y \wedge y = Y\}}{\text{y} := \text{x} - \text{y}; \text{x} := \text{x} - \text{y}}{\{x = Y \wedge y = X\}}}{\cfrac{\{x = X \wedge y = Y\}}{\text{x} := \text{x} + \text{y}; \text{y} := \text{x} - \text{y}; \text{x} := \text{x} - \text{y}}{\{x = Y \wedge y = X\}}}$$

Fig. 10 A derivation in Hoare's $\mathcal{HK}$

ditions/postconditions in Hoare's $\mathcal{HK}$. After all, fallacious reasoning begins with the use of formulæ that are, from the point of view of the candidate calculus, already logically problematic.

The moral of the story is that we should exercise great care in the formulation of safety claims, and what they are about, to avoid the kind of problems mentioned above, or others of a similarly problematic logical nature. As a first step, we may choose to restrict the formulation of safety claims to fragments of ordinary languages, such as English, that are expressive enough to capture the safety claims that we need, but that maintain a reasonable degree of logical tractability. This would provide us with a basis on which to engineer a formal language for s-formulæ, and a corresponding formal semantics, which caters for a faithful translation of safety claims, catering for a thorough and systematic understanding of the sort of claims involved in safety reasoning and how to reason about them.

3.3.2 Some Thoughts on Toulmin

The definition of a s-inference figure given in Fig. 6 provides a necessary level of technicality for putting in context and discussing the appeal to Toulmin's argument patterns (see [34]) in the formulation of a safety argument, an important topic at the heart of notations for safety arguments.

Let us recall some of the basics of Toulmin's notion of an argument pattern. In [34], Toulmin asks himself: How should we lay an argument out, if we want to show the sources of its validity? In answering this question, Toulmin identifies the following elements: claim (C), data (D), warrant (W), qualifier (Q), rebuttal (R), and backing (B). Resorting to this basic stock of concepts, Toulmin lays out his famous notion of an argument pattern as in Fig. 11.

A very simple explanation of these concepts is offered in [19]. There, it is explained that Toulmin articulates his argument patterns in the context of justifying an assertion in response to a challenge. The challenge starts with the assertion of a claim (C), of which we may be asked: What have we got to go on? To which we would offer the data (D). Upon offering the data (D), we may be asked: How do you get there? (How

Fig. 11 Toulmin's argument pattern

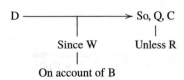

do we get from D to C?) To which we would present the warrant (W). The warrant is, thus, what allows us to infer the claim from the data. Warrants may be qualified by modalities (Q), e.g., 'probably', 'generally', 'necessarily, or 'presumably'. If the warrant is defeasible, i.e., open to revision or annulment, then, we would state the conditions of rebuttal (R). Lastly, we may also be asked for a justification of the warrant itself, to which we would present the backing (B).

Setting aside Toulmin's notion of a backing (B), it is not difficult to see that, though with some restrictions, our formulation of a s-inference figure schema in Fig. 6 borrows elements from Toulmin's argument patterns and articulates them in Gentzen's terminology. More precisely, incorporating the modalities (Q) into the logical connectives of the language of s-formulæ, the set of A_is, **B**, and **R** in Fig. 6 can be viewed as standing in analogy with Toulmin's triple of data (D), claim (C), and rebuttal (R) in the obvious way, i.e., D relates to the conjunction of the A_is, C relates to **B**, and R relates to **R** (this is the reason why we named the instances of **R** rebuttals). Toulmin's notion of a warrant can be viewed as standing in analogy with the s-inference figure schema in Fig. 6.

The restrictions that we referred to above are linguistic and logical constraints on the kind of rebuttals allowed. According to Toulmin, rebuttals indicate circumstances in which the general authority of the warrant would have to be set aside (see [34, p. 94]). There are, at least, two possible ways in which Toulmin's view of a rebuttal can be understood. First, (i) as indicating a set of circumstances in which the *claim* licensed by the warrant would have to be set aside. Second, (ii) as indicating a set of circumstances in which the *warrant itself* would have to be set aside. The analogy between a warrant and a s-inference figure schema allows for the following clarification: (i) Implies that an instance of the s-inference schema cannot be used in a particular s-derivation; (ii) Implies that the s-inference schema cannot be part of the s-inference figure schemata defining the $\mathcal{SK}$. When understood in this sense, (i) speaks to the defeasible aspect of s-derivations, whereas (ii) results in a denial of the proposed calculus (intuitionistic reasoning, arising as a result of rejecting the principle of the excluded middle, see [8]; or the various systems of Deontic logic, arising in view of the so-called paradoxes of obligations and contrary-to-duties, see [26], are examples of the second kind of rebuttals). In defining rebuttals as s-formulæ, and under the proviso that the language for s-formulæ cannot refer to properties of the $\mathcal{SK}$, we preempt the formulation of rebuttals of the second kind. Including rebuttals of the second kind makes room for paradoxes of language, as they can refer to s-derivations. Paradoxes of language are something that we clearly wish to steer away from. We consider that this restriction presents a firmer basis on which to start building the $\mathcal{SK}$.[7]

The relation between Toulmin's argument patterns and s-inference figures places the work of Toulmin in the context of safety reasoning: Toulmin's argument patterns present a framework in which to formulate what s-inference figures, or s-figure schemata, may look like. However, Toulmin's argument patterns are not s-inference figures, nor s-figure schemata. This means that Toulmin's argument patterns do not define, at least not obviously, a calculus for safety reasoning, our sought after $\mathcal{SK}$. Such a calculus, which we view as a fundamental tool for analysing the logical well-formedness of safety arguments, is only defined by the provision and justification of a sensible set of s-inference figures via a set of s-inference figure schemata. In other words, the appeal to Toulmin's argument patterns in the context of safety reasoning is rather limited; it serves as a way of showing the sources of validity of a safety argument, but it does not propose a way of assessing the validity of said sources.

3.3.3 Regarding s-Derivations

Two immediate questions may be asked about the $\mathcal{SK}$: (i) Are s-derivations suitable as the formal counterpart of safety arguments? (ii) Are s-derivations suitable for supporting the logical analyses needed to establish the well-formedness and the soundness of the inference licenses used in safety arguments?

Our answer to question (i) is, at this point, an expression of desire. Evidently, we do consider that s-derivations present a suitable framework for incorporating safety arguments. This view is partly justified by the intent of notations such as the GSN or the CAE. Whether this view is fully justified is debatable. So far, we have been unable to produce an example of the incorporation of a safety argument as a s-derivation. This is, partly, due to our own limitations, to the lack of a language for s-formulæ, and to the logical rigour that we intend to put in place in the integration of a safety argument into a s-derivation (something that we hope to improve on); but this is also due to the logical havoc reigning over the handful of examples of safety arguments that we have inspected in detail (something that we expect to shed some light on).

Our answer to question (ii) is, even at this point, more satisfactory. In particular, the concept of an s-derivation enables us to discuss some basic notion of well-formedness. Let π_1 and π_2 be two s-derivations; if the s-premisses of π_2 are a subset of the s-premisses of π_1 and the s-endformula of π_2 belongs to the set of rebuttals of the s-inference figures in π_1, then, we call π_2 a rebutting derivation for π_1. We call a s-derivation internally coherent in the absence of a rebutting s-derivation for it.[8,9] The

---

[7]If we really wish to revise a logic supporting safety reasoning by revising its inference rules, then this is a logic engineering job and we have not thought about what this may entail.

[8]We are calling a s-derivation internally coherent in the absence of a rebutting s-derivation for it, not in the presence of a proof that such a rebutting s-derivation does not exist; the latter is far more difficult to establish.

[9]Internally coherent s-derivations also make precise the role of rebuttals. They are not negated premisses, nor premisses of any kind; they are a source of defeasibility. To consider rebuttals as

second footnote is in response to the comment that rebuttals are negated assumptions. Obviously, s-derivations that are internally incoherent are logically ill-formed.

The concept of internal coherence makes precise in which sense a s-derivation is defeasible, i.e., open to revision or annulment. To illustrate this point, let π be the s-derivation in Fig. 7. As it stands, π is internally coherent. It remains so even if we extend its set of s-premisses. However, if we are, from the extended set of s-premisses, able to construct a s-derivation with $\neg v$ as its s-endformula, then, π is no longer internally coherent. Losing this status is a direct result of the s-inference figure in Fig. 9. Intuitively, this s-inference figure may be read as: If $\Diamond g$ has been established by means of s-premiss r, we have $r \supset \Diamond g$, now without the s-premiss r. The rebuttal $\neg v$ states the conditions under which this inference license is locally inapplicable, i.e., in those situations in which there is also a derivation of $\neg v$, i.e., in those situations in which it is possible to establish that 'the go signal is not visible'. The s-inference figure in Fig. 9 is 'locally inapplicable' for there are situations in which its use is perfectly permissible (e.g., in Fig. 7).

To be noted, the discovery of a rebutting s-derivation π_1 for a s-derivation π calls for a revision of π as a *whole*, and *possibly* establishes its annulment; as a 'whole', because s-inference figures in s-derivations are not, in general, localized to parts of the s-derivation; and, 'possibly', and not 'necessarily', because even in the presence of a rebutting s-derivation, we may still be able to 'repair' the original s-derivation, e.g., by resorting to s-inference figures not affected by the rebutting s-derivation.

The discussion on internal coherence is important for two main reasons. First, because it sets apart safety reasoning from mathematical reasoning; the former is defeasible while the latter is not. Second, because it has a bearing on compositional safety reasoning. Let us explain this with an example. Suppose that the s-derivation (i) $\{r \wedge g \supset \Diamond d, (s \cup g) \wedge (g \cup s), r \supset (r \mathsf{W} d)\} \vdash_{SK}^{\{\neg v\}} r \supset \Diamond d$ in Fig. 7 results from composing two s-derivations, (ii) $\{(s \cup g) \wedge (g \cup s)\} \vdash_{SK}^{\{\neg v\}} r \supset \Diamond g$, and (iii) $\{r \wedge g \supset \Diamond d, r \supset \Diamond g, r \supset (r \mathsf{W} d)\} \vdash_{SK} r \supset \Diamond d$, in the obvious way, i.e., by gluing (ii) and (iii) together at $r \supset \Diamond g$. This composition is reminiscent of Gentzen's Cut Rule.[10] In this case, two internally coherent s-derivations, i.e., (ii) and (iii), are composed into an internally coherent s-derivation, i.e., (i). But this is not the general case, as there is no guarantee that we will not be able to obtain the rebuttal of one of the s-inference figures in one of the composing s-derivations from the joint

---

premisses has some drawbacks. Suppose that π is an s-derivation with rebuttal **R**. Let π be internally coherent, i.e., we do not have an s-derivation π' of **R** from the premisses of π. Adding **R** to the premisses of π implies that we need to discharge it in some conditional; otherwise, we are changing the set of premisses from which we argue. Not discharging an added rebuttal has the unwanted consequence that it makes any s-derivation incoherent in the presence of the rule of reflexivity, i.e., we can always conclude what is in the premisses. Adding the $\neg$**R** also has unwanted effects. If it so happens that from the premisses of π we can prove **R**, but we have not done so yet, e.g., because we have not found such a s-derivation, adding $\neg$**R** to the premiss set of π means that we now have to deal with a premiss set that involves a glaring contradiction, i.e., **R** and $\neg$**R**. The moral of the story is: rebuttals occupy a special place in s-derivations as sources of defeasibility; considering them as part of the premisses of a s-derivation needs to be done with extreme care.

[10]Gentzen's Cut Rule: If $\{A_i\} \vdash_{N\!K} B$ and $\{B_j\} \cup \{B\} \vdash_{N\!K} C$, then, $\{A_i\} \cup \{B_j\} \vdash_{N\!K} C$ (see [14]).

set of premisses of the composed s-derivations. In other words, the composition of internally coherent s-derivations to form a larger s-derivation may result in the larger s-derivation being internally incoherent; and in order to establish whether the larger s-derivation is internally coherent or not, we may have to revise this s-derivation as a whole. In any case, compositionality is lost.

The preceding discussion on internal coherence also allows us to discuss technically the use of a form of *eliminative induction* in safety reasoning. As we mentioned in Sect. 2, eliminative induction was first developed by Bacon, and taken up by philosophers such as Mill, Keynes, Popper, et al. A reference to eliminative induction in safety reasoning is [15]. Briefly, eliminative induction works like this: Let us suppose that we conclude a property P and that, at the same time, we identify that this may not be the case in the presence of one or more properties P_i. The set of P_is associates some uncertainty to P. If none of the P_is can be concluded, then, the uncertainty associated with P is reduced. In the context of s-derivations and internal coherence, eliminative induction takes the following form: For a s-derivation π, the property P corresponds to the s-endformula of π. Each of the P_is corresponds to a rebuttal of π. Let us now suppose that π is internally coherent, i.e., that we have not found a rebutting s-derivation for it. Internal coherence associates some uncertainty to π, i.e., that expressed by its rebuttals. Since internal coherence alone does not establish that there are no rebutting s-derivations for π, simply that we have not found them, and since establishing that there are no rebutting s-derivations for π is non-trivial, instead, as a form of eliminative induction, we can attempt to construct s-derivations whose s-enformulæ are the negations of the rebuttals of π. The latter s-derivations enable us to reduce the uncertainty associated with π, and thus with P. This form of eliminative induction involves the presentation of a set of s-derivations. In this set, one s-derivation is designated as a main s-derivation. The assumption is that there is some uncertainty associated with this main s-derivation, as indicated by its rebuttals. The remaining s-derivations in the set are intended to reduce this uncertainty. This form of reasoning is an example of a form of probabilistic reasoning related to confidence, a topic that we discuss in more detail in Sect. 4.2.3. Let us illustrate this view of eliminative induction with a simple example. As it stands, the s-derivation $\{r \wedge g \supset \Diamond d, (s \cup g) \wedge (g \cup s), r \supset (r \mathbf{W} d)\} \vdash_{SK}^{\{\neg v\}} r \supset \Diamond d$ in Fig. 7 is internally coherent. There is, associated to this s-derivation, some uncertainty, namely, that indicated by $\neg v$. Recall that this s-derivation corresponds to an argument which concludes that 'if the train is *r*eady, then, it eventually *d*eparts', that one of the s-inference figures is contingent on the go signal being visible, and that the s-formula $\neg v$ corresponds to the property 'the go signal is not *v*isible'. To reduce this uncertainty, we can focus on constructing a s-derivation having v as its s-endformula, i.e., an argument which concludes that 'the go signal is *v*isible'.

3.4   Some Final Remarks on Safety Arguments

There are some final remarks about the difference between s-derivations and GSN diagrams that we can only elucidate at this point.

First, GSN strategies have no concept analogous to that of discharging an initial s-formula. This limitation severely restricts most forms of conditional reasoning. Is conditional reasoning forbidden in safety cases? How are we to reason conditionally without suitable mechanisms for introducing and discharging conditionals?

Let us digress for a moment to the issue of an initial formula being discharged to explain its ramifications in some detail. We begin by discussing what is the case in Gentzen's $\mathcal{NK}$. In Gentzen's $\mathcal{NK}$, discharging an initial formula means: (i) incorporating said formula into the lower formula of some inference figure in the derivation and (ii) eliminating said formula from the set of premises of the derivation. Though (ii) is not necessary, keeping initial formulæ that have been discharged as part of the premises of a derivation is superfluous; and this is something that we wish to avoid (see [8]). In fact, what Gentzen is after with his $\mathcal{NK}$ is a derivation that is *logistic*, i.e., one in which all initial formulæ in a derivation are discharged (see [14, p. 295]). To achieve this, Gentzen proposes to convert any non-logistic derivation π_1 into a logistic derivation π_2 whose endformula is an instance of $\mathbf{A} \supset \mathbf{B}$; in this instance of $\mathbf{A} \supset \mathbf{B}$, the instance of $\mathbf{B}$ is the endformula of π_1, and the instance of $\mathbf{A}$ is the conjunction of the formulæ in the set of premises of π_1. In a more general setting, Gentzen's proposal requires the use of the Compactness and Deduction Theorems for the logical calculus in which derivations are formulated (see [8]). It is not clear to us whether or not such (meta) theorems hold for safety reasoning, i.e., whether they hold in our sought after $\mathcal{SK}$ (and we are inclined to believe that they do not). In other words, it seems that in safety reasoning we are required to be able to deal with genuine premises, i.e., premises that cannot be discharged. To have at hand suitable mechanisms for dealing with such premises is of utmost importance. In addition, it is well known that different discharge policies give rise to different conditionals. For example, in the s-inference figure in Fig. 7 we allowed for the s-premiss r to be discharged vacuously (as is usually done in the introduction of the material conditional). If we forbid this, then, we obtain a form of a relevant conditional. Alternatively, if we allow for r to be discharged only once, then, we obtain a form of linear conditional (see [30]). It is clear to us that safety reasoning involves different kinds of conditionals. Discussing what discharge policies are allowed in safety reasoning may shed some light on which conditional we are referring to. How are we supposed to do this without proper mechanisms for tracking which initial s-formulæ correspond to which conditional? These issues are not at all properly dealt with in GSN diagrams.

Second, GSN diagrams have no concept analogous to rebuttals. In this sense, they are more limiting than goal decomposition structures in KAOS, which incorporate the notion of an *obstacle* to a goal (see [23]). Without rebuttals, the defeasible aspects of safety arguments are left implicit or are simply ignored.

In summary, the discussion that we have presented in this section is not a matter of logical pedantry. Instead, our discussion pinpoints some important issues to be addressed if safety reasoning is meant to be grounded on logical principles, and it exposes the leading causes of fallacies and the challenges in safety reasoning by bringing them into the foreground with the use of appropriate logical machinery.

4 Evidence in Safety Reasoning

As we mentioned in Sect. 3, safety cases are a cornerstone of safety reasoning. In addition to structured arguments, a defining characteristic of safety cases is the use of evidence as a grounding mechanism for safety arguments. In this section, we pay close attention to the concept of evidence, to how it can be incorporated into our program for formalizing safety reasoning in the form of a logical calculus, and to some of the challenges that it brings with it.

4.1 Evidence in Safety Cases

To provide some context for discussion, let us recall some basic facts about the role of initial formulæ in Gentzen's NK. Gentzen mentions in [14, p. 292] that a distinguishing feature of his NK is that derivations start from what he calls *assumptions*, to which logical deductions are then applied. Gentzen's assumptions are the initial formulæ of a derivation. As we have noted in Sect. 3.1, in Gentzen's formulation of the NK, it is possible for the initial formulæ of a derivation to be discharged or not. We have called initial formulæ that are not discharged *premisses*. An important characteristic of the premisses of a derivation is that they are, in a sense, given *deus ex machina*. This is not the case in safety reasoning, where the safety claims from which a safety argument is built need to be provided with a rationale which justifies their postulation. In other words, the s-premisses in a s-derivation cannot be taken as being given *deus ex machina*. This is reminiscent of the notion of justified belief in studies in epistemology or scientific explanation. It is at this point that evidence makes an appearance.

The definition of a s-derivation given in Sect. 3.2 enables us to discuss the use of evidence in safety cases in technical terms. However, in order to do so, we need, first and foremost, to be (i) precise about what we mean by evidence and to be (ii) able to refer to evidence in the language of s-formulæ.

As to (i), the uses of 'evidence' that we have observed in safety arguments, in particular in those referred to as *solutions* in GSN diagrams, refer to results obtained via testing, simulation, model analyses, or other observation-based mechanisms, including past experiences. These uses regard 'evidence' as some kind of data. This view of evidence is problematic for data does not, and cannot, in and of itself, be used as a basis for constructing a safety argument. To explain this issue, we take an example presented in [38, p. 195]. In a court of law, a bloodied knife, i.e., a piece of data, can be used both by the prosecution or the defense in their respective cases.

The use of the bloodied knife in court, i.e., the use of this piece of data in court, may involve claims such as: 'the bloodied knife was found at the crime scene', 'the bloodied knife was used by the accused to stab the victim', 'the bloodied knife was planted at the crime scene', etc. The bloodied knife is a source of many such claims (some of which may be incompatible with others). What this example shows is that, in isolation, a piece of data is not a truth bearer, i.e., it cannot be assigned a truth value; a truth bearer is a claim about it. In other words, data becomes evidence, in the epistemological or scientific sense of the term, when it stands in a precisely defined testing relationship with some claims postulated about it. To avoid any confusion, we will refer to a piece of data as *evidence*, and to a claim about a piece of data as an *evidential claim*.

Let us now turn our attention to (ii). Immediately from the distinction between evidence and evidential claim, we would need *evidence terms* and *evidence formulæ* in the language of s-formulæ. Evidence terms would include, at least, constants for concrete pieces of evidence, and variables for arbitrary pieces of evidence. Evidence formulæ would include, at least, quantifiers binding variables in evidence terms. An evidence formula is *grounded* if it has no free variables (where 'free variable' has the usual meaning). An evidence formula is said to be *ground atomic* if it has no quantifiers and if its testing relationship with its evidence term is self-evident (intersubjectively agreed).

Evidence terms and evidence formulæ can be understood by drawing an analogy between terms and formulæ in the language of FOL. More precisely, in the language of FOL, terms denote objects; formulæ are the formal counterpart of claims about objects. For instance, in their standard interpretation, the terms $S(n)$, 0, denote the successor of a natural number, and the natural number zero, respectively. In these terms, the variable n is used to indicate an arbitrary natural number, and the constant 0 to indicate the number zero. In turn, the formula $\neg \exists n (S(n) = 0)$ is the formal counterpart of 'there is no natural number whose successor is zero'. In this formula, the existential quantifier binds the variable n. If we understand evidence terms and evidence formulæ in this way, the former serve as a way to denote pieces of evidence, while the latter are the formal counterpart of claims about evidence.

We are now in a position to make precise in which sense a safety argument is to be taken as being grounded on evidence. We do this in relation to s-derivations. Namely, we define a s-derivation as grounded on evidence if its s-premisses, i.e., its undischarged initial s-formulæ, are ground atomic evidence formulæ. In consequence, a safety argument is grounded on evidence if its incorporation into a s-derivation results in the latter being grounded on evidence.

4.2 Some Comments on Evidence in Safety Arguments

As with safety arguments, a significant part of the definitions of an evidence term and an evidential s-formula needs to be completed and fully worked out. Nevertheless, evidence terms and evidential s-formulæ allow us to discuss technically some aspects of the use of evidence in safety cases.

4.2.1 Regarding Ground Atomic Evidence Formulæ

We have defined a safety argument as being grounded on evidence if its incorporation into a s-derivation results in the latter being grounded on evidence. The first part of this definition corresponds to our program of making precise what is a safety argument via its incorporation into a logical calculus (our sought after $\mathcal{SK}$). The second part of this definition corresponds to our view of the use of evidence in safety cases and its logical characterization. The idea is that a ground atomic evidence formula plays a role similar to an *axiom* of a classical logical theory, i.e., a formula that is regarded as accepted or self-evident. This is precisely what a ground atomic evidence formula aims to capture. More elaborate evidence formulæ, e.g., those that are not ground atomic, must perforce involve some reasoning.

To illustrate the points above, let us suppose that the go signal example in Sects. 2 and 3, indicating whether the train can depart or not, consists, among other things, of a piece of software toggling the light from red to green, and from green to red. Let us suppose further that this piece of software is proven correct in Hoare's $\mathcal{HK}$, i.e., that there is, for this piece of software, a derivation π akin to that in Fig. 10. Technically speaking, π is not a proof that the piece of software itself is correct, but rather a proof of the correctness of a (syntactical) model of the piece of software in Hoare's $\mathcal{HK}$ in relation to some specification. But the piece of software itself and its (syntactical) model in Hoare's $\mathcal{HK}$ are different things. Now, let us suppose that we use the proof of correctness of the (syntactical) model of the piece of software in Hoare's $\mathcal{HK}$ to argue that the piece of software itself is dependable (in a more general sense than 'correct'). In this context, the former is a piece of evidence and the latter is an evidence claim. In the language of s-formulæ, we would then have an evidence term to denote the piece of evidence, i.e., π, and an evidence formula as the formal counterpart of the evidence claim, i.e., that the piece of software itself is dependable, respectively. A question that we could ask ourselves at this point is: Would this evidence formula meet the criterion of being ground atomic? The answer is *no*. The problem is that the testing relationship between the evidence term and the evidence formula is not self-evident, i.e., it already involves some reasoning, e.g., about the adequacy of the proof of correctness of the (syntactical) model of the piece of software in relation to a claim about the dependability of the piece of software itself. For this reason, the evidence formula cannot be used as a premiss in a s-derivation. What can be used as a ground atomic evidence formula is the formal counterpart of a claim along the lines of 'the (syntactical) model of the piece of software meets its specification'. It is the role of a safety argument to take us from this basic claim about evidence (possibly in conjunction with other basic claims about evidence), to the claim that the piece of software itself is dependable.

The preceding discussion shows that the burden is on finding ground atomic evidence formulæ, i.e., evidence terms and evidence formulæ whose testing relationship is self-evident. These ground atomic evidence formulæ serve as the basis on which we would construct the s-derivation that would take us to a s-endformula. The danger is that, without a proper formulation of a ground atomic evidence formula, or set thereof, a significant amount of effort needs to be devoted to eliciting in which sense a piece of evidence relates to an evidence claim, something that is prone to error. An open question is whether the testing relationship between evidence terms and evidence formulæ is part of the $\mathcal{SK}$ or is external to it.

4.2.2 Regarding Multiple Atomically Grounded Evidence Formulæ

The discussion about ground atomic evidence formulæ raised to the surface the use of multiple pieces of evidence in relation to a single evidence claim.

To illustrate this phenomenon, let us take up again the go signal example in Sects. 2 and 3. Namely, let us suppose that the go signal, indicating whether the train can depart or not, consists, among other things, of a piece of software toggling the light from red to green, and from green to red. In addition, let us suppose that this piece of software is proven correct in Hoare's $\mathcal{HK}$, i.e., that there is, for this piece of software, a derivation π akin to that in Fig. 10. Let us suppose further that we use π to argue that the piece of software itself is dependable (in a more general sense than 'correct'). Repeating ourselves, in this context, the former is a piece of evidence and the latter is an evidence claim; which would cause us to have, in the language of s-formulæ, an evidence term to denote the piece of evidence, and an evidence formula as the formal counterpart of the evidence claim, respectively. In Sect. 4.2.1 we discussed that this evidence formula is not ground atomic for it already involves some reasoning, e.g., about the adequacy of π in relation to a claim about the dependability of the piece of software itself. Among other things, the adequacy of π in relation to a claim about dependability hinges on how faithful the model of the piece of software in Hoare's $\mathcal{HK}$ is to the piece of software itself, something which depends, in turn, on some assumptions on the piece of software itself, e.g., that arithmetic computations do not result in an overflow. The use of input/output testing data on the piece of software itself presents an interesting use of evidence to validate this kind of assumption. Moreover, this leads in a more or less natural way to the use of different input/output testing data, e.g., obtained from different testing methods, to validate the same assumption, e.g., because the different testing methods cover different aspects of the assumption. In technical terms, we are in a scenario in which a s-formula, i.e., the formal counterpart of one of the assumptions, is the s-endformula of various s-derivations, each of which has as its premises ground atomic evidence formulæ whose evidence terms denote the different input/output testing data. To put all these different s-derivations together in one single s-derivation, we need to relax the definition of a s-derivation to allow for the upper s-formula of a s-inference figure to be the lower s-formula of more than one s-inference figure. Such a relaxation has no analogy in Gentzen-like derivations, for in traditional logical calculi, one derivation of an endformula is as good as any other. However, the situation is different in safety arguments due to the *confidence* value that we tend to associate with them. We have here an example of what has sometimes been referred to as a *multi-legged* argument (see [2]). The idea is that each leg is logically sufficient, but the legs taken together provide greater confidence in the logical result. We discuss this in Sect. 4.2.3, after we introduce some basics on *confidence measures*.

4.2.3   Evidence and Confidence

As previously mentioned, there is an inherent uncertainty associated with safety arguments. To begin with, the use of data in evidential reasoning naturally involves uncertainty. There is uncertainty in gathering data, in the processes of observation and measurement from which we obtain the data, in how the data is used, in the claims that we formulate about data, etc. For example, we often use some form of inductive reasoning to assert a universal conclusion from a finite number of observations, e.g., in testing of programs, and this inherently involves some lack of certainty in the universal conclusion, e.g., whether the test cases are sufficient to justify the conclusion. Second, multi-legged arguments are often used to reduce the uncertainty (i.e., increase confidence) in an argument, and this clearly means that we are not entirely sure of the conclusion of some arguments, no matter how stringent we might have been in developing the argument. Thirdly, the use of eliminative induction involves uncertainty of various kinds. For example, we cannot be certain that all the possibilities for issues to be examined have been discovered. We may also not be able to positively eliminate all possible cases, leaving some open as the risk involved is deemed too low to worry about. Modelling this uncertainty is key to evaluating the confidence we place in the safety argument, and in the claim that it establishes. In the previous section, we focused on rebuttals as a source of uncertainty. In this section, we focus on evidence.

The kind of uncertainty associated with evidence that we have in mind goes beyond uncertainty associated with statistical values (e.g., test cases returned the expected result 8 out of 10 times). It also includes the uncertainty associated with the way in which the evidence is obtained (e.g., the test cases are devised properly, they are executed in the right environment, the results are repeatable, etc). The latter kind of uncertainty associated with evidence is typically systematized by using acceptance criteria for the inclusion of certain data as evidence. The various confirmation measures for work products proposed in ISO 26262 provide an example of such acceptance criteria (see [21]).[11] In the cases where we have a pass/fail acceptance criterion, things are relatively simple, i.e., the data item either gets included in the safety argument, or not, i.e., the data is accepted as evidence or not. However, if there is a degree of acceptability, and a certain threshold that needs to be met in order for the data to be acceptable as evidence, we would like to have acceptability values at hand when evaluating the confidence we may place in the safety argument. For example, consider the following confirmation measure found in ISO 26262: 'The work products referenced in the safety case are available and sufficiently complete' (see [21, pt. 2, p. 21]). Checking whether the work products are *available* is not difficult, but measuring whether they are *sufficiently complete* is not trivial, as this needs to be precisely defined, so as not to be open to arbitrary interpretation. Unfortunately, this kind of precision is often missing.

---

[11] ISO 26262 is a safety standard developed to fit the needs of the automotive domain. The standard applies to electrical and/or electronic (E/E) systems within road vehicles (see [21]).

The above focuses our attention on one of the biggest issues that plague confidence modelling and evaluation. There is a lack of precise definitions, benchmarks, and evaluation techniques, all of which hinder the possibility of defining meaningful acceptance criteria, i.e., ones to which we can assign values. This is very similar to the situation in quality management, where vague definitions are common. Let us draw an analogy to elaborate on this point. In quality management, some concepts are associated with an abundance of definitions and quality measures. This has meant that something as simple as the efficiency of a computer program might mean completely different things to different stakeholders. In addition, since there are various models and methods for product quality assurance, e.g., various ways of measuring the efficiency of a computer program (memory usage and/or speed), any value associated with one of the quality characteristics of a product has to be accompanied by additional information in order for this value to be meaningful (see [17]). The same is clearly true for confidence modelling in the safety domain. To refer back to the example above, if an expert has stated that the safety case references work products that are sufficiently complete, this expert needs to provide a definition of *sufficiently complete* in measurable terms, and explain the measurement procedure used to arrive at this conclusion. If done in this way, the claim made by the expert can be reviewed and potentially compared on a more objective basis, an otherwise well-nigh impossible task. Of course, objective measures are difficult to come by, and sometimes relying on subjective measures is the only practical approach. But these subjective measures still ought to be given explicit definitions to enable results to be reproduced. This is crucial for the validation of confidence metrics and the measures associated with them.

In addition to providing a precise definition of the confidence metrics and the measures associated with them, we need rules for combining and decomposing those measures. One example of the latter in the safety-critical domain is the decomposition of *Automotive Safety Integrity Levels*, ASILs, in ISO 26262 (see [21, pt. 9]). In more detail, in ISO 26262, each safety requirement of the E/E system being considered has an ASIL associated with it. The stringency of the ASIL depends on the criticality of the safety requirement. Note that ISO 26262 allows for ASILs to be weakened during the decomposition of a safety requirement. In this respect, the obvious question that arises is: What is needed to guarantee that the weakened ASILs associated with the decomposed safety requirements guarantee the ASIL of the original safety requirement? The problem is that the decomposition of ASIL levels suggested in ISO 26262 has not been explicitly justified. Instead, it rests on domain experience, as does its validation. This is clearly a shortcoming. One way in which the combination of different measures is addressed in the field of quality management is through utility functions. Though the definition of such utility functions may be subjective, when explicitly defined, they enable us to reproduce results and to interpret these results in a repeatable and objective fashion.

In light of the above, and in order to make some progress, we need to start with widely agreed upon definitions of confidence and what its evaluation entails. Without this, the production of confidence measures becomes very vague and is largely based on the opinions and prejudices of experts. In [16], we find a survey of confidence modelling approaches suggested for use in the safety-critical domains. All the various approaches are illustrated by means of examples. Borrowing from them, though we are not yet ready to provide a more complete working definition for this framework, we can outline some of its key elements. First, each evidence term has to be associated with a confidence value (or a tuple of values), produced as a reflection of its acceptance criteria as well as additional sources of uncertainty. The measures for these confidence values have to be precise and to meet the representation condition of measurement theory (see [11]), namely, that the mapping from the empirical domain of attributes to the formal domain of measures is a homomorphism (i.e., that the assignment of measures to attributes does not violate properties of attributes, e.g., that height does not make a baby's height bigger than a grown person's). In addition, the measurement statements must be meaningful, i.e., the truth value of the measurement statements must remain invariant under all admissible scale transformations (e.g., 'the temperature in Toronto is 20 C and is twice as much as in Buenos Aires, where it is 10 C' is not a meaningful statement as the transformation of Celsius to Fahrenheit, an admissible transformation, does not preserve the truthfulness of the statement). The measurement scales might exhibit different properties (being classified, being ordered, having quantified differences, etc.) based on the scale used. This would in turn depend on the property (subject to uncertainty) being modelled. To again make a parallel with quality management, we know that quality measures can take a number of forms (non-numeric, or quantitative, both of which are further subdivided and correspond to different scales) depending on the domain-specific content they model. In fact, due to the fact that the different types of measures sometimes cannot be meaningfully combined, it is possible that instead of a single confidence value, we end up with a tuple of confidence values. Transforming between different types of measures is not impossible, but it might not bring any added benefit, and might instead obscure some valuable information (e.g., through transforming a precise value into a range one). The difference between confidence modelling and quality management, and indeed the biggest issue, lies in the propagation of confidence measures associated with uncertainty. One possible way in which this issue may be addressed is through the use of Jøsang's Subjective Logic (see [22]). After introducing well-defined confidence measurement scales and procedures, and utility functions for combining the confidence values as well as a logic for propagating them, we should proceed to empirically validate our framework and make any necessary adjustments.

An additional challenge to modelling confidence is what some people call the three Ps: Process, Product, and People. For example, ISO 26262 states that we need to explicitly note the qualifications and level of independence of the people tasked with carrying out the confirmation measures (see [21, pt. 2, p. 12]). This may be construed as a form of multi-legged argument. Each leg of the argument corresponds to how an independent team arrives at a conclusion, the premises of each leg would then correspond to items of evidence that have been independently obtained, or, alternatively, independently vetted. Let us illustrate this by extending our train example. Suppose that we recognized the visibility of the go sign as one of the sources of uncertainty. In other words, for whatever reason, we cannot be totally sure whether the go sign is visible or not. This said, we do want to establish, with some level of confidence, that the go sign is visible. To reach this conclusion, we approach two experts in the field of vision inspection, who will conduct independent experiments. Each expert starts out with the same set of experiment participants (the train operators), and the same experimental environment (riding in the train alongside the train operators). Though both experts start with the same premises, i.e., in the same setting, the individual experiments that they conduct might be devised in a different way, e.g., based on the different impediments to visibility that they might have thought of and decided to check against. For example, both experts might take into account the fact that one of the train operators uses allergy medication, reported to cause blurry vision as a side effect, but only one of them considers the use of sunglasses, and that their lens hue and tint density might negatively affect visibility in certain conditions (pink, blue, and green lenses can make red lights indistinguishable). Having the experts design their experiments independently guards against confirmation bias and leads to increased confidence in the final result, i.e., that the go signal is visible. The results, observations, claims, and the like, made by each of the experts would then be included in their own leg of the overall safety argument. Each leg by itself may not provide sufficient confidence, but when put together they reduce the uncertainty and increase the confidence in the claim that 'the go signal is visible'.

5 Discussion

In Sects. 3 and 4, we discussed some various bits and pieces of the puzzle of safety reasoning independently from each other. In this section, we put these bits and pieces together in an attempt to present a coherent picture.

We frame our discussion in the context of our running example: a train departing from a station. Our goal is to establish that this is done safely, for which we would like to build a safety case, i.e., a structured argument. This structured argument is our claim of safety. The structured argument itself corresponds to a s-derivation π in the $\mathcal{SK}$. As a first step in the construction of π, we need to determine its s-endformula. This s-endformula is the property which, if established, via a structured argument, assures safety. (It is not directly a claim of 'safety', for such a claim of 'safety' is outside the structured argument and the logical calculus in which we state it, and it is associated with our conception of what it means for the train to depart safely.) Let us suppose that this property is 'the train $d$eparts iff it is $r$eady'. To this property, there corresponds, in the language of s-formulæ, the s-formula $(r \supset \Diamond d) \wedge (\neg r \supset \neg \Diamond d)$. Given the structure of the s-endformula, we can think of proceeding with the construction of the s-derivation which establishes separately in two s-derivations, π_1 and π_2; π_1 would have $r \supset \Diamond d$ as its s-endformula; π_2 would have $\neg r \supset \neg \Diamond d$ as its s-endformula; π would obtain by combining π_1 and π_2. π_1 would correspond to a structured argument establishing that 'if the train is $r$eady then it eventually $d$eparts'. π_2 would correspond to a structured argument establishing that 'if the train is not $r$eady then it does not eventually $d$epart'. We have shown what π_1 may look like in Fig. 7. We have also shown that there is some uncertainty associated with π_1, namely, that indicated by $\neg v$. The latter s-formula corresponds to the property 'the go signal is not visible'. As we have said, in the absence of a s-derivation which has $\neg v$ as its s-enformula, π_1 is internally coherent. Since finding that there are no such derivations, what we can do instead, is to construct a s-derivation π_3, which has v as its s-endformula. This s-derivation corresponds to a structured argument making a case for 'the go signal is visible'. In this way, π_1 would be accompanied by a π_3, with π_3 being there to reduce the uncertainty associated with π_1. This is an application of eliminative induction. The case with π_2 would be similar. We are now in a situation in which we have two internally coherent s-derivations, π_1 and π_2, which we want to combine in a single internally coherent s-derivation, π. A priori, we could glue π_1 and π_2 together to form π by introducing the missing logical connective, $\wedge$. However, to guarantee that π is internally coherent, we would have to inspect π as a whole. This is a necessary step to eliminate the possibility of one of the rebuttals of π_1 being established from the combination of the s-premises of π_1 in combination with those of π_2, and similarly with the rebuttals of π_2.

Thus far, nothing has been said about π being grounded on evidence, i.e., the s-premises of π could be arbitrary s-formulæ. This is a situation that we would wish to remedy. For this, we would have to show how the s-premises of π that are not ground atomic evidence s-formulæ can be obtained from ground atomic evidence s-formulæ. This involves an extension of π. This extension is also a s-derivation, let us call it π'. In contrast to π, the s-premises in π' are ground atomic evidence s-formulæ. It would seem that π' contains a distinguished part that is 'purely evidential', i.e., obtained through evidential reasoning (this distinction is, although with a different flavour, also noted in [32]). It is open to debate where this 'purely evidential' part ends. Perhaps Carnap's distinction between the observable and the theoretical in the language of science (see [5, ch. 23]) provides a foundation on which to settle this debate. But this thesis needs further investigation. It should be noted that 'purely evidential' reasoning needs not appear solely when there is a need to make a s-derivation grounded on evidence. It may also appear while attempting to remove the uncertainty associated with a s-derivation. For example, in the example above, it is possible, perhaps even natural, for π_3, i.e., the s-derivation corresponding to a structured argument making a case for 'the go signal is visible', to be 'purely evidential'.

Emerging from our discussion in Sect. 4.2.3, we would associate with each s-derivation a confidence value. In order to assign a confidence value (or a tuple of values) to each evidence term, we shall start by reviewing the sources of uncertainty associated with it, including any acceptance criteria that have been specified. If the acceptance criteria have been properly defined, they would provide the scale of measurement and the measurement procedure to be used. However, for the sources of uncertainty that have not been explicitly considered we would need to add two steps. Firstly, based on the property being modelled (availability, objectivity, independence, etc.), we would select a scale for its measurement such that we can formulate useful and truth-preserving measurement statements. Then, we would select and describe a measurement procedure, which is practical and reliable (it should return the same result under the same conditions). After defining the confidence measurement scales and procedures, and obtaining precise confidence values for the evidence terms, we would introduce utility functions for combining them (these might vary across products and companies), and a logic for propagating them. Lastly, the framework shall be empirically validated and adjusted so as to make sure that the representation condition still holds after the use of our chosen utility functions and logic. Resuming the example above, we would associate with π, i.e., the s-derivation corresponding to a structured argument making a case for 'the train $d$eparts iff it is $r$eady', some confidence value. This value will, in turn, be the value associated with our claim of safety.

6   Conclusions

The present practice of safety cases, recorded in some notation, is the result of over 25 years of work. However, to date, notations for safety cases have no semantics. This makes their understanding and assessment difficult, if not well nigh impossible, and prone to error, with the apparent negative consequences. In this work, we have started to travel the long road to providing a semantics for safety cases. Our work builds on the idea that the semantics for what is a structured argument should be based on a logical calculus. We have discussed the main ingredients of such a logical calculus, as well as the challenges that its development represents.

The situation with notations for safety cases is not new. Immediately coming to mind is the Unified Modelling Language (UML) (see [4]). This language underwent a similar historical development over a similar period of time. In both cases it has become clear that simply providing a loose syntax is not enough. Engineering disciplines rely on scientific theories and mathematics to enable precision in design and analyses to support sound engineering decisions. This was acknowledged by the OO community, who started to incorporate mathematical precision into its notations some years ago, not without its hurdles and sometimes against the protests of the notation's inventors! The safety case community is slowly awakening to this. The increasing complexity of safety-critical systems, and the recognition that relying on the informal understanding and intuition of individuals, regardless of their experience, is not only unscientific, but a historic invitation to disaster, have been the major forces pushing the need for proper engineering guarantees about safety; notations for safety cases are no exception.

We propose to develop a proper scientific and engineering basis for safety case understanding and construction on logical grounds. To this end, we have introduced a working definition of a safety case via its incorporation in a precisely defined calculus. In line with other researchers in the area (see [25]), we observe that assurance case reasoning is more akin to the argument based reasoning ideas of Toulmin than to the conventional deductive logic reasoning well known to mathematicians and software engineers (or computer scientists). This form of reasoning is already known in domains such as legal reasoning and scientific reasoning/explanation (from which we have taken some of our ideas). The logical roots of our proposal are based on Gentzen's program for formalizing mathematical reasoning in terms of a logical language, inference rules to support reasoning steps, and proofs to capture the 'informal' notion of argument used by mathematicians. One can debate about the adequacy of Gentzen's formalization, but if one accepts it, and most mathematicians have, then one can make remarkable progress in analysing mathematical reasoning, including developing automated tools such as theorem provers and model checkers. Though safety reasoning is very different in character from mathematical reasoning, we can use an analogous approach to that of Gentzen. In particular, we can focus on the same ingredients, i.e., a formalized logical language for expressing safety claims, a well-defined notion of inference step (enlarged by incorporating some of the ideas of Toulmin's definition of an argument pattern), a well-defined notion of derivation (capturing what is a safety argument), and a new ingredient, grounded proofs, i.e., the idea that all initial formulæ in a derivation cannot be taken for granted but that they need to be justified by evidence. The latter enables a proper understanding of the notion of evidence and the role it plays in safety arguments. We hope to have taken some steps in the right direction.

References

1. Adelard, Claim, Argument, Evidence Notation. Adelard (2015), http://www.adelard.com/asce/choosing-asce/cae.html
2. R. Bloomfield, B. Littlewood, Multi-legged arguments: the impact of diversity upon confidence in dependability arguments, in *International Conference on Dependable Systems and Networks (DSN'03)* (IEEE, 2003), pp. 25–34
3. A. Bochman, Non-monotonic reasoning, in *Handbook of the History of Logic: The Many Valued and Nonmonotonic Turn in Logic*, vol. 8, ed. by D. Gabbay, J. Woods (North-Holland, Amsterdam, 2007), pp. 555–632
4. G. Booch, J. Rumbaugh, I. Jacobson, *The Unified Modeling Language User Guide*, 2nd edn. (Addison-Wesley Professional, Boston, 2005)
5. R. Carnap, *An Introduction to the Philosophy of Science*, 5th edn. (Dover, Mineola, 1966)
6. V. Cassano, T. Maibaum, S. Grigorova, A (proto) logical basis for the notion of a structured argument in a safety case. In: *18th International Conference on Formal Engineering Methods (ICFEM'16)*. LNCS, vol. 10009 (2016), pp. 1–17
7. W. Cooper, The propositional logic of ordinary discourse. Inquiry **11**(1–4), 295–320 (1968)
8. D. van Dalen, *Logic and Structure*, 5th edn. (Springer, Berlin, 2013)
9. R. Darimont, A. van Lamsweerde, Formal refinement patterns for goal-driven requirements elaboration, in *4th ACM SIGSOFT Symposium on Foundations of Software Engineering (SIGSOFT'96)* (ACM, 1996), pp. 179–190
10. J. van Eijck, M. Stokhof, The gamut of dynamic logics (2011), in [13], pp. 499–600
11. N.E. Fenton, Software measurement: a necessary scientific basis, in *Predictably Dependable Computing Systems* (Springer, Berlin, 1995), pp. 67–78
12. D. Gabbay, J. Woods (eds.), *Handbook of the History of Logic: Inductive Logic*, vol. 10 (North-Holland, Amsterdam, 2011)
13. D. Gabbay, J. Woods (eds.), *Handbook of the History of Logic: Logic and the Modalities in the Twentieth Century*, vol. 7 (North-Holland, Amsterdam, 2011)
14. G. Gentzen, Investigations into logical deduction. Am. Philos. Q. **1**(4), 288–306 (1964)

15. J. Goodenough, C. Weinstock, A. Klein, Eliminative induction: a basis for arguing system confidence, in *35th International Conference on Software Engineering (ICSE'13)* (2013), pp. 1161–1164
16. P. Graydon, M. Holloway, An investigation of proposed techniques for quantifying confidence in assurance arguments. Saf. Sci. **92**, 53–65 (2017)
17. S. Grigorova, The elusive quest: software product quality evaluation. Master's thesis, McMaster University, Canada, 2009
18. C. Hempel, *Aspects of Scientific Explanation: And Other Essays in the Philosophy of Science* (Free Press, New York, 1965)
19. D. Hitchcock, Toulmin's warrants, in *Anyone Who Has a View: Theoretical Contributions to the Study of Argumentation*, ed. by F. van Eemeren et al. (Springer, Berlin, 2003), pp. 69–82
20. C. Hoare, An axiomatic basis for computer programming. Commun. ACM **12**(10), 576–580 (1969)
21. International Organization for Standardization, ISO 2626: Road Vehicles – Functional Safety. Version 1 (2011)
22. A. Jøsang, *Subjective Logic: A Formalism for Reasoning Under Uncertainty* (Springer International Publishing, Berlin, 2016)
23. A. van Lamsweerde, *Requirements Engineering: From System Goals to UML Models to Software Specifications* (Wiley, Hoboken, 2009)
24. Z. Manna, A. Pnueli, *The Temporal Logic of Reactive and Concurrent Systems: Specification* (Springer Science+Business Media, Berlin, 1991)
25. J. McDermid, Safety arguments, software and system reliability, in *2nd International Symposium on Software Reliability Engineering (ISSRE'91)* (IEEE, 1991), pp. 43–50
26. P. McNamara, Deontic logic (2011), in [13], pp. 197–288
27. P. Øhrstrøm, P. Hasle, Modern temporal logic: the philosophical background (2011), in [13], pp. 447–498
28. G. Pólya, *How to Solve It*, 2nd edn. (Princeton University Press, Princeton, 2004)
29. K. Popper, *An Introduction to the Philosophy of Science* (Routledge, Abingdon, 2002)
30. G. Restall, *Proof Theory and Philosophy*. Draft Book (2006), http://consequently.org/writing/ptp
31. J. Rushby, Logic and epistemology in safety cases, in *32nd International Conference on Computer Safety, Reliability, and Security (SAFECOMP'13)*. LNCS, vol. 8153 (2013), pp. 1–7
32. J. Rushby, On the interpretation of assurance case arguments, in *New Frontiers in Artificial intelligence - JSAI-ISAI 2015 Workshops, LENLS, JURISIN, AAA, HAT-MASH, TSDAA, ASD-HR, and SKL (Revised Selected Papers)*. LNCS, vol. 10091 (2015), pp. 331–347
33. The GSN Working Group, Goal Structuring Notation. The GSN Working Group (2011), http://www.goalstructuringnotation.info/
34. S. Toulmin, *The Uses of Argument* (Cambridge University Press, Cambridge, 2003)
35. UK Ministry of Defense, Defence standard 00-56 issue 4: safety management requirements for defence systems (2007)
36. S. Veloso, P. Veloso, R. de Freitas, An application of logic engineering. Log. J. IGPL **13**(1), 29–46 (2005)
37. W. Vincenti, *What Engineers Know and How They Know It: Analytical Studies from Aeronautical History* (Johns Hopkins University Press, Baltimore, 1993)
38. T. Williamson, *Knowledge and Its Limits* (Oxford University Press, Oxford, 2000)

The Indefeasibility Criterion for Assurance Cases

John Rushby

Abstract Ideally, assurance enables us to know that our system is safe or possesses other attributes we care about. But full knowledge requires omniscience, and the best we humans can achieve is well-justified belief. So what justification should be considered adequate for a belief in safety? We adopt a criterion from epistemology and argue that assurance should be "indefeasible," meaning that we must be so sure that all doubts and objections have been attended to that there is no (or, more realistically, we cannot imagine any) new information that would cause us to change our evaluation. We explore application of this criterion to the interpretation and evaluation of assurance cases and derive a strict but practical characterization for a sound assurance case.

1 Introduction

One widely quoted definition for a safety case comes from the UK Ministry of Defence [1]:

> A safety case is a structured argument, supported by a body of evidence that provides a compelling, comprehensible, and valid case that a system is safe for a given application in a given operating environment.

An *assurance case* is simply the generalization of a safety case to properties other than safety (e.g., security) so, *mutatis mutandis*, we can accept this definition as a basis for further consideration.

Key concepts that we can extract from the definition are that an assurance case uses a *structured argument* to derive a *claim* or goal (e.g., "safe for a given application in a given operating environment") from a body of *evidence*. The central requirement is for the overall case to be "compelling, comprehensible and valid"; here, "compelling" and "comprehensible" seem to be subjective judgments, so I will focus on the notion

J. Rushby (✉)
Computer Science Laboratory, SRI International, 333 Ravenswood Avenue, Menlo Park, CA 94025, USA
e-mail: Rushby@csl.sri.com

© Springer Nature Singapore Pte Ltd. 2021
Y. Ait-Ameur et al. (eds.), *Implicit and Explicit Semantics Integration in Proof-Based Developments of Discrete Systems*,
https://doi.org/10.1007/978-981-15-5054-6_12

of a "valid" case and, for reasons I will explain later, I prefer to use the term *sound* as the overall criterion.

There are two ways one might seek a definition of "sound" that is appropriate to assurance cases: one would be to fix the notion of "structured argument" (e.g., as classical deduction, or as defeasible reasoning, or as Toulmin-style argumentation) and adopt or adapt its notion of soundness; the other is to look for a larger context in which a suitable form of soundness can be defined that is independent of the style of argument employed. I will pursue the second course and, in Sect. 3, I will argue for the *indefeasibility criterion* from epistemology. I will apply this to assurance case arguments in Sect. 4 and argue for its feasibility in Sect. 5. Then, in Sect. 6, I will consider how the indefeasibility criterion applies in the evaluation of assurance case arguments.

Surrounding these sections on assurance are sections that relate assurance to system behavior and to certification. The top-level claim of an assurance case will generally state that the system satisfies some critical property such as safety or security. Section 2 relates confidence in the case, interpreted as a subjective probabilistic assessment that its claim is true, to the likelihood that critical system failures will be suitably rare—which is the basis for certification. Section 7 considers probabilistic assessments of an assurance case in support of this process. Section 8 presents brief conclusions and speculates about the future.

2   Assurance and Confidence in Freedom from Failure

A sound assurance case should surely allow—or even persuade—us to accept that its claim is true. There are many different words that could be used to describe the resulting mental state: we could come to *know* that the claim is true, or we could *believe* it, or have *confidence* in it. I will use the term "belief" for this mental state and will use "confidence" to refer to the strength of that belief.

So an assurance case gives us confidence in the belief that its claim is true. For a system-level assurance case, the top-level claim is generally some critical property such as safety (i.e., a statement that nothing really bad will happen), but we may also have "functional" claims that the system does what is intended (so it is useful as well as safe). A system-level assurance case will often be decomposed into subsidiary cases for its subsystems, and the functional and critical claims will likewise be decomposed. At some point in the subsystem decomposition, we reach "widgets" where the claims are no longer decomposed and we simply demand that the subsystem satisfies its claims.

Software assurance cases are generally like this: software is regarded as a widget and its local claim is correctness with respect to functional requirements, which then ensure the critical requirements of its parent system; of course there is a separate assurance task to ensure that the functional requirements really do ensure the critical requirements and hence the top-level claim. This division of responsibility is seen most clearly and explicitly in the guidelines for commercial aircraft certification, where DO-178C [2] focuses on correctness of the software and ARP 4754A [3] provides safety assurance for its requirements. If we assume the requirements are good and focus strictly on software assurance, any departure from correctness constitutes a *fault*, so a software assurance case gives us confidence that the software is fault-free. Confidence can be expressed numerically as a subjective probability so, in principle, a software assurance case should allow us to assess a probability p_{nf} that represents our degree of confidence that the software is free of faults (or nonfaulty).

What we really care about is not freedom from faults but the absence of failure. However, software can fail only if it encounters a fault, so software that is, with high probability, free of faults will also be free of failures, with high probability. More particularly, the probability of surviving n independent demands without failure, denoted $p_{srv}(n)$, is given by

$$p_{srv}(n) = p_{nf} + (1 - p_{nf}) \times (1 - p_{F|f})^n, \tag{1}$$

where $p_{F|f}$ is the probability that the software Fails, if faulty.[1] A suitably large n can represent the system-level assurance goal. For example, "catastrophic failure conditions" in commercial aircraft ("those which would prevent continued safe flight and landing") must be "so unlikely that they are not anticipated to occur during the entire operational life of all airplanes of one type" [5]. If we regard a complete flight as a demand, then "the entire operational life of all airplanes of one type" can be satisfied with n in the range 10^8–10^9.

The first term of (1) establishes a lower bound for $p_{srv}(n)$ that is independent of n. Thus, if assurance gives us the confidence to assess, say, $p_{nf} \geq 0.9$ (or whatever threshold is meant by "not anticipated to occur") then it seems we have sufficient confidence to certify the aircraft software. However, we also need to consider the case where the software does have faults.[2] We need confidence that the system will not suffer a critical failure despite those faults, and this means we need to be sure that the second term in (1) will be well above zero even though it decays exponentially.

This confidence could come from prior failure-free operation. Calculating the overall $p_{srv}(n)$ can then be posed as a problem in Bayesian inference: we have assessed a value for p_{nf}, have observed some number r of failure-free demands, and want to predict the probability of seeing $n - r$ future failure-free demands. To do this, we need a prior distribution for $p_{F|f}$, which may be difficult to obtain and difficult to justify. However, Strigini and Povyakalo [4] show there is a distribution that delivers *provably worst-case* predictions; using this, we can make predictions that are guaranteed to be conservative, given only p_{nf}, r, and n. For values of p_{nf} above 0.9, their results show that $p_{srv}(n)$ is well above the floor given by p_{nf}, provided $r > \frac{n}{10}$.

---

[1] I am omitting many details here, such as the interpretation of subjective probabilities, and the difference between aleatoric and epistemic uncertainty. The model and analysis described here are due to Strigini and Povyakalo [4], who give a comprehensive account.

[2] Imagine using this procedure to provide assurance for multiple aircraft types; if $p_{nf} = 0.9$ and we assure 10 types, then one of them may be expected to have faults.

Thus, in combination with prior failure-free experience (which is gained incrementally, initially from tests and test flights, and later from regular operation), an assessment $p_{nf} > 0.9$ provides adequate assurance for extremely low rates of critical failure, and hence for certification. I have presented this analysis in terms of software (where the top claim is correctness) but, with appropriate adjustments to terminology and probabilities, it applies to assurance of systems and properties in general, even autonomous systems. (It also applies to subsystems; one way to mitigate faults and failures in low-assurance subsystems is to locate them within a suitable architecture where they can be buttressed with high-assurance monitors or other mechanisms for fault tolerance; Littlewood and Rushby [6] analyze these cases.) This analysis is the only one I know that provides a credible scientific account for how assurance and certification actually work in practice. Those who reject probabilistic reasoning for critical properties need to provide a comparably credible account based on their preferred foundations.

Failures of the assurance process do not invalidate this analysis. For example, the Fukushima nuclear meltdown used inappropriate assessment of hazards, and the Boeing 737Max MCAS appears to have violated every principle and process of safety engineering and assurance. Sections 3–6 consider how to structure and evaluate an assurance case so that aberrations such as Fukushima and the 737Max MCAS are reliably detected and rejected. In the remainder of this section and in Sect. 7, I focus on how a probabilistic assessment such as $p_{nf} \geq 0.9$ can be derived from a successful assurance case.

One approach would be to give a probabilistic interpretation of the argument of the case. It is certainly reasonable to assess evidence (i.e., the leaves of the argument) probabilistically, and I will discuss this in Sect. 4. However, a fully probabilistic interpretation requires the interior of the argument to be treated this way, too, which will take us into probability logics or their alternatives such as fuzzy set "possibility theory" or the Dempster–Shafer "theory of evidence." Unfortunately, despite much research, there is no generally accepted interpretation for the combination of logic and probability. Furthermore, it is not clear that any proposed interpretations deliver reliable conclusions for assurance case arguments. Graydon and Holloway [7, 8] examined 12 proposals for using probabilistic methods to quantify confidence in assurance case arguments: 5 based on Bayesian Belief Networks (BBNs), 5 based on Dempster–Shafer or similar forms of evidential reasoning, and 2 using other methods. By perturbing the original authors' own examples, they showed that all the proposed methods can deliver implausible results.

An alternative approach is to revert to the original idea that the overall case should be sound in some suitable sense, and the probabilistic assessment is a measure of our confidence in that soundness. So now we need a suitable interpretation for the soundness of an assurance case. The intent is that a sound case should lead us, collectively, to believe its claim, and that claim should be true. The means by which the case induces belief is by providing *justification*, so it looks as if soundness should involve these three notions: belief, justification, and truth. As it happens, epistemology, the branch of philosophy concerned with knowledge, has traditionally (since Plato) combined these three terms to interpret *knowledge* as Justified True Belief

(JTB), so we may be able to draw on epistemology for a suitable characterization of a sound assurance case. This idea is developed and explored in the following four sections; we then return, in Sect. 7, to consider probabilistic assessment of confidence in the resulting process.

3   Epistemology and the Indefeasibility Criterion

Few philosophers today accept the basic version of JTB due to what are called "Gettier cases"; these are named after Edmund Gettier who described two such cases in 1963 [9]. Gettier's is the most widely cited modern work in epistemology with over 3,000 citations, many of which introduce new or variant cases. However, these all follow the same pattern, which had previously been exemplified by the "stopped clock case" introduced by Bertrand Russell in 1912 [10, p. 170]:

> Alice sees a clock that reads two o'clock, and believes that the time is two o'clock. It is in fact two o'clock. However, unknown to Alice, the clock she is looking at stopped exactly twelve hours ago.

The general pattern in these cases is "bad luck" followed by "good luck"; in the stopped clock case, Alice believes that it is two o'clock and her belief is justified because she has looked at a clock. But the clock is stopped ("bad luck") so her belief could well be false; however, the clock stopped *exactly* twelve hours ago ("good luck") so her belief is in fact true. Thus, Alice has a belief that is justified and true— but the case does not seem to match our intuitive concept of knowledge, so there must be something lacking in the JTB criterion.

Those interested in assurance will likely diagnose the problem as weakness in Alice's justification: if this were an assurance case it would be criticized for not considering the possibility that the clock is wrong or faulty. Many epistemologists take the same view and seek to retain JTB as the definition of knowledge by tightening the notion of "justification." For example, Russell's student Ramsey proposed that the justification should employ a "reliable process" [11], but this just moves the problem on to the definition of reliable process. A more widely accepted adjustment of this kind is the *indefeasibility* criterion [12–14]. A justified belief is indefeasible if it has no defeaters, where a *defeater* is a claim which, if we were to believe it, would render our original belief unjustified. (Thus, a defeater to an argument is like a hazard to a system.)

There are difficulties even here, however. A standard example is the case of Tom Grabit [12]:

> We see someone who looks just like Tom Grabit stealing a book from the library, and on this basis believe that he stole a book. Unbeknownst to us, Tom's mother claims that he is away on a trip and has an identical twin who is in the library. But also unbeknownst to us, she has dementia: Tom is not away, has no brother, and did steal a book.

The problem is that the claim by Tom's mother is a defeater to the justification (we saw it with our own eyes) for our belief that Tom stole a book. But this defeater is itself defeated (because she has dementia). So the indefeasibility criterion needs to be amended so that there are no *undefeated* defeaters to our original belief, and this seems to invite an infinite regress. Some current work in epistemology attempts to repair, refute, or explore this and similar difficulties [15, 16], but at this point I prefer to part company with epistemology.

Epistemology seeks to understand knowledge, and one approach is to employ some form of justified true belief. But truth is known only to the omniscient; as humans, the best we can aspire to is "well-justified" belief. Much of the inventiveness in Gettier's examples is in setting up a poorly justified belief (which is defeated by the "bad luck" event) that is nonetheless true (due to the second, "good luck," event). For assurance, we are not interested in poorly justified beliefs that turn out to be true, and many of the fine distinctions made by epistemologists are irrelevant to us. We are interested in well-justified beliefs (since that is our best approach to truth) and what we can take from epistemology is indefeasibility as a compelling criterion for adequately justified belief.[3]

Observe that there are two reasons why an assurance case might be flawed: one is that the evidence is *too weak* to support the claim (to the extent we require) and this is managed by our treatment of the *weight of evidence*, as will be discussed in Sect. 4.1; the other is that there is something logically *wrong* or *missing* in the case (e.g., we overlooked some defeater), and these are eliminated by the notion of indefeasible justification.

Hence, the combination of justification and indefeasibility is an appropriate criterion for soundness in assurance cases. To be explicit, I will say that an assurance case is *justified* when it is achieved by means of a valid argument (and I will explain validity in Sect. 4), and I will say that an assurance case is justified *indefeasibly* when there is no (or, more realistically, we cannot imagine any) new information that would cause us to retract our belief in the case (i.e., no defeaters). A *sound* case is one that is justified indefeasibly and whose weight of evidence crosses some threshold for credibility.

In addition to contributing to the definition of what it means for a case to be sound, another attractive attribute of indefeasible justification is that it suggests how reviewers can challenge an assurance case: search for defeaters (flaws in the valid argument providing justification are eliminated by checking its logic, which can be automated). I discuss this in more detail in Sect. 6.

---

[3] When I said "truth is known only to the omniscient" I was implicitly employing the *correspondence* criterion for truth, which is the (commonsense) idea that truth is that which accords with reality. There are other criteria for truth, among which Peirce's *limit* concept is particularly interesting: "truth is that concordance of a …statement with the ideal limit towards which endless investigation would tend to bring …belief" [17, Vol 5, para 565]. Others paraphrase it as that which is "indefeasible— that which would not be defeated by inquiry and deliberation, no matter how far and how fruitfully we were to investigate the matter in question" [18]. Russell criticized Peirce's limit concept on the grounds that it mixes truth with epistemology, but I think it is interesting for precisely this reason: independent inquiries, performed 50 years apart, converge on indefeasibility as the fundamental basis for justification, knowledge, and truth.

There are two immediate objections to the indefeasibility criterion. The first is that to establish indefeasibility we must consider all potential defeaters, and that could be costly as we might spend a lot of resources checking potential defeaters that are subsequently discarded (either because they are shown not to defeat the argument or because they are themselves defeated). However, I believe that if a case is truly indefeasible, then potential defeaters can either be quickly discarded (because they are not defeaters, for reasons that were already considered and recorded in justifying the original case), or themselves quickly defeated (for similar reasons). The second objection is that indefeasibility is unrealistic: how can we know that we have thought of all the "unknown unknowns"? I address this objection in Sect. 5, but note here that the demanding character of indefeasibility is precisely what makes it valuable: it raises the bar and requires us to make the case that we have, indeed, thought of everything.

A variant on both these objections is the concern that indefeasibility can provoke overreaction that leads to prolix arguments, full of material included "just in case" or in anticipation of implausible defeaters. A related concern is that indefeasibility gives reviewers license to raise numerous imagined defeaters. The first of these must be excluded by good engineering management: proposed defeaters, or proposed counterevidence for acknowledged defeaters, must first be scrutinized for relevance, effectiveness, and parsimony. For the second, note that rather than inviting "nuisance" defeaters during development or review, indefeasibility is a tool for their exclusion. An indefeasible case anticipates, refutes, and records all credible objections that might be raised by its reviewers. So as a case approaches completion and we become more confident that all defeaters have been recognized, so it becomes easier to discard proffered "nuisance" defeaters—because either they are not new or not defeaters, for reasons that have already been considered, or because they can themselves be defeated (for similar reasons).

4   Interpretation and Application of Indefeasibility

An assurance case justifies its claim by means of a *structured argument*, which is a hierarchical collection of individual *argument steps*, each of which justifies a *local claim* on the basis of *evidence* and/or lower level local *subclaims*. A trivial example is shown on the left in Fig. 1, where a top claim C is justified by an argument step AS_1 on the basis of evidence E_3 and subclaim SC_1, which itself is justified by argument step AS_2 on the basis of evidence E_1 and E_2.

Assurance cases often are portrayed graphically, as in the figure, and two such graphical notations are in common use: Claims-Argument-Evidence, or CAE [19], and Goal Structuring Notation, or GSN [20] (the notation in Fig. 1 is generic, although its element shapes are those of GSN). In a real assurance case, the boxes in the figure will contain, or reference, descriptions of the artifacts concerned: for evidence (circles) this may be substantial, including results of tests, formal verifications, etc.; for claims and subclaims (rectangles) it will be a careful (natural language or formal)

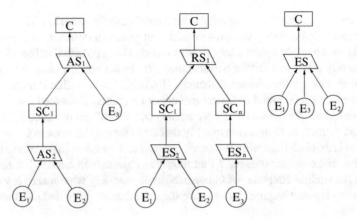

Fig. 1 A structured argument in free (left) and simple form (center) and refactored (right)

statement of the property claimed; and for argument steps (parallelograms) it will be a detailed justification or "warrant" why the cited subclaims and evidence are sufficient to justify the local parent claim.

It is important to note that this interpretation of assurance case arguments applies to CAE, for example, but that GSN, although it appears similar, uses a very different interpretation. What I call argument steps (pictured as parallelograms) are called "strategies" in GSN and their purpose is to describe how the argument is being made (e.g., as an enumeration over components or over hazards), rather than to state an inference from subclaims to claim. In fact, GSN strategies are often omitted and sets of "subgoals" (i.e., subclaims) are connected directly to a "goal" (i.e., claim), and the implicit argument is taken to be some obvious decomposition. I do not attempt to provide an interpretation for GSN strategies. In the interpretation described here, and in CAE "blocks" [21], an argument step that employs a decomposition must provide a narrative justification (i.e., warrant) and possibly some supporting evidence for the decomposition employed (e.g., why it is necessary and sufficient to enumerate over just *these* hazards, or why the claim distributes over the components).

As a concrete example of our interpretation, let us suppose that the left side of Fig. 1 is a (trivialized) software assurance case, where the claim C concerns software correctness. Evidence E_1 might then be test results, and E_2 a description of how the tests were selected and the adequacy of their coverage, so that SC_1 is a subclaim that the software is adequately tested and argument step AS_2 provides a warrant or justification for this. In addition, we need to be sure that the deployed software is the same as that tested, so E_3 might be version management data to confirm this and argument step AS_1 provides a warrant that the claim of software correctness follows if the software is adequately tested, and the tested software is the deployed software. Of course, a real assurance case will concern more than testing and even testing will require additional items of supporting evidence (e.g., the trustworthiness of the test oracle), so real assurance cases are large. On the other hand, evidence must support a specific claim, and claims must contribute to an explicit argument, so there is hope that assurance cases can be more focused and therefore more succinct than current processes driven by guidelines such as DO-178C that require large quantities of evidence with no explicit rationale.

Observe that the argument step AS_1 on the left of Fig. 1 uses both evidence E_3 and a subclaim SC_1. Later, in Sect. 5, I will sketch how to interpret such "mixed" argument steps, but it is easier to understand the basic approach in their absence. By introducing additional subclaims where necessary, it is straightforward to convert arguments into *simple form* where each argument step is supported either by subclaims (boxes) or by evidence (circles), but not by a combination of the two. The mixed or free form argument on the left of Fig. 1 is converted to simple form in the center by introducing a new subclaim SC_n and a new argument step ES_n above E_3.

The benefit of simple form is that argument steps are now of two kinds: those supported by subclaims are called *reasoning steps* (in the example, argument step AS_1 is relabeled as reasoning step RS_1), while those supported by evidence are called *evidential steps* (in the example, these are the relabeled step ES_2 and the new step ES_n) and the key to our approach is that the two kinds of argument step are interpreted differently.

Specifically, evidential steps are interpreted "epistemically," while reasoning steps are interpreted "logically." The idea is that evidential steps whose "weight of evidence" (as described below) crosses some threshold are treated as premises in a conventional logical argument in which the reasoning steps are treated as axioms. This is a systematic version of "Natural Language Deductivism" (NLD) [22], which interprets informal arguments as attempts to create deductively valid arguments. NLD differs from deductive proof in formal mathematics and logic in that its premises are "reasonable or plausible" rather than certain, and hence its conclusions are likewise reasonable or plausible rather than certain [23, Sect. 4.2]. Our requirement that the weight of each evidential step must cross some threshold systematizes what it means for the premises to be reasonable or plausible or, as we often say, credible. (Hence, there is no conceptual problem with evidence based on expert opinion, or incomplete testing, provided these are buttressed by warrants, and possibly additional evidence, for their credibility.)

Our treatment of reasoning steps shares with NLD the requirement that these should be deductively valid (i.e., the subclaims must imply or entail the parent claim); this differs from other interpretations of informal argumentation, which adopt criteria that are weaker (e.g., the subclaims need only "strongly suggest" the parent claim) [24], or different (e.g., the Toulmin style of argument) [25]. Weaker (or different) criteria may be appropriate in other argumentation contexts: indeed, the very term "natural language deductivism" was introduced by Govier [26] as a pejorative to stress that this style of argument does not adequately represent "informal argument." However, our focus is not informal arguments in general, but the structured arguments of assurance cases, where deductive validity is a natural counterpart to the requirement for indefeasibility, and so we can adopt the label NLD with pride. We consider the case of those who assert the contrary in Sect. 4.2.

Because our treatment is close to that of formal logic, we adopt its terminology and say that an argument is *valid* if its reasoning steps are logically so (i.e., true in all interpretations) and that it is *sound* if, in addition, its evidential steps all cross their thresholds for credibility.[4] Thus, our requirement for a sound assurance case is that its argument is sound in the sense just described (which we also refer to as a *justified* argument), and indefeasible.

We now consider the two kinds of argument steps in more detail.

4.1 Evidential Steps

My recommended approach for evidential steps is described in a related paper [27]; here, I provide a summary and connect it to the indefeasibility criterion.

When we have an evidential step with some collection of evidence E, our task is to decide if this is sufficient to accept its local claim C as a premise. We cannot expect E to prove C because the relation between evidence and claims is not one of logic but of epistemology (i.e., it concerns knowledge and belief). Thus, when an evidential step uses two or more items of evidence to support a subclaim (as, for example, at the lower left of the arguments in Fig. 1), the interpretation is not that the conjunction of the evidence logically supports the subclaim, but that each supports it to some degree and together they support it to a greater degree. The reason we have several items of evidence supporting a single claim is that there are rather few claims that are directly observable. Claims like "correctness" can only be inferred from indirect and partial observations, such as testing and reviews. Because these observations provide indirect and incomplete evidence, we combine several of them, in the belief that, together, their different views provide an accurate evaluation of that which cannot be observed directly. Furthermore, an observation may provide valid evidence only in the presence of other evidence: for example, testing is credible only if we have a trustworthy way of assessing test results (i.e., an oracle), so an evidential step concerning testing must also include evidence for the quality of the oracle employed.

Thus, as previously noted, the assessment of evidential steps is not a problem in logic (i.e., we are not *deducing* the claim from the evidence) but in epistemology: we need to assess the extent to which the evidence allows us to *believe* or *know* the truth of the subclaim. Subjective probabilities provide a basis for assessing and reporting confidence in the various beliefs involved and we need to combine these in some way to yield a measure for the "weight" of the totality of evidence E in support of claim C. This topic has been studied in the field of Bayesian confirmation theory [28] where suitable *confirmation measures* have been proposed. The crucial idea is that E should not only support C but should discriminate between C and other claims, and the negation $\neg C$ in particular. This suggests that suitable measures will concern the

---

[4]It is because these usages are standard in logic that we prefer *sound* to *valid* in [1].

difference or ratio of the conditional probabilities $P(E \mid C)$ and $P(E \mid \neg C)$.[5] There are several such measures but among the most recommended is that of Kemeny and Oppenheim [29]

$$\frac{P(E \mid C) - P(E \mid \neg C)}{P(E \mid C) + P(E \mid \neg C)};$$

this measure is positive for strong evidence, near zero for weak evidence, and negative for counterevidence.

When an evidential step employs multiple items of evidence $E_1, \ldots, E_i$, which may not be independent of one another, we need to estimate conditional probabilities for the individual items of evidence and combine them to calculate the overall quantities $P(E_1, \ldots, E_i \mid C)$ and $P(E_1, \ldots, E_i \mid \neg C)$ used in the chosen confirmation measure; Bayesian Belief Nets (BBNs) and their tools provide ways to do this ([27] gives an example).

This probabilistic model, supported by suitable BBN tools, can be used to calculate a confirmation measure that represents the weight of evidence in support of an evidential claim, and a suitable threshold on that weight (which may differ from one claim to another) can be used to decide whether to accept the claim as a premise in the reasoning steps of the argument. I concede that it is difficult to assign credible probabilities to the estimations involved, so in practice the determination that evidence is sufficient to justify a claim will generally be made by (skilled) human judgment, unassisted by explicit probabilistic calculations. However, I believe that judgment can be improved and honed by undertaking numerical examples and "what if" experiments using the probabilistic model described here. And I suggest that assurance templates that may be widely applied should be subjected to quantitative examination of this kind. The example in [27] provides an elementary prototype for this kind of examination.

The probabilistic model helps us understand how the various items of evidence in an evidential step combine to lend weight to belief in its claim. Applying the model to a specific evidential step, whether this is done formally with BBNs or informally by human judgment, involves determination that the collection of evidence is "valid" (e.g., does not contain contradictory items) and credible (i.e., its weight crosses our threshold for acceptance). The indefeasibility criterion comes into play when we ask whether the evidence supplied is also "complete." Specifically, indefeasibility requires us to consider whether any *defeaters* might exist for the evidence supplied. For example, testing evidence is defeated if it is not for exactly the same software as that under consideration, and formal verification evidence is defeated if its theorem prover might be unsound.

It might seem that since testing merely samples a space, it must always be incomplete and therefore vulnerable to defeat. This is true, but I maintain that this kind of "graduated" defeat is different in kind and significance to true "noetic" defeat.

---

[5]It might seem that we should be considering $P(C \mid E)$ and its variants rather than $P(E \mid C)$; these are related by Bayes' rule but it is easier to estimate the likelihood of concrete observations, given a claim about the world, than vice versa.

Almost all evidence is imperfect and partial; that is why evidential steps are evaluated epistemically and why we use probabilities (either formally or intuitively) to record our confidence. Testing is no different than other forms of evidence in this regard. Furthermore, we can choose how partial is our testing: depending on the claim, we can target higher levels of "coverage" for unit tests, or higher levels of statistical validity for random system tests. Some other kinds of evidence share this "graduated" character: for example, we can choose how much effort to devote to human reviews. Thus, the potential for defeat in graduated forms of evidence is acknowledged and managed. It is managed through the "intensity" of the evidence (e.g., effort applied, as indicated by hours of human review, or coverage measures for testing) and probabilistic assessment of its resulting "weight." If that weight is judged insufficient, then evidence that is vulnerable to graduated defeat might be buttressed by additional evidence that is strong on the graduated axis, but possibly weaker on others. Thus testing, which considers interesting properties but for only a limited set of executions, could, for suitable claims, be buttressed by static analysis, which considers *all* executions, but only for limited properties.

"Noetic" defeat is quite different from graduated defeat: it signifies something is wrong or missing and undermines the whole basis for given evidence. For example, if our test oracle (the means by which we decide whether or not tests are successful) could be faulty, or if the tested components might not be the same as those in the actual system, then our tests have no evidential value.

The indefeasibility criterion requires us to eliminate noetic defeaters and to manage graduated ones. Consideration of potential noetic defeaters may lead us to develop additional evidence or to restrict the claim. According to the dependencies involved, additional evidence can be combined in the same evidential step as the original evidence or it can be used in dedicated evidential steps to support separate subclaims that are combined in higher level reasoning steps. For example, in the center of Fig. 1, evidence E_3 might concern version management (to counter the noetic defeater that the software tested is not the same as that deployed), and it supports a separate claim that is combined with the testing subclaim higher up in the argument. On the other hand, if this were evidence for quality of the oracle (the means by which test results are judged) it would be better added directly to the evidential step ES_2 since it is not independent of the other evidence in that step, leading to the refactored argument on the right of Fig. 1.

We now turn from evidential steps to reasoning steps.

4.2   Reasoning Steps

Evidential steps are the bridge between epistemology and logic: they establish that the evidence is sufficient, in its context, to treat their subclaims as premises in a logical interpretation of the reasoning steps. That logical interpretation is a "deductive" one, meaning that the conjunction of subclaims in a reasoning step must imply or entail its claim. This interpretation is not the usual one: most other treatments of assurance case arguments require only that the collection of subclaims should "strongly suggest" the claim, a style of reasoning generally called "inductive" (this is a somewhat unfortunate choice as the same term is used with several other meanings in mathematics and logic). The deductive interpretation is a consequence of our requirement for indefeasibility: if a reasoning step is merely inductive, we are admitting a "gap" in our reasoning that can be filled by a defeater.

Some authors assert that assurance case arguments cannot be deductive due to complexity and uncertainty [30, 31]. I emphatically reject this assertion: the whole point of an assurance case is to manage complexity and uncertainty. In the interpretation advocated here, all uncertainty is confined to the evaluation of evidential steps, where (formal or informal) probabilistic reasoning may be used to represent and estimate uncertainty in a scientific manner. In the inductive interpretation, there is no distinction between evidential and reasoning steps so uncertainty can lie anywhere, and there is no requirement for indefeasibility so the argument can be incomplete as well as unsound.

Nonetheless, the requirement for indefeasibility, and hence for deductive reasoning steps, strikes some as an unrealizable ideal—a counsel of perfection—so in the following section I consider its feasibility and practicality.

5 Feasibility of Indefeasibility

One objection to the indefeasibility criterion for assurance cases is that it sets too high a bar and is infeasible and unrealistic in practice. How can we ever be sure, an objector might ask, that we have thought of all the "unknown unknowns" and truly dealt with all possible defeaters? My response is that there are systematic ways to develop deductive reasoning steps and techniques that shift the doubt into evidential steps where it can be managed appropriately.

Many reasoning steps represent a decomposition in some dimension and assert that if we establish some claim for each component of the decomposition then we can conclude a related claim for the whole. For example, we may have a system X that is composed of subsystems $X1, X2, \ldots, Xn$ and we argue that X satisfies claim C, which we denote $C(X)$, by showing that each of its subsystems also satisfies C: that is, we use subclaims $C(X1), C(X2), \ldots, C(Xn)$. We might use this reasoning step to claim that a software system will generate no runtime exceptions by showing it to be true for each of its software components. However, this type of argument is not always deductively valid—for example, we cannot argue that an airplane is safe by arguing that its wheels are safe, its rudder is safe, ...and its wings are safe. Deductive validity is contingent on the property C, the nature of the system X, and the way in which the subsystems $X1, X2, \ldots, Xn$ are composed to form X. Furthermore, claim $C(X)$ may not follow simply from the same claim applied to the subsystems, but from different subclaims applied to each: $C1(X1), C2(X2), \ldots, Cn(Xn)$. For example, a system may satisfy a timing constraint of 10ms if its first subsystem satisfies a constraint of 3ms, its second satisfies 4ms, and its third and last satisfies

2ms (together with some assumptions about the timing properties of the mechanism that binds these subsystems together).

I assert that we can be confident in the deductive character of systematically constructed reasoning steps of this kind by explicitly stating suitable assumptions or side conditions (which are simply additional subclaims of the step) to ensure that the conjunction of component subclaims truly implies the claim. In cases where the subclaims and claim concern the same property C, this generally follows if C distributes over the components and the mechanism of decomposition, and this would be an assumption of the template for this kind of reasoning step. In more complex cases, formal modeling can be used to establish deductive validity of the decomposition under its assumptions. Bloomfield and Netkachova [21] provide several examples of templates for reasoning steps of this kind, which they call "decomposition blocks."

Deductiveness in these steps derives from the fact that we have a definitive enumeration of the components to the decomposition and have established suitable assumptions. A different kind of decomposition is one over hazards or threats. Here, we do not have a definitive enumeration of the components to the decomposition: it is possible that a hazard might be overlooked. In cases such as this, we transform concerns about deductiveness of the reasoning step into assessment of evidence for the decomposition performed. For example, we may have a general principle or template that a system is safe if all its hazards are eliminated or adequately mitigated. Then we perform hazard analysis to identify the hazards—and that means *all* the hazards—and use a reasoning step that instantiates the general principle as a decomposition over the specific hazards that were identified and attach the evidence for hazard analysis as a side condition. Thus our doubts about deductiveness of the reasoning step that enumerates over hazards are transformed into assessment of the credibility of the evidence for the completeness of hazard analysis (e.g., the method employed, the diligence of its performance, historical effectiveness, and so on).

This is not a trick; when reasoning steps are allowed to be inductive, there is no requirement nor criterion to justify how "close" to deductive (i.e., indefeasible) the steps really are. Under the indefeasibility criterion, we need to justify the deductiveness of each reasoning step, either by reference to physical or logical facts (e.g., decomposition over enumerable components or properties) or to properly assessed evidence, such as hazard analysis, and this is accomplished by the method described.

Both kinds of decomposition discussed above employ assumptions or side conditions (or as will be discussed below, "provisos") to ensure the decomposition is indefeasible. Assumptions (as we will call them here) are logically no different than other subclaims in an argument step. That is, an argument step

$$p_1 \text{ AND } p_2 \text{ AND } \cdots \text{ AND } p_n \text{ IMPLIES } c, \text{ ASSUMING } a$$

is equivalent to

$$a \text{ AND } p_1 \text{ AND } p_2 \text{ AND } \cdots \text{ AND } p_n \text{ IMPLIES } c. \qquad (2)$$

If the original is an evidential step (i.e., $p_1, p_2, \ldots p_n$ are evidence) and a is a subclaim, then (2) is a mixed argument step involving both evidence and subclaims. In Fig. 1 of Sect. 4, we explained how such arguments could be converted to simple form. By that method we might obtain

$$p_1 \text{ AND } p_2 \text{ AND } \cdots \text{ AND } p_n \text{ IMPLIES } c_1 \tag{3}$$

$$a \text{ AND } c_1 \text{ IMPLIES } c \tag{4}$$

and an apparent problem is that the required assumption has been lost from (3). However, this is not a problem at all. The structure of an assurance case argument (as we have defined it) is such that every subclaim must be true. Hence, it is sound to interpret (3) under the assumption a even though it is established elsewhere in the tree of subclaims. In the same way, evidence E_3 in the left or center of Fig. 1 can be interpreted under the assumption of subclaim SC_1. This treatment can lead to circularity, and checks to detect it could be expensive. A sound and practical restriction is to stipulate that each subclaim or item of evidence is interpreted on the supposition that subclaims appearing earlier (i.e., to its left in a graphical presentation) are true. Thus, mixed argument steps like (2) are treated as reasoning steps subject to the evidentially supported assumptions represented by a and this interpretation can be applied either directly or via the conversion to simple form.

Beyond the objection, just dismissed, that the indefeasibility criterion is unrealistic or infeasible in practice is the objection that it is the *wrong* criterion—because science itself does not support deductive theories.

This contention derives from a controversial topic in the philosophy of science concerning "provisos" (sometimes spelled "provisoes") or *ceteris paribus* clauses (a Latin phrase usually translated as "other things being equal") in statements of scientific laws. For example, we might formulate the law of thermal expansion as follows: "the change in length of a metal bar is directly proportional to the change in temperature." But this is true only if the bar is not partially encased in some unyielding material, and only if no one is hammering the bar flat at one end, and…. This list of provisos is indefinite, so the simple statement of the law (or even a statement with some finite set of provisos) can only be inductively true. Hempel [32] asserts there is a real issue here concerning the way we understand scientific theories and, importantly, the way we attempt to confirm or refute them. Others disagree: in an otherwise sympathetic account of Hempel's work in this area, his student Suppe describes "where Hempel went wrong" [33, pp. 203, 204], and Earman and colleagues outright reject it [34].

Rendered in terms of assurance cases, the issue is the following. During development of an assurance case argument, we may employ a reasoning step asserting that its claim follows from some conjunction of subclaims. The assertion may not be true in general, so we restrict it with additional subclaims representing necessary assumptions (i.e., provisos) that are true (as other parts of the argument must show) in the context of this particular system. The "proviso problem" is then: how do we know that we have not overlooked some necessary assumption? I assert that this is just a variant on the problem exemplified by hazard enumeration that was discussed earlier, and is solved in the same way: we provide explicit claims and suitable evidence that the selected assumptions are sufficient. Unlike inductive cases, where assumptions or provisos may be swept under the rug, in deductive cases we must identify them explicitly and provide evidentially supported justification for their correctness and completeness.

Some philosophers might say this is hubris, for we cannot be sure that we do identify all necessary assumptions or provisos. This is, of course, true in the abstract but, just as we prefer well-justified belief to the unattainable ideal of true knowledge, so we prefer well-justified assumptions to the limp veracity of inductive arguments. With an inductive reasoning step, we are saying "this claim holds under these provisos, but there may be others," whereas for a deductive step we are saying "this claim holds under these assumptions, and this is where we make our stand." This alerts our reviewers and raises the stakes on our justification. The task of reviewers is the topic of the following section.

6 Challenges and Reviews

Although reasoning steps must ultimately be deductive for the indefeasible interpretation, I recommend that we approach this via the methods and tools of the inductive interpretation. The reason for this is that assurance cases are developed incrementally: at the beginning, we might miss some possible defeaters and will not be sure that our reasoning steps are deductive. As our grasp of the problem deepens, we may add and revise subclaims and argument steps and only at the end will we be confident that each reasoning step is deductive and the overall argument is indefeasible. Yet even in the intermediate stages, we will want to have some (mechanically supported) way to evaluate attributes of the case (e.g., to check that every subclaim is eventually justified), and an inductive interpretation can provide this, particularly if augmented to allow explicit mention of defeaters.

Furthermore, even when we are satisfied that the case is deductively sound, we need to support review by others. The main objection to assurance cases is that they are prone to "confirmation bias" [35]: this is the human tendency to seek information that will confirm a hypothesis, rather than refute it. The most effective counterbalance to this and other fallibilities of human judgment is to subject assurance cases to vigorous examination by multiple reviewers with different points of view. Such a "dialectical" process of review can be organized as a search for potential defeaters. That is, a reviewer asks "what if this happens," or "what if that is not true."

The general idea of a defeater to a proposition is that it is a claim which, if we were to believe it, would render our belief in the original proposition unjustified. Within argumentation, this general idea is refined into specific kinds of defeaters. Pollock [36, p. 40] defines a *rebutting defeater* as one that (in our terminology) contradicts the claim to an argument step (i.e., asserts it is false), while an *undercutting defeater* merely doubts it (i.e., doubts that the claim really does follow from the proffered subclaims or evidence); others subsequently defined *undermining defeaters* as those that doubt some of the evidence or subclaims used in an argument step. This taxonomy of defeaters can be used to guide a systematic critical examination of an assurance case argument.

For an elementary example, we might justify the claim "Socrates is mortal" by a reasoning step derived from "all men are mortal" and an evidential step "Socrates is a man." A reviewer might propose a rebutting defeater to the reasoning step by saying "I have a CD at home called 'The Immortal James Brown,'[6] so not all men are mortal." The response to such challenges may be to adjust the case, or it may be to dispute the challenge (i.e., to defeat the defeater). Here, a proponent of the original argument might rebut the defeater by observing that James Brown is dead (citing Google) and therefore indubitably mortal. An undercutting defeater for the same reasoning step might assert that the claim cannot be accepted without evidence, and an adjustment might be to interpret "mortal" as "lives no more than 200 years" and to supply historical evidence of human lifespan. An undermining defeater for the evidential step might challenge the assumption that Socrates was a historical figure (i.e., a "real" man).

I think the record of such challenges and responses (and the narrative justification that accompanies them) should be preserved as part of the assurance case to assist further revisions and subsequent reviews. The fields of defeasible and dialectical reasoning provide techniques for recording and evaluating such "disputed" arguments. For example, *Carneades* [37] is a system that supports dialectical reasoning, allowing a subargument to be *pro* or *con* its conclusion: a claim is "in" if it is not the target of a *con* that is itself "in" unless …(the details are unimportant here). Weights can be attached to evidence, and a *proof standard* is calculated by "adding up" the *pro*s and *con*s supporting the conclusion and their attendant weights. For assurance cases, we ultimately want the proof standard equivalent to a deductive argument, which means that no *con* may be "in" (i.e., every defeater must be defeated). Takai and Kido [38] build on these ideas to extend the *Astah GSN* assurance case toolset with support for dialectical reasoning [39].

7 Probabilistic Interpretation

In Sect. 2, we explained how confidence in an assurance case, plus failure-free experience, can provide assurance for extremely low rates of critical failure, and hence for certification. Sections 3–6 have described our approach to interpretation and evaluation of an assurance case, so we now need to put the two pieces together. In particular, we would like to use the determination that a case is sound (i.e., its argument is valid, all its evidential steps cross the threshold for credibility, it is indefeasible, and all

---

[6]The CD in question is actually called "Immortal R&B Masters: James Brown.".

these assessments have withstood dialectical challenge) to justify expressions of confidence such as $p_{nf} \geq 0.9$ in the absence of faults. This is a subjective probability, but one way to give it a frequentist interpretation is to suppose that if 10 systems were successfully evaluated in the same way, at most one of them would ever suffer a critical failure in operation.

This is obviously a demanding requirement and not one amenable to definitive demonstration. One possibility is to justify $p_{nf} \geq 0.9$ for this assurance case by a separate assurance case that is largely based on evidential steps that cite historical experience with the same or similar methods (for example, no civil aircraft has ever suffered a catastrophic failure condition attributed to software assured to DO-178B/C Level A[7]). For this reason among others, I suggest that assurance for really critical systems should build on successful prior experience and that templates for their assurance cases should be derived from existing guidelines such as DO-178C [2] rather than novel "bespoke" arguments.

Different systems pose different risks, and not all need assurance to the extreme level required for critical aircraft software. Indeed, aircraft software itself is "graduated" according to risk. So a sharpened way to pose our question is to ask how a given assurance case template can itself be graduated to deliver reduced assurance at correspondingly reduced cost or, dually, how our overall confidence in the case changes as the case is weakened. Eliminating or weakening subclaims within a given argument immediately renders it defeasible, so that is not a viable method of graduation. What remains is lowering the threshold on evidential steps, which may allow less costly evidence (e.g., fewer tests), or the elimination or replacement of some evidence (e.g., replace static analysis by manual review). When evidence is removed or changed, some defeaters may be eliminated too, and that can allow the removal of subclaims and their supporting evidence (e.g., if we eliminate static analysis we no longer need claims or evidence about its soundness).

It is difficult to relate weakened evidence to explicit reductions in the assessment of p_{nf}. Again, we could look to existing guidelines such as DO-178C, where 71 "objectives" (essentially items of evidence) are required for Level A software, 69 for Level B, 62 for Level C, and 26 for Level D. Alternatively, we could attempt to assess confidence in each evidential step (i.e., a numerical value for $P(C \mid E)$) and assess p_{nf} as some function of these (e.g., the minimum over all evidential steps). The experiments by Graydon and Holloway mentioned earlier [7, 8] suggest caution here, but some conservative approaches are sound. For example, it follows from a theorem of probability logic [40] that *doubt* (i.e., 1 minus probabilistic confidence) in the claim of a reasoning step is no worse than the sum of the doubts of its supporting subclaims.

---

[7]This remains true despite the 737Max MCAS crashes; as far as we know, the MCAS software satisfied its requirements; the flaws were in the requirements, whose assurance is the purview of ARP 4754A [3], which Boeing apparently failed to apply with any diligence.

It has to be admitted that quantification of this kind rests on very subjective grounds and that the final determination to accept an assurance case is a purely human judgment. Nonetheless, the model of Sect. 2 and the interpretation suggested here do establish a probabilistic approach to that judgment, although there is clearly opportunity for further research.

8 Conclusion

I have reviewed the indefeasibility criterion from epistemology and argued that it is appropriate for assurance case arguments. I also proposed a systematic version of Natural Language Deductivism (NLD) as the basis for judging soundness of assurance case arguments: the interior or reasoning steps of the argument should be deductively valid, while the leaf or evidential steps are evaluated epistemically using ideas from Bayesian confirmation theory and are treated as premises when their evaluation crosses some threshold of credibility. NLD ensures correctness or soundness of the argument, while indefeasibility ensures completeness. I derived requirements for the evidential and reasoning steps in such arguments and argued that they are feasible and practical, and that postulating defeaters provides a systematic way to challenge arguments during review.

I propose that assurance case templates satisfying these criteria and derived from successful existing assurance guidelines (e.g., DO-178C) can provide a flexible and trustworthy basis for assuring future systems.

The basis for assurance is systematic consideration of every possible contingency, which requires that the space of possibilities is knowable and enumerable. This is true at design time for conventional current systems such as commercial aircraft, where conservative choices may be made to ensure predictability. But more recent systems such as self-driving cars and "increasingly autonomous" (IA) aircraft pose challenges, as do systems that are assembled or integrated from other systems while in operation (e.g., multiple medical devices attached to a single patient). Here, we may have software whose internal structure is opaque (e.g., the result of machine learning), an imperfectly known environment (e.g., a busy freeway where other road users may exhibit unexpected behavior), and interaction with other systems (possibly due to unplanned stigmergy via the plant) whose properties are unknown. These challenge the predictability that is the basis of current assurance methods. I believe this basis can be maintained and the assurance case framework can be preserved by shifting some of the gathering and evaluation of evidence, and assembly of the final argument, to integration or run time [41–43], and that is an exciting topic for future research.

Acknowledgements This work was partially funded by SRI International and builds on previous research that was funded by NASA under a contract to Boeing with a subcontract to SRI International.

I have benefited greatly from extensive discussions on these topics with Robin Bloomfield of Adelard and City University. Patrick Graydon of NASA provided very useful comments on a previous iteration of the paper, as did the editors and reviewers for this volume.

References

1. UK Ministry of Defence: Defence Standard 00-56, Issue 4: Safety Management Requirements for Defence Systems. Part 1: Requirements (2007)
2. Requirements and Technical Concepts for Aviation (RTCA) Washington, DC: DO-178C: Software Considerations in Airborne Systems and Equipment Certification (2011)
3. Society of Automotive Engineers: Aerospace Recommended Practice (ARP) 4754A: Certification Considerations for Highly-Integrated or Complex Aircraft Systems (2010). Also issued as EUROCAE ED-79
4. L. Strigini, A. Povyakalo, Software fault-freeness and reliability predictions, in *SafeComp 2013: Proceedings of the 32nd International Conference on Computer Safety, Reliability, and Security*. Lecture Notes in Computer Science, vol. 8153, Toulouse, France (Springer, 2013), pp. 106–117
5. Federal Aviation Administration: System Design and Analysis (1988). Advisory Circular 25.1309-1A
6. B. Littlewood, J. Rushby, Reasoning about the reliability of diverse two-channel systems in which one channel is "possibly perfect". IEEE Trans. Softw. Eng. **38**, 1178–1194 (2012)
7. P.J. Graydon, C.M. Holloway, An investigation of proposed techniques for quantifying confidence in assurance arguments. Saf. Sci. **92**, 53–65 (2017)
8. P.J. Graydon, C.M. Holloway, An investigation of proposed techniques for quantifying confidence in assurance arguments. Technical Memorandum NASA/TM-2016219195, NASA Langley Research Center, Hampton VA (2016)
9. E.L. Gettier, Is justified true belief knowledge? Analysis **23**, 121–123 (1963)
10. B. Russell, *Human Knowledge: Its Scope and Limits* (George Allen & Unwin, London, England, 1948)
11. F.P. Ramsey, Knowledge, in *Philosophical Papers of F. P. Ramsey*, ed. by D.H. Mellor. (Cambridge University Press, Cambridge, UK, 1990), pp. 110–111 (original manuscript, 1929)
12. K. Lehrer, T. Paxson, Knowledge: undefeated justified true belief. J. Philos. **66**, 225–237 (1969)
13. P.D. Klein, A proposed definition of propositional knowledge. J. Philos. **68**, 471–482 (1971)
14. M. Swain, Epistemic defeasibility. Am. Philos. Q. **11**, 15–25 (1974)
15. J. Turri, Is knowledge justified true belief? Synthese **184**, 247–259 (2012)
16. J.N. Williams, Not knowing you know: a new objection to the defeasibility theory of knowledge. Analysis **75**, 213–217 (2015)
17. C. Hartshorne, P. Weiss, A.W. Burks (eds.), *Collected Papers of Charles Sanders Peirce*, vols. 1–8 (Harvard University Press, Cambridge, MA, 1931–1958)
18. C. Misak, Review of "Democratic Hope: Pragmatism and the Politics of Truth" by Robert B. Westbrook. Trans. Charles S. Peirce Soc. **42**, 279–282 (2006)
19. Adelard LLP London, UK: ASCAD: Adelard Safety Case Development Manual (1998). https://www.adelard.com/resources/ascad.html
20. T. Kelly, Arguing Safety—A Systematic Approach to Safety Case Management. DPhil thesis, Department of Computer Science, University of York, UK (1998)
21. R. Bloomfield, K. Netkachova, Building blocks for assurance cases, in *ASSURE: Second International Workshop on Assurance Cases for Software-Intensive Systems, Naples, Italy, IEEE International Symposium on Software Reliability Engineering Workshops* (2014), pp. 186–191
22. L. Groarke, Deductivism within pragma-dialectics. Argumentation **13**, 1–16 (1999)
23. L. Groarke, Informal logic, in *The Stanford Encyclopedia of Philosophy*, ed. by E.N. Zalta, Spring 2017 edn. (Metaphysics Research Lab, Stanford University, 2017)

24. J.A. Blair, What is informal logic? in *Reflections on Theoretical Issues in Argumentation Theory*. Argumentation Library, vol. 28, ed. by F.H. van Eemeren, B. Garssen (Springer, 2015), pp. 27–42
25. S.E. Toulmin, *The Uses of Argument* (Cambridge University Press, 2003) Updated edition (the original is dated 1958)
26. T. Govier, *Problems in Argument Analysis and Evaluation*. Studies of Argumentation in Pragmatics and Discourse Analysis, vol. 5 (De Gruyter, 1987)
27. J. Rushby, On the interpretation of assurance case arguments, in *New Frontiers in Artificial Intelligence: JSAI-isAI 2015 Workshops, LENLS, JURISIN, AAA, HAT-MASH, TSDAA, ASD-HR, and SKL*, Revised Selected Papers. Lecture Notes in Artificial Intelligence, vol. 10091, Kanagawa, Japan (Springer, 2015), pp. 331–347
28. J. Earman, *Bayes or Bust? A Critical Examination of Bayesian Confirmation Theory* (MIT Press, 1992)
29. K. Tentori, V. Crupi, N. Bonini, D. Osherson, Comparison of confirmation measures. Cognition **103**, 107–119 (2007)
30. V. Cassano, T.S. Maibaum, S. Grigorova, *Towards Making Safety Case Arguments Explicit, Precise, and Well Founded* (This volume)
31. M. Chechik, R. Salay, T. Viger, S. Kokaly, M. Rahimi, Software assurance in an uncertain world, in *International Conference on Fundamental Approaches to Software Engineering (FASE)*. Lecture Notes in Computer Science, vol. 11424, Prague, Czech Republic (Springer, 2019), pp. 3–21
32. C.G. Hempel, Provisoes: a problem concerning the inferential function of scientific theories. Erkenntnis **28**, 147–164 (1988). Also in conference proceedings "The Limits of Deductivism," ed. by A. Grünbaum, W. Salmon (University of California Press, 1988)
33. F. Suppe, Hempel and the problem of provisos, in *Science, Explanation, and Rationality: Aspects of the Philosophy of Carl G. Hempel*, ed. by J.H. Fetzer (Oxford University Press, 2000), pp. 186–213
34. J. Earman, J. Roberts, S. Smith, Ceteris Paribus lost. Erkenntnis **57**, 281–301 (2002)
35. N. Leveson, The use of safety cases in certification and regulation. J. Syst. Saf. **47**, 1–5 (2011)
36. J.L. Pollock, *Cognitive Carpentry: A Blueprint for How to Build a Person* (MIT Press, 1995)
37. T.F. Gordon, H. Prakken, D. Walton, The Carneades model of argument and burden of proof. Artif. Intell. **171**, 875–896 (2007)
38. T. Takai, H. Kido, A supplemental notation of GSN to deal with changes of assurance cases, in *4th International Workshop on Open Systems Dependability (WOSD), Naples, Italy, IEEE International Symposium on Software Reliability Engineering Workshops* (2014), pp. 461–466
39. Astah: (Astah GSN home page). http://astah.net/editions/gsn
40. E.W. Adams, *A Primer of Probability Logic* (Center for the Study of Language and Information (CSLI), Stanford University, 1998)
41. J. Rushby, Trustworthy self-integrating systems, in *12th International Conference on Distributed Computing and Internet Technology, ICDCIT 2016*. Lecture Notes in Computer Science, Bhubaneswar, India, vol. 9581, ed. by N. Bjørner, S. Prasad, L. Parida (Springer, 2016), pp. 19–29
42. J. Rushby, Automated integration of potentially hazardous open systems, in *Sixth Workshop on Open Systems Dependability (WOSD)*, ed. by M. Tokoro, R. Bloomfield, Y. Kinoshita (Keio University, Tokyo, Japan, DEOS Association and IPA, 2017), pp. 10–12
43. J. Rushby, Assurance and assurance cases, in *Dependable Software Systems Engineering* (Marktoberdorf Summer School Lectures, 2016). NATO Science for Peace and Security Series D, vol. 50, ed. by A. Pretschner, D. Peled, T. Hutzelmann (IOS Press, 2017), pp. 207–236

Refinement Based Modelling

An Event-B Development Process for the Distributed BIP Framework

Badr Siala, Jean-Paul Bodeveix, Mamoun Filali, and Mohamed Tahar Bhiri

Abstract We present a refinement-based methodology to design correct by construction distributed systems specified as Event-B models. Our approach makes explicit the transition from formal requirements to a distributed executable model. Starting from an Event-B machine, we propose successive steps in order to split and schedule the computation of complex events and then to map them on subcomponents. The specification of these steps is done through two domain-specific languages. From these specifications, two refinements are generated. Eventually, a distributed code architecture is also generated. The correctness of the process relies on the correctness of the refinements and the translation. We target the distributed BIP framework.

1 Introduction

Usually, the development of distributed systems starts with some requirements and leads to the proposal of a set of distributed components. As a matter of fact, the BIP framework is proposed to support the corresponding deployment which can be executed. Continuing the work outlined in [27], we propose here to make explicit the transition from formal requirements to distributed executable models. It relies on Domain-Specific Languages (DSL)-guided step by step refinements supported by the Event-B method. More precisely, we are concerned with providing tool support to

---

B. Siala · M. T. Bhiri
Université de Sfax, Sfax, Tunisia
e-mail: siala@irit.fr

M. T. Bhiri
e-mail: tahar_bhiri@yahoo.fr

B. Siala · J.-P. Bodeveix · M. Filali (✉)
Université de Toulouse IRIT CNRS UPS, Toulouse, France
e-mail: filali@irit.fr

J.-P. Bodeveix
e-mail: bodeveix@irit.fr

© Springer Nature Singapore Pte Ltd. 2021
Y. Ait-Ameur et al. (eds.), *Implicit and Explicit Semantics Integration
in Proof-Based Developments of Discrete Systems*,
https://doi.org/10.1007/978-981-15-5054-6_13

assist system design using a safe refinement-based process. The considered systems will be seen as a collection of interacting actors. The first levels of the process provide a centralized view of the system behavior. It will be built by taking into account system requirements incrementally, in the form of a series of abstract machines written in Event-B [5, 12]. Then, we propose dedicated, user guided through several DSLs, refinement generators to take into account the distributed nature of the designed system. As a result, we obtain a set of interacting machines of which composition is proven to conform to the abstract levels. The system can then be executed on a distributed platform via a translation to the BIP (Behavior, Interaction, Priority) language [8]. By now, it should be clear that our aim is not to fully automate the distribution process but to assist it. While keeping modest, the difference is similar to that between a model checker where the proof of a judgement is automatic and a theorem proving assistant where the user has to compose basic strategies in order to make his proof. Actually, while a theorem proving assistant helps to construct the proof of a goal, we intend to help in the elaboration of a distributed model through refinement patterns [25]. Moreover, our objective is not to address the development of distributed algorithms as [7] but to help in decomposing a centralized specification over a fixed number of subcomponents during a deployment step.

The semantics of Event-B and BIP are based on labeled transition systems thereby promoting their coupling. Event-B is used for the formal specification and the decomposition of initially centralized reactive systems. BIP is used for the implementation and the deployment of distributed systems specified and verified in Event-B. The skeleton of the BIP code is automatically generated from Event-B.

Sections 2 and 3 present Event-B composition/decomposition techniques and the component-based model BIP. Section 4 presents Event-B patterns modeling BIP connectors and associated problems. Section 5 proposes our development process of distributed systems by coupling Event-B and BIP. Section 6 relates our distributed systems development approach to existing work. We conclude the paper in Sect. 7 and present some perspectives.

2   Event-B

In this section, we introduce the Event-B language and method and take the electronic hotel key system as an illustrative key study. Then we present extensions of Event-B for model composition and decomposition and the limits of the shared event decomposition tooling.

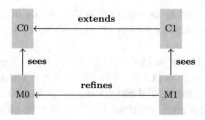

Fig. 1 Event-B development step

2.1 Introduction to Event-B

The Event-B method allows the development of correct by construction systems and software [5]. It supports a formal development process based on a refinement mechanism with mathematical proofs. Static data models are introduced incrementally through a chain of context extensions. Dynamic data updated by events are introduced in machines and subsequently refined. Each machine can access context data through the sees link (Fig. 1).

Contexts define abstract data types through sets, constants, and axioms while machines define symbolic labeled transition systems. The state of a transition system is defined as the value of machine variables. Labeled transitions are defined by events specifying the new value of variables while preserving invariants. Moreover, the **theorem** clause expresses facts that should be satisfied. Proof obligations for well-formedness, invariant preservation, and theorems are automatically generated by the Rodin tool [3]. They can be discharged thanks to automatic proof engines (CVC4, Z3 …) or through human-assisted proofs.

For the most part, Event-B uses standard set theory and its usual set notation. As a matter of fact, in Event-B, arrays and functions are both considered as sets of couples. Some notations are specific to Event-B:

- **pair construction:** pairs are constructed using the maplet operator $\mapsto$. A pair is thus denoted $a \mapsto b$ instead of (a, b). The set of pairs $a \mapsto b$ where $a \in A$ and $b \in B$ is denoted $A \times B$.
- A subset of $A \times B$ is a *relation*. The set of relations from A to B is denoted $A \leftrightarrow B = \mathcal{P}(A \times B)$. A relation $r \in A \leftrightarrow B$ has a domain: **dom**(r) and a codomain: **ran**(r). When a relation r relates an element of **dom**(r) with at-most one element, it is called a function. The set of partial functions from A to B is denoted $A \nrightarrow B$, the set of total functions is denoted $A \rightarrow B$. The image of a set A by a relation r is denoted $r[A]$.
- The composition of two relations $r_1 \in A \leftrightarrow B$ and $r_2 \in B \leftrightarrow C$ is denoted as $r_1; r_2$.

- The **direct product** $r_1 \otimes r_2$ of relations $r_1 \in A \leftrightarrow B_1$ and $r_2 \in A \leftrightarrow B_2$ is the relation containing the pairs $(x \mapsto (y_1 \mapsto y_2))$ where $x \mapsto y_1 \in r_1$ and $x \mapsto y_2 \in r_2$.
- **domain restriction:** $D \lhd r = \{x \mapsto y \mid (x \mapsto y) \in r \wedge x \in D\}$
- **range restriction:** $r \rhd D = \{x \mapsto y \mid (x \mapsto y) \in r \wedge y \in D\}$
- **overwrite:** $f \Leftarrow g = ((\mathbf{dom}(f) \backslash \mathbf{dom}(g)) \lhd f) \cup g$. For instance, such a notation is used to denote a new array obtained by changing the element of an array a at index i: $a \Leftarrow \{i \mapsto e'\}$.

As already said, Event-B machines specify symbolic transitions through events. An event has three optional parts: parameters (**any** p1 ...pn), guards (**where** ...) specifying constraints to be satisfied by parameters and state variables, and actions (**then** ...) specifying state variables updates. Guards are defined in set-based predicate logic. Concurrent updates of distinct variables may be deterministic ($x := e$), non-deterministic ($x :\in E$ or $x :\mid P(x, x')$). In $x :\in E$, x takes any value belonging to the set E. In $x :\mid P(x, x')$, the new value x' of x is specified by the predicate P.

2.2 Case Study

As a running example, we will consider the electronic hotel key system case study [24]. As said in [5], *the purpose of such a system is to guarantee that between your checking and checkout of the hotel, you can enter the room you booked and no-one else can do so.* The behavior of the considered system is not exactly the intended one: the previous owner of a room can enter the room while the next one has not introduced his card in the room card reader. It means that this action ends *checkin* and starts *checkout*. Given this event mapping, the informal specification is fulfilled.

The context (Listing 1) introduces basic data structures: guests, rooms, and cards defined as ordered pairs of keys. State variables (Listing 2) declare the current key of a room (currk), the rooms owned by a guest (owns), the cards issued by the hotel, and cards owned by a guest.

```
context chotel
sets ROOM GUEST KEY
constants CARD
axioms
      @crd CARD = KEY × KEY
end
```

```
machine hotel sees chotel
variables currk owns issued cards
invariants
      @currk_ty currk ∈ ROOM → KEY
      @owns_ty owns ∈ ROOM → ℙ(GUEST)
      @issued_ty issued ∈ ℙ(CARD)
      @cards_ty cards ∈ GUEST → ℙ(CARD)
```

Listing 1 Hotel context

Listing 2 Hotel state variables

The dynamics of the system is described by events, one of which, named register being given in Listing 3.

This is a non-deterministic event, parametrized by the variables g for the incoming guest, r for the room to be chosen, and c for the card to be issued. The `where` part specifies which of these triples are allowed: the room should be free (g1), the card should not have been issued (g2), and the card should open the door (g3). The `then` part specifies how the state space is updated: the current key of the room will be the second key of the card (a1), the card has been issued (a2), which is owned by the guest (a3) who owns the room (a4).

```
event register
any g r c where
    @tg g ∈ GUEST
    @tr r ∈ ROOM
    @tc c ∈ CARD
    @g1 owns(r) = ∅
    @g2 c ∉ issued
    @g3 prj1(c) = currk(r)
then
    @a1 currk(r) := prj2(c)
    @a2 issued := issued∪{c}
    @a3 cards(g) := cards(g)∪{c}
    @a4 owns(r) := {g}
end
```

Listing 3 Hotel register event

2.3 *Event-B Extensions*

Recently, Event-B has been enhanced by reuse techniques such as genericity [28], abstraction [19], composition and decomposition [6, 29]. In this paper, we are mainly concerned by composition, and decomposition. They allow the formal combination of specifications through the refinement mechanism. Two methods of composition/decomposition were identified for Event-B: shared variable [30] and shared event [29]. Shared variable composition/decomposition is suitable for shared-memory parallel systems whereas shared event composition/decomposition is suitable for message-passing distributed systems. In this paper, we limit ourselves to the shared event composition/decomposition approach inspired by CSP where processes synchronize on the same event and may exchange messages. In Event-B, subcomponents (sub-specifications) can synchronize through shared events and exchange data specified by the common value of their parameters.

2.3.1 Shared Event Composition

The shared event composition of Event-B machines is represented by a new construct called **composed machine** [29]. This operation requires the disjointness of the sets of state variables of the machines to be composed. It is defined as a machine merging subcomponents' properties: conjunction of invariants, union of variables, and parallel synchronization of events. The composition of two events which have common parameters p is defined as follows [29]:

```
E1 ≜ any p,x where G(p,x,u) then S(p,x,u) end
E2 ≜ any p,y where H(p,y,v) then T(p,y,v) end
E1 ‖ E2 ≜ any p,x,y where G(p,x,u)∧H(p,y,v) then S(p,x,u) ‖ T(p,y,v) end
```

where x, y, p are sets of parameters from the events E1 and E2 and u and v are the variables of the two subcomponents. Sending a value d can be modeled by using a guard of the form $p = d$ while a receipt will be modeled by an action $u:=p$. The other guards will constrain the sent value either at the sending point or at the receiving point. This design pattern originating from CSP has been proposed by Butler for action systems [11] and in [29] for Event-B.

The **composed machine** is supposed to satisfy the Event-B standard Proof Obligations (POs) related to invariants and refinements. Moreover, during the composition of several subcomponents, it is possible to add a composition invariant relating the states of subcomponents.

Like CSP parallel composition, Event-B shared event composition is monotonic under refinement [29]. Actually, the composition of refined subcomponents is a refinement of the composition of initial subcomponents.

2.3.2 Shared Event Decomposition

Decomposition is a mean to master the complexity (divide and conquer) or to introduce architectural aspects (see Sect. 5). It can be seen as the inverse of composition where an Event-B model is split into several simpler subcomponents. Concretely, decomposition is specified by a set of subcomponent names and a partition of variables, each class being mapped to a subcomponent. An important point is that the composition of subcomponents refines the initial centralized model. However, decomposition fails if a guard or an action refers to variables mapped to different locations. Within the scope of distributed systems, we propose a support to help solving these problems. Decomposition can also fail if the synthesized typing invariant is not strong enough. It could lead to a badly formed expression where some partial functions are applied outside their definition domain. We do not consider this problem here. It is a real challenge to address this problem automatically. Experiments have been done to keep stronger invariants in the projection [26].

2.3.3 Shared Event Decomposition Limits

The shared event decomposition plugin has strong restrictions. First, it does not apply on models where guards or actions access variables mapped on different components as the tool would not know how to split them. Moreover, even if each guard or action refers to only one variable, the resulting components produced by this tool could be unusable. Consider two variables a and b mapped on components C_1 and C_2 and the event ev (Listing 4):

Applying [29] is possible: each of C_1 and C_2 gets a copy of ev with respectively g_1 and g_2 as their unique guard, but this leads to another problem: we get two synchronized events specifying constraints over the parameter p. Their separate refinement could lead to incompatible choices and thus to a deadlock resulting from the assembly. The transformations we propose in Sect. 5 will allow the user to avoid these problems by guiding the refinement process.

```
event  ev
any  p  where
  @g1  a  >  p
  @g2  p  <  b
then
  @a  p1  :=  p
end
```

Listing 4 Event split

2.3.4 Shared Event Composition/Decomposition Tool

The Rodin platform provides an interactive tool [30] as a plugin allowing the shared event composition/decomposition of Event-B specifications. Composition is defined by editing a *composed machine* which designates the subcomponents and defines synchronization events as a product of subcomponent events. Conversely, decomposition is built by naming subcomponents and mapping variables on them. In case of success, the tool generates a machine for each subcomponent and a composed machine. Given that the decomposition of the invariants depends on the scope of the variables, invariants containing variables distributed over several subcomponents are discarded.

3 The BIP Component-Based Model

The BIP language [8] allows to build component-based systems. To achieve this, it offers a means to describe atomic components and composition operators describing composite components. In BIP, an architecture is a hierarchical model consisting of a structured collection of components obtained by composition of atomic components which represent the leaves of the hierarchical model.

3.1 Atomic Components

An atomic BIP component declares data, ports, and a behavior. Data variables (**data**) are typed. Ports (**port**) give access to some variables and constitute the component **interface**. The behavior is defined by a port, a guard, and a variable update function. According to the component-based paradigm, a BIP component is a design-time concept (a type) and a runtime concept (an instance). This is also true for ports. Listings 5 and 6 present, respectively, the port types and an atomic component `ty_Desk` produced by our BIP code generator (see Sect. 5.4).

```
port type ty_empty_port ()
port type ty_register_Desk (INT register_g , INT  register_c)
port type ty_register_Guest (INT register_g , INT register_c )
```

Listing 5 Port types

```
atom type ty_Desk ()
  /* state variables */
  data INT currk ...
  /* temporary variables */
  data INT register_g
  data INT register_c
  /* port instances */
  export port ty_empty_port compute_register_r ()
  export port ty_register_Desk register (register_g ,register_c )
  place P0
  initial to P0 do /* initialize variables */
  /* transitions */
  on compute_register_c from P0 to P0 provided register_g_computed
  on register from P0 to P0 provided register_g_computed do /* action */
end
```

Listing 6 Atomic component ty_Desk

3.2 Coordination Between BIP Components

The component-based model BIP has three layers called Behavior, Interaction, and Priority. The *Behavior* layer describes the behavior of atomic components (see Sect. 3.1) while layers *Interaction* and *Priority* describe the architectural aspects of a component-based system. This separation between behavioral and architectural aspects is an asset in BIP [8]. The synchronization constraints between BIP components are expressed through interactions defined by the **connector** construct whereas scheduling constraints between these interactions are expressed through the *Priority* concept.

BIP connectors. A connector is a *design-time* concept when declared as a type and can be instantiated to a *runtime* concept. A BIP connector is defined by

- a set of ports $\{p_1, ..., p_n\}$ of subcomponents involved in an interaction.
- an optional port p with variables exported by the connector allowing to compose the connectors.
- a set of interactions which are subsets of $\{p_1, ..., p_n\}$. Every interaction can be annotated by a guard, an *upstream transfer functions* (up) and *downstream transfer functions* (down). The guards of the interactions involve variables in the scope of ports and connector variables. Here, we limit ourselves to simple connectors restricted to data transfer (Sect. 4).

For example, Listing 7 defines two connector types.[1] The first one denotes a pure synchronization and the second one a synchronization with data exchange.

---

[1]Produced by our BIP code generator in Sect. 5.4.

```
connector type ty_compute_register_r
    (ty_empty_port Desk,ty_empty_port Guest)
  define Desk Guest
  on Desk Guest down {}
end
connector type ty_register
    (ty_register_Desk Desk,ty_register_Guest Guest)
  define Desk Guest
  on Desk Guest down {
    Guest.register_c=Desk.register_c;Desk.register_g=Guest.register_g;
  }
end
```

Listing 7 Connector types

Composite component. In BIP, a composite component is both present at design-time (as a type) and at runtime (as an instance). It includes the following elements:

- atomic or composite components declared by the keyword **component**;
- connectors which connect the components forming the composite component declared by the keyword **connector**;
- priority rules declared by the keyword **priority**;
- exported ports that define the interface of the composite component.

Listing 8 presents a composite component. It contains two atomic components and a connector for coordinating them.

```
compound type ty_hotel_decomposition ()
  component ty_Desk Desk()
  component ty_Guest Guest()
  connector ty_register register(Desk.register , Guest.register)
  ...
end
```

Listing 8 The Hotel root component type

3.3 BIP Execution and Operational Semantics

The BIP execution engine starts with the computation of executable interactions (Interaction layer). Then, it schedules these interactions, taking into account the priority constraints (Priority layer). Finally, the transitions of the atomic components involved in the interaction are executed (Behavior layer).

Step 1 (Initialization): each component executes its initialization transition in its own thread.

Step 2 (Stable Situation): each thread runs until it reaches a port synchronization.

Step 3 (Executable Interactions): this step aims at computing the set of legal interactions where components are ready to synchronize on interaction ports and satisfy the associated guard.

Step 4 (Selection of an executable interaction): selection takes into account priority constraints: an interaction can be elected if no interaction of greater priority is legal. The selection is non-deterministic.

Step 5 (Execution of the elected interaction): the execution of an interaction comes to running its **down** statement which performs data transfer between port variables of subcomponents.[2]

Step 6 (Evolution of atomic components): the transitions of atomic components concerned by the interaction are executed.

Step 7: jump to Step 2.

3.4 The BIP Tool-Chain

The BIP tool-chain includes translators from other languages to BIP, formal verification tools, and code generators from a BIP model. The BIP language features a static checker called D-Finder [8]. It is a compositional verification tool (invariants, deadlock). Likewise, the BIP language has a runtime verification tool [17]. The code generators take the BIP model and generate single-threaded or multi-threaded code that can be executed and analyzed [21].

4 Distributed Event-B and BIP

In this section, we present Event-B event definition schemes that model the synchronization of events to be executed on subcomponents. They must match the BIP connectors that perform only data transfer. These schemas will be seen as elements of an Event-B0 language targeted by automatic or manual refinement steps. They can be seen as describing a shallow embedding of BIP connectors within Event-B and are supposed be the source of the *shared event decomposition* plugin [29] introduced in Sect. 2.3.1 to build the specification of BIP subcomponents and connectors.

4.1 Decomposable Machine and Synchronization Event

Our aim being the development of systems implemented as a set of communicating distributed components made executable via their BIP transcription, we must be able to represent distribution in Event-B. This goes through the notion of decomposition with synchronization on shared events as introduced by Butler [29] and presented in Sect. 2.3.1. The mapping of variables to subcomponents makes it possible to associate with a decomposable machine a set of BIP machines composed via connectors expressing the semantics of the synchronization events. Each event thus corresponds to a connector connected to the components synchronized via ports giving read/write

---

[2]We have omitted to speak about the **up** transfer function which is used to transfer data upwards in the connector hierarchy.

access to some of their variables. Projected events become synchronized transitions on these same ports. The guards and actions of the Event-B event are checked/executed by the component or by the connector according to the location of the accessed data. The parameters represent either a data transfer or non-determinism which must be eliminated in order to obtain an executable model. After raising the problem of the decomposition of non-deterministic events, we detail the authorized synchronization schemes.

4.2 Decomposition of a Non-deterministic Event

One of the characteristics of an Event-B machine is its ability to express non-determinism. This non-determinism derives either from the existence of several triggerable events in a given state, or from a non-determinism internal to an event. Let us consider this second case and the Listing 9.

```
variables
    c1  // on C1
    c2  // on C2
event ndet
    any i where
        @i1  i < c1  // on C1
        @i2  i < c2  // on C2
    end
```

Listing 9 Non-deterministic event

```
variables
    c1
event ndet
    any i where
        @i1  i < c1
    end
```

Listing 10 Projection on C1

```
variables
    c2
event ndet
    any i where
        @i2  i < c2
    end
```

Listing 11 Projection on C2

This machine is decomposable on components C1 and C2. The event ndet then becomes a non-deterministic synchronization event. Its two projections produce a non-deterministic choice of value for the parameter i. As in CSP, if the choices are identical, the synchronized event will be triggerable. However, as the projections are non-deterministic, the a priori separate development of each component consists in restricting the choice of values for i and proposing an algorithm determining a local solution. Synchronization can then become impossible and lead to a deadlock. To avoid this problem, we impose the determinism of synchronization events and provide a tool named *event splitting* to help eliminate non-determinism (Sect. 5.2). It will remove the parameters of events entended to be synchronized.

4.3 Data Transfer

This first pattern represents the transfer of data stored in the variable x of the component C0 to the variables xi of the components Ci. The centralized event exports the value of x via the parameter vx. This pattern is represented in BIP by an n-ary connector (Fig. 2) whose down action copies the value of x read via the port p0 in each of the variables xi modified via the ports pi. The shape of the event makes

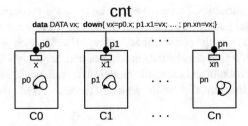

Fig. 2 N-ary connector `cnt`

explicit the transfer of the data and the mapping of actions to components, the right part of the assignments not referring to remote variables.

```
event cnt
  any vx where
    @g1 x = vx // on C0
  then
    @a1 x1 := vx // on C1
       ...
    @an xn := vx // on Cn
end
```

Listing 12 Event `cnt`

4.4 Using a Remote Variable in an Action

This schema (Fig. 3) generalizes the previous one by modeling the use of the received value in any expression in order to update the state of the receiver components. On the Event-B side, the synchronization parameter appears in the right-hand expression. On the BIP side, an action must be added to each receiver subcomponent to exploit the transferred value. It is important here to note that the actions of the subcomponents are performed after the data transfers. The semantics of the BIP model is therefore consistent with that of the Event-B model.

```
event cnt
  any vx where
    @g1 x = vx // on C0
  then
    @a1 a1 := a1+vx // on C1
      .
      .
      .
    @an an := an+vx // on Cn
end
```

Listing 13 Event `cnt`

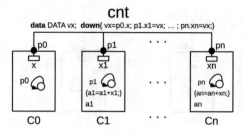

Fig. 3 N-ary connector cnt

4.5 Using a Remote Variable in a Guard

The next pattern models an n-ary synchronization in which each component compares the transmitted value (vx) with local data (ai).

```
event cnt
  any vx where
    @g x=vx  // on C0
    @g1 vx > a1  // on C1
    ...
    @gn vx > an  // on Cn
  end
```

Listing 14 Event cnt

First attempt in BIP. A direct translation in BIP (Figs. 4) inspired from the previous paragraph would be incorrect since the transfer of data is carried out after the test of the guards. The test would, therefore, be performed with the old values of the xi variables receiving a copy of the remote variable x.

In order to take into account the transmitted value, the guards should be tested by the connector which should therefore make a copy of the ai and then the test of the n guards (Figs. 5). However, in order to preserve the usual meaning of a connector, that is a synchronization and data transfer mean, we forbid computations inside connectors and thus consider this schema as incorrect. The Event-B model is therefore not part of the Event-B0 subset and must be transformed.

The transformed model. Since guards can only use local data, copies of the remote data should be performed. Here we declare the xi variables located on the Ci components and intended to receive a copy of x. The Boolean variable x_fresh indicates whether the copies are up to date. This update is performed by the share_x event. The cnt event now compares each local copy with the local data of the component when the copies are up to date. The Sect. 5.3 will generalize this transformation and introduce it as a refinement of the initial model in order to establish its correctness.

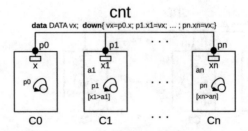

Fig. 4 Incorrect n-ary connector cnt

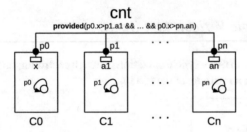

Fig. 5 Forbidden n-ary connector cnt

```
variables
  xi // copy of x on Ci
  x_fresh // on C0
invariants
  @invi x_fresh = TRUE ⇒ xi = x
events
  event share_x
  any vx where
    @vx vx = x // on C0
    @nfr x_fresh = FALSE // on C0
  then
    @xi xi := x // on Ci
    @fri x_fresh := TRUE // on C0
  end
  event cnt
  any vx where
    @fr x_fresh = TRUE // on C0
    @gi xi > ai // on Ci
  end
```

Listing 15 Refinement introducing local copies

The BIP model (Fig. 6) introduces two n-array connectors, one per event. The first one, share_x copies the variable x on each component, while cnt synchronizes the tests of now local guards.

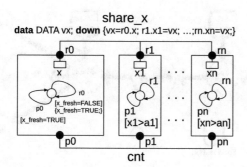

Fig. 6 The two n-ary connectors `cnt` and `share_x`

5   Toward a Distribution Process

Our goal is to provide a process for guiding the user refinements in order to map an initial "centralized" design (as explained in Sect. 2) on a distributed architecture. The proposed process can be seen as a continuation of the basic methodology which captures requirements as successive refinements of an initial specification. However, as we target a system engineering process, our aim is not to propose a fully automatic distribution tool. For example, in the hotel case study, the behavior of the guest should be mapped on a `Guest` component. Figure 7 illustrates the proposed process. It is based on three steps: a splitting step which splits events in order to allow the incremental and local resolution of non-determinism, a mapping step which introduces *components*, and *mappings* of variables over these components and a distributed code generation step.

5.1   The Decomposition Process

In order to support such a process, we consider two domain-specific languages (DSL), one for specifying event parameters computation order and the other for specifying the mapping of machine variables and possibly the location of guard computations. The transformation steps are explicitly specified through the proposed DSLs. These two specifications are used to generate refined models and projections for subcomponents automatically. The correctness of the refinements ensures the correctness of the development. Our process, applied to the hotel case study, is illustrated by Fig. 8.

Fig. 7   Process steps

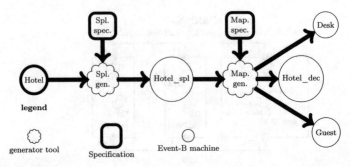

Fig. 8 Hotel transformations

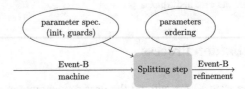

Fig. 9 Event splitting step

5.2 The Event Splitting Step

The splitting step allows the user to inject heuristics for computing event parameters specified by a set of constraints: an event can be split in order to allow the incremental resolution of its non-determinism. This transformation can be useful if the event is non-deterministic and intended to be shared by several subcomponents. Non-determinism will be constrained to occur on local events so that data exchanged will be locally computed before. This step is guided by the user as he may want to control the order in which non-determinism is resolved.[3]

Figure 9 illustrates the profile of the transformation implemented as a Rodin plugin. It takes as input an Event-B machine and a *splitting specification*, whose structure is described by a domain-specific language.

event ev **when** p_1 ... p_n **parameter** p **init** v **with** g_1 ... g_m
            when ...   **parameter**   ...

We specify for some of the model events, e.g., *ev*, the parameters (p) to be computed, the parameters on which it depends (p_i), the default value v of p (for typing purposes) and the guards (g_i) acting as the specification of the value of p. The plugin generates a refinement of the input machine.

Such a specification provides a partial order on event parameters. It is used to schedule newly introduced events aiming at computing and storing in a state variable

---

[3]We consider here that non-determinism is only introduced through event parameters.

the value of their associated parameter. Ordering constraints are implemented through the introduction of one boolean variable for each parameter, its *computed state*. The machine invariant is extended by the properties of the newly introduced variables: if a variable has been computed, its specification, given by its guards, is satisfied. When all the parameters of an event have been computed as state variables, the event itself can be fired. The progress of parameters computation is ensured by a *variant* defined as the number of parameters remaining to be computed. More precisely, the previous specification for parameter p of event ev will produce the following machine contents:

```
machine generated refines input_machine
variables
   ev_p  ev_p_computed //witness and status for param. p of ev
invariants
   @ev_gi ev_p_computed ⇒ gi // where p is replaced by ev_p
variant  // count of the remaining parameters to compute
   {FALSE ↦ 1, TRUE ↦ 0}(ev_p_computed) + ...
events
   event INITIALISATION extends INITIALISATION
   then
      @ev_p ev_p := v
      @ev_p_comp ev_p_computed := FALSE
   end

   convergent event compute_ev_p // computes parameter p of event ev
   any p where
      @gi gi // guards acting as p specification
      @pi ev_pi_computed = TRUE //parameters, p depends on, are computed
      @p ev_p_computed = FALSE // p remains to be computed
   then
      @a ev_p := p  //computed value stored in state variable ev_p
      @computed ev_p_computed := TRUE // makes the variant decrease
   end

   event ev refines ev
   when
      @p_comp ev_p_computed = TRUE
   with
      @p p = ev_p // parameter p of inherited event is refined to ev_p
   then
      @pi ev_pi_computed := FALSE // for all ev_pi with updated guards
      ...    // replace p by ev_p in actions of the refined event
   end
end
```

Listing 16 Generated machine for the splitting refinement

An important point is that we get a refinement of the input machine. It should be proved by the user by discharging the standard proof obligations generated by Rodin and has actually been proved for the hotel example. Three main properties should be established: convergent events refine skip as they do not modify inherited state variables and preserve the invariant. They cannot be launched indefinitely as they make the variant (a natural number) decrease. Lastly, the event ev is refined as new state variables which take place of the parameters of the inherited event to satisfy their guards. The refined invariant is also preserved thanks to the reset of the *computed state* of parameters which depend on guards using updated variables. However, it has to be noted that this transformation can introduce deadlocks. Consider, for example, an event which assigns x and y such that $x \geq 0, y > 0, y < x$. If two events are introduced in sequence, one for computing x and one for y, the second event will be blocking if 0 is chosen for x. In such a case, the user should introduce by himself $x > 0$ as a new constraint for x. This strengthening does not introduce deadlocks as it is a consequence of the initial guard. This knowledge will be automatically added by the plugin as a new invariant. Thanks to this enriched invariant, the absence of deadlock can be established.

To sum up, in the methodology we propose, the user is expected to add derived properties (annotated as theorems) to guards. These properties are transferred to the invariant section by the plugin and can be used to prove the absence of deadlock.

Application to our example. With respect to our example, the register event (see Listing 3) has three parameters: g, r, c. We specify that the parameter g should be computed first as the arrival of a guest is supposed to trigger the various actions. Then, a room is chosen in r and its associated card is computed in c. For each parameter, we specify its initial value and the name of guards which constitute its specification. The dependencies for the register event (see Listing 3) are specified as follows:

```
splitting hotel_splitted refines hotel
events
  event register
    parameter g init g0 with tg // tg does ¬depend on r,c
    when g parameter r init r0 with tr g1 // fired after computation of g
    when g r parameter c init c0 with tc g2 g3 // fired after g,r
end
```

Listing 17 Splitting specification

An executable refinement of the register event should find a value for each of its parameters so that the guards are satisfied. First, a value for g can be selected so that the guard @tg g ∈ Guest is satisfied. This will be the goal of the event associated to g introduced by the splitting plugin. Once g has been chosen, the parameter r should be selected so that guards @tr r ∈ ROOM and @g1 owns (r) = ∅. Last, an event is created to compute the parameter c specified by the guards @tc c ∈ CARD @g2 c ∉ issued and @g3 prj1 (c) = currk (r). These three events will be scheduled in this order to use the value selected for one parameter to compute the next one.

It has to be noted that such a transformation can introduce a deadlock: let us consider the room entering event and the same parameter splitting: if the chosen guest has no valid card for the chosen room, the event in charge of selecting the card cannot be fired. This problem can be solved by allowing to fire again the events associated to the selection of g, r, or c until a solution satisfying all the constraints is found. As a consequence, the variant will not decrease but it concerns environment events, not system events. The solution we have adopted is to tag system parameters. Only system parameters should make the variant decrease, which means that responses to external events should take finite time.

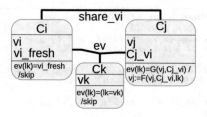

Fig. 10   Local copies and distant access

5.3   The Mapping Step

The aim of this step is to generate a distributed implementation over subcomponents of an Event-B centralized model. As for the splitting step, the mapping step
takes as input a machine and a *mapping specification* described using a dedicated
domain-specific language. The user can thus provide a set of subcomponent names
and declare a mapping from machine variables and possibly event guards to subcomponents. Then, the tool generates a refinement of the input machine and one
projection machine for each subcomponent. This step has two phases: the first one,
called the *replication phase*, replicates the variables over the components in order
to allow a local access to remote variables; the second one, called the *projection
phase*, isolates each component as such. The first phase generates a refinement of
the input machine which is in turn refined by the product of its projections, thanks
to the shared event decomposition mechanism [30].

The replication phase. Given the mapping of machine variables to subcomponents,
this phase builds a refinement of the input machine by introducing local copies of
distant variables accessed by guards. It maps each guard or action to a component
and performs some renaming.

Figure 10 presents a component-based view of the transformed model. The focus
is put on event ev of component C_j:

$$ev \triangleq G(v_i) \Longrightarrow v_j := F(v_j, v_i, v_k)$$

Its guard reads the local copy of v_i while the action has remote access to v_k. Event
synchronization ensures that the local copy of v_i is up-to-date and gives access to
v_k by constraining the event parameter (lk in the figure, `local_vk` in the code
pattern).

Listing 18 presents the transformation pattern focused on component C_i. The
resulting machine should refine the input machine. This is for the moment verified
by discharging the proof obligations generated by Rodin. As previously, we plan to
establish this result at the meta-level and the arguments will be very similar to those
given for the splitting transformation.

```
machine generated refines input_machine variables
  vi // inherited variables, on Ci
  Cj_vi // copy of vi mapped on Cj (used by a Cj guard)
  vi_fresh // true if vi has been copied, on Ci
invariants
  @Cj_vi_f vi_fresh = TRUE ⇒ Cj_vi = vi // copy is synchronized
variant {FALSE ↦ 1, TRUE ↦ 0}(vi_fresh) + ...
events
  convergent event share_vi // shared by Ci and Cj
  any local_vi where
    @g vi_fresh = FALSE // on Ci
    @l local_vi = vi // on Ci
  then
    @to_Cj Cj_vi := local_vi // on Cj
    @done vi_fresh := TRUE // on Ci
  end
  event ev refines ev // shared by Ci, Cj, Ck
  any local_vk where
    @vj_access local_vk = vk // on Ck, access to remote variables
    @vi_fresh vi_fresh = TRUE // on Ci, copy to Cj has been done
    @g [vi:=Cj_vi]g // inherited guard on Cj, access to local copy of vi
  then
    @a vj := [vi := Cj_vi || vk := local_vk]e // on Cj
  end
end
```

Listing 18 Replication phase

Furthermore, as for the splitting plugin, the freshness of copies is reset when the source variable is updated by an action.

The projection phase. It generates a machine for each component, as would do the *shared event decomposition* plugin [30]. However, thanks to the replication phase, guards and actions over remote variables are now accepted. For component C_j, we get the following code template:

```
machine Cj
variables vj Cj_vi
invariants ... // keep only those referring vj and Cj_vi
events
  event share_vi // sync with Ci event, import vi
  any local_vi then
    @to_Cj Cj_vi := local_vi
  end
  event share_vj // sync with Cl event, export vj
  any local_vj then
    @to_Cl local_vj := vj
  end
  event ev
  any local_vk // read by some Cj action
  where
    @vj_fresh vj_fresh = TRUE // needed by Cl, vj has been exported
    @g [vi := Cj_vi]g // mapped on Cj, access to copy of vi
  then
    @a vj := [vi := Cj_vi;vk := local_vk]e
  end
end
```

Listing 19 Projection phase

We have to note that some invariants may be lost: we only keep those who refer variables local to the considered component. It means that the correctness of the resulting machines (i.e., the fact that events preserve the remaining invariants) is not guaranteed and should be proven. If this is not possible, invariants should be added by the user. However, the composition of the projections, as defined in [30], to which lost invariants are added is, by construction, the machine we had before decomposition. As a consequence, thanks to the monotony of composition, the design process can be pursued on each component machine.

Application to our example. Listing 20 specifies hotel subcomponents and the mapping of the variables currk owns issued on the component Desk and the variable cards on the component Guest.

```
mappings
   variables currk owns issued ↦ Desk;
   variable cards ↦ Guest;
   variables register_r register_c ↦ Desk;
```

Listing 20 Hotel components and mapping specification

5.4 The Code Generation Step

This step assumes that the input Event-B model conforms to a subset of Event-B, we called Event-B0, which plays the role of the subset B0 of the B language that is translated to C or Ada. In the considered subset, shared events should be those resulting from the application of the replication phase of the mapping step. Thus parameterized events should conform to the send/receive pattern used to represent the access to remote variables. Furthermore, we suppose that subcomponent machines do not need to be refined. Events should be deterministic (constraints on parameters present in guards should be in solved form) and use a subset of the Event-B expression and predicate languages for which there exists a direct mapping to their BIP counterparts, i.e., quantifications should not be used. For this purpose, we require that set expressions and predicates have been refined to calls to a set library [15] of which signature has a C implementation within the BIP framework. Here, we present how the architectural part of the BIP code is generated. The generator takes as input the mapping specification (subcomponent names, variable, and guards mappings) and the refined machine produced by the mapping step.

Port type generation. For each shared event and each component of which variables are referenced by this event, we generate a port type taking as parameter the type of exported variables (variables mapped to this component and used by guards or actions mapped to other components). A port type for synchronization purpose only is generated for all events that do not export variables. Listing 6 provides port types generated for our example.

Connector type generation. For each event which uses variables of several components, we generate a connector type taking as parameters ports specified by the previously introduced port types. They are supposed to be synchronous. They define a down action which copies (via the ports) variables of one component to their copies located in components which need them. Listing 7 illustrates the application of this rule in our example.

Subcomponent skeleton generation. For each subcomponent, we generate an atomic BIP component. It contains

- variables mapped to this component as well as variables of other components referenced by guards or actions mapped to this component.
- instances of the port types associated to this component
- for each event, a transition synchronized on the corresponding port instance, and the BIP translation of guards and actions mapped to this component.

As an illustration, Listing 6 gives an extract of the atomic component type ty_Desk generated by our plugin.

Composite component generation. The root component contains an instance of each subcomponent and connector. Each connector instance takes as parameter a port instance defined in one of the concerned subcomponents. Listing 8 provides the code of our example resulting from this step.

The generated BIP architecture should for now be completed manually by the data types and behaviors of atomic components. To achieve this, we envision to use the Theory component [15] of the Rodin platform. Indeed, the Theory component allows to develop proved mathematical theories (datatypes, operators, rewrite rules, inference rules). This allows the extension of Event-B by useful data structures such as arrays, linked lists, and hash tables.

6 Related Work

Several formalisms such as process algebra, input/output automata, UNITY and TLA$^+$ have been proposed to model and mostly to reason over concurrent and distributed systems. However, to the best of our knowledge, their effective use within development frameworks leading to a distributed implementation has not yet been a general tendency. We can also cite Action Systems adapted to the specification and design of distributed systems (Disco [23]), but it does not support vertical refinement and, to the best of our knowledge, does not offer any verification environment.

Modeling distributed systems in Event-B has been addressed for a long time. We can cite the Leader Election Algorithm [4]. However, the deployment on distributed platforms was not their main concern. The automatic generation of source code from formal specifications is supported by few formal methods such as B and Event-B. In [9], an approach is developed allowing the generation of efficient code from B formal developments by using an imperative intermediate language B0. Several Event-B source code generators have been proposed [14, 18, 31, 32]. Indeed, an

Event-B model can represent sequential, concurrent, or distributed code as well as reactive, distributed, or hybrid systems. The work described in [31] proposes a set of plugins for the Rodin development tool that automatically generates imperative sequential code from an Event-B formal specification. In [32], Java code is extracted from an Event-B model structured to represent a distributed system. These works do not take into account Event-B composition. Whereas the works described in [16] generate concurrent Ada code restricted to binary synchronization. The automatic refinement of B machines is also possible thanks to the Bart tool [13]. Also, in Event-B, the atomicity decomposition plugin [25] defines a DSL to parametrize the refinement generator. However, the refinement pattern is dedicated to event splitting and does not apply to our problem.

7 Conclusion

In this paper, we have presented a distribution process for system design formally expressed as Event-B models. Starting from an Event-B machine, the studied process proposes successively the splitting step and the mapping step. The specification of these two steps is done through two domain-specific languages. Eventually, a distributed Event-B model and a distributed BIP code architecture are also automatically generated. In our work, we make explicit the distribution of Event-B models through domain-specific languages. As we said in the introduction, our primary aim is to assist the user in the design of distributed systems. Providing a fully automatic (implicit) process is not in our objectives as we target system engineering and requirements may provide constraints in functions/data to component mapping. Each proposed step generates refinements. The proof obligations generated by Rodin for these refinements remain to be discharged in order to assert the correctness of the developed model. Our experiments show that these proof obligations concern essentially the well-definedness of some generated expressions and are easy to discharge.

The correctness of the process relies on the correctness of the refinements and of the machines defining the introduced subcomponents. These correctness properties are thus implicitly stated by the usual Event-B proof obligations. Thus, the user is asked to establish the correctness of the plugin applications. However, the final translation to BIP is not verified. This could rely on translation validation or transformation verification and is left to future work.

As future work, we also envision to enhance the tooling of our process. Currently, the splitting and mapping steps have been implemented with the xtext [2] language infrastructure, and the refinements and the BIP code have been generated with the accompanying xtend language [1] which provides support for writing code generators. We are interested in achieving a distributed code generator plugin for the Rodin platform by taking into account types and the translation of Event-B expression and predicate languages.

With respect to the code generation step, we are also interested in targeting kernels for critical applications. We are especially interested in the so-called Ravenscar [33] and RTSJ [10, 20] profiles. The challenge here will be to consider how to integrate high-level applicative properties with low-level properties pertaining to the underlying physical architecture. We are also interested in studying how the proof obligations generated by the refinements can be discharged definitively at the meta-level. In the long term, we seek to enrich the set of transformations and to provide a library of certified transformations dedicated to the development of distributed systems for various architectures. Finally, our framework could be extended to support parameterized distributed systems (token ring, spanning tree, …) and could benefit from the recently proposed BIP extension [22].

Acknowledgements We thank the anonymous reviewers for their constructive and helpful comments.

References

1. Java 10, today! http://www.eclipse.org/xtend/. Accessed 16 Jan 2006
2. Language engineering for everyone! https://eclipse.org/Xtext. Accessed 16 Jan 2006
3. J. Abrial, M.J. Butler, S. Hallerstede, T.S. Hoang, F. Mehta, L. Voisin, Rodin: an open toolset for modelling and reasoning in Event-B. STTT **12**(6), 447–466 (2010)
4. J. Abrial, D. Cansell, D. Méry, A mechanically proved and incremental development of IEEE 1394 tree identify protocol. Formal Asp. Comput. **14**(3), 215–227 (2003)
5. J.-R. Abrial, *Modeling in Event-B: System and Software Engineering*, 1st edn. (Cambridge University Press, New York, 2010)
6. J.-R. Abrial, S. Hallerstede, Refinement, decomposition, and instantiation of discrete models: application to Event-B. Fundam. Inf. **77**(1–2), 1–28 (2007). Jan
7. R. Back, R. Kurki-Suonio, Decentralization of process nets with centralized control. Distrib. Comput. **3**(2), 73–87 (1989)
8. A. Basu, S. Bensalem, M. Bozga, J. Combaz, M. Jaber, T.-H. Nguyen, J. Sifakis, Rigorous component-based system design using the BIP framework. IEEE Softw. **28**(3), 41–48 (2011)
9. D. Bert, S. Boulmé, M.-L. Potet, A. Requet, L. Voisin, Adaptable translator of B specifications to embedded C programs, in *FME*, LNCS, vol. 2805 ed. by K. Araki, S. Gnesi, D. Mandrioli (Springer, Berlin, 2003), pp. 94–113
10. J. Bodeveix, R. Cavallero, D. Chemouil, M. Filali, J. Rolland. A mapping from AADL to Java-RTSJ, in *Proceedings of the 5th International Workshop on Java Technologies for Real-time and Embedded Systems, JTRES 2007, Institute of Computer Engineering, Vienna University of Technology, 26-28 September 2007, Vienna, Austria*, ACM International Conference Proceeding Series ed. by G. Bollella (ACM, 2007), pp. 165–174
11. M. Butler, A CSP Approach to Action Systems. Ph.D. thesis, Oxford University (1992)
12. D. Cansell, D. Méry, Formal and incremental construction of distributed algorithms: On the distributed reference counting algorithm. Theor. Comput. Sci. **364**(3), 318–337 (2006)
13. Clearsy. Bart (B automatic refinement tool), http://tools.clearsy.com/wp-content/uploads/sites/8/resources/BART_GUI_User_Manual.pdf
14. A. Edmunds, M. Butler, Tasking Event-B: an extension to Event-B for generating concurrent code. Event Dates: 2nd April 2011, February 2011
15. A. Edmunds, M.J. Butler, I. Maamria, R. Silva, C. Lovell, Event-B code generation: Type extension with theories, in *ABZ proceedings* (2012), pp. 365–368
16. A. Edmunds, A. Rezazadeh, M.J. Butler, Formal modelling for ada implementations: tasking event-b, in *Reliable Software Technologies - Ada-Europe 2012 - 17th Ada-Europe International Conference on Reliable Software Technologies, Stockholm, Sweden, June 11-15, 2012.*

Proceedings, Lecture Notes in Computer Science, vol. 7308, ed. by M. Brorsson, L.M. Pinho (Springer, 2012), pp. 119–132

17. Y. Falcone, M. Jaber, T.-H. Nguyen, M. Bozga, S. Bensalem, Runtime verification of component-based systems in the BIP framework with formally-proved sound and complete instrumentation. Softw. Syst. Model. **14**(1), 173–199 (2015)
18. A. Fürst, T. Hoang, D. Basin, K. Desai, N. Sato, K. Miyazaki, Code generation for Event-B, in *IFM*, LNCS, vol. 8739 ed. by E. Albert, E. Sekerinski (Springer, 2014), pp. 323–338
19. A. Fürst, T. S. Hoang, D. Basin, N. Sato, K. Miyazaki, Formal system modelling using abstract data types in Event-B, in *ABZ proceedings*, LNCS, vol. 8477 (Springer, 2014), pp. 222–237
20. J. Gosling, G. Bollella, *The Real-Time Specification for Java* (Addison-Wesley Longman Publishing Co. Inc, Boston, 2000)
21. M. Jaber, *Centralized and Distributed Implementations of Correct-by-construction Component-based Systems by using Source-to-source Transformations in BIP*. Theses, Université Joseph-Fourier - Grenoble I, 2010
22. I. Konnov, T. Kotek, Q. Wang, H. Veith, S. Bliudze, J. Sifakis, Parameterized systems in BIP: design and model checking, in *Proceedings of the 27th International Conference on Concurrency Theory (CONCUR 2016)* (2016), pp. 30:1–30:16
23. R. Kurki-Suonio, H.-M. Järvinen, Action system approach to the specification and design of distributed systems, in *Proceedings of the 5th International Workshop on Software Specification and Design*, IWSSD '89, New York, NY, USA (1989). ACM, pp. 34–40
24. T. Nipkow, Verifying a hotel key card system, in *Theoretical Aspects of Computing (ICTAC 2006)*, vol. 4281 ed. by K. Barkaoui, A. Cavalcanti, A. Cerone (2006), pp. 1–14
25. A. Salehi Fathabadi, M. Butler, A. Rezazadeh, *A Systematic Approach to Atomicity Decomposition in Event-B* (Springer, Berlin, 2012), pp. 78–93
26. B. Siala, *Décomposition formelle des spécifications centralisées Event-B : application aux systèmes distribués BIP*. Ph.D. thesis, Université Paul Sabtier. Toulouse (2017)
27. B. Siala, M.T. Bhiri, J. Bodeveix, M. Filali, An event-b development process for the distributed BIP framework, in *Formal Methods and Software Engineering - 18th International Conference on Formal Engineering Methods, ICFEM 2016, Tokyo, Japan, November 14-18, 2016, Proceedings*, Lecture Notes in Computer Science, vol. 10009 ed. by K. Ogata, M. Lawford, S. Liu (2016), pp. 313–328
28. R. Silva, M. Butler, Supporting reuse of Event-B developments through generic instantiation, in *ICFEM '09 Proceedings* (2009), pp. 466–484
29. R. Silva, M. Butler, Shared event composition/decomposition in Event-B, in *FMCO Formal Methods for Components and Objects* (2010)
30. R. Silva, C. Pascal, T.S. Hoang, M. Butler, Decomposition tool for Event-B. Softw. Pract. Exper. **41**(2), 199–208 (2011)
31. N.K. Singh, Eb2all: an automatic code generation tool, in *Using Event-B for Critical Device Software Systems* (Springer, London, 2013), pp. 105–141
32. M. Tounsi, M. Mosbah, D. Méry, From event-b specifications to programs for distributed algorithms, in *2013 Workshops on Enabling Technologies: Infrastructure for Collaborative Enterprises* (2013), pp. 104–109
33. T. Vardanega, J. Zamorano, J.A. de la Puente, On the dynamic semantics and the timing behavior of Ravenscar kernels. Real-Time Syst. **29**(1), 59–89 (2005)

Explicit Exploration of Refinement Design in Proof-Based Approach: Refinement Engineering in Event-B

Fuyuki Ishikawa, Tsutomu Kobayashi, and Shinichi Honiden

Abstract Control of abstraction levels is key to tackling the increasing complexity of emerging systems such as cyber-physical systems. Formal methods for dependability assurance have been used to explore this point by using refinement mechanisms, with which complex models are gradually constructed and verified. However, refinement mechanisms to derive the whole specification of systems are relatively new, as with the Event-B method, and refinement design is still an "art." In this chapter, we discuss the problem of refinement design and present our approach for explicitly exploring and manipulating possible refinement designs. Specifically, we report our experiences on refinement planning and refactoring to support engineering activities on refinement.

1 Introduction

One of the key challenges for system dependability is how to deal with the increasing complexity in system modeling and verification. Event-B is a formal method for tackling this challenge with its flexible refinement mechanism [3]. With this refinement mechanism, we can gradually introduce and verify concepts and constraints in a system while moving from abstract prescriptive representations into concrete realizable ones. Due to its flexibility, this refinement mechanism requires designing the refinement steps. In other words, we need to examine how symbols, predicates,

F. Ishikawa (✉) · T. Kobayashi · S. Honiden
National Institute of Informatics, Tokyo, Japan
e-mail: f-ishikawa@nii.ac.jp

T. Kobayashi
e-mail: t-kobayashi@nii.ac.jp

S. Honiden
e-mail: honiden@nii.ac.jp

© Springer Nature Singapore Pte Ltd. 2021
Y. Ait-Ameur et al. (eds.), *Implicit and Explicit Semantics Integration in Proof-Based Developments of Discrete Systems*,
https://doi.org/10.1007/978-981-15-5054-6_14

and their proofs in the whole specification are decomposed and modularized into refinement steps. This type of problem was not dominant in classical refinement mechanisms for transforming a specification to a code. For example, data refinement, as with the classical B-Method [1], can be supported using automated tools using typical patterns, e.g., converting a set-based representation to an array-based one with a loop [6].

Refinement design can affect various quality attributes on an Event-B model. For example, it can affect the verifiability of the model. A "bad" design leads to large refinement steps, which are difficult for engineers or automated provers to verify. Another example is reusability. As each refinement step depends on the preceding steps, reuse of some steps succeeding from the most abstract step is very straight-forward by just throwing away the unnecessary concrete steps. However, reusing aspects scattered throughout various steps of the model is not trivial.

Nevertheless, refinement design has been an "art." In other words, the "design space" of refinement and the process to explore it has been implicit. Although there are models that are believed to be "good" according to textbooks or case studies, only the resulting models are presented with a few words on the design. Current approaches for supporting refinement design are based on experiences such as guidelines or patterns [7, 20, 24, 26]. There has been little effort in supporting methods or tools for explicitly manipulating the refinement design (space) to support engineers.

In this chapter, we present our approach for explicitly exploring refinement designs and their spaces on the basis of our experiences [15–17]. This explicit approach enables us to provide supporting methods and tools by manipulating refinement designs and spaces. Specifically, we tackled the problem of planning, i.e., exploration of the possible design space. We also tackled the problem of refactoring, i.e., changing the design to another one in an existing model. We discuss quality attributes of Event-B models, specifically, comprehensibility, verifiability, and reusability. These attributes can be clearly examined with our approach.

In the remainder of this chapter, we first introduce the refinement mechanism of Event-B for system modeling and verification (Sect. 2). We then discuss the impact of a refinement design (Sect. 3). We present our investigations in refinement planning and refactoring (Sect. 4). Finally, we give an overview of the relevant literature (Sect. 5) before concluding remarks (Sect. 6).

2 Preliminary: Refinement in Event-B

In this section, we briefly give an overview of Event-B [3], focusing on its refinement mechanism.

2.1 Characteristics of Refinement Mechanisms

Event-B is a method that emerged after the success of its ancestor B-Method. B-Method targets software components and their correct implementation [1, 2]. By contrast, Event-B primarily targets modeling and verification of system models that can contain not only software components but also physical elements controlled by them and environmental elements that interact with them. The difference in the focus leads to different definitions of refinement mechanisms, the common core feature of the two methods.

A refinement mechanism generally allows for gradual stepwise transition from abstract representations to concrete representations. Thus, a model consists of multiple steps with different abstraction levels. Concrete steps inherit a certain set of aspects and their constraints from their preceding steps in a manner defined by the refinement mechanism of the target method. This transformation from abstract representations to concrete ones is generally the core of software development. This is enhanced in the context of formal methods by rigorous definitions and mechanisms for consistency between steps.

With B-Method, the refinement mechanism aims at gradually obtaining the code proved to be correct in terms of the specification. A model in B-Method defines states and operations in each software component. The interface is defined in the most abstract step and strictly inherited in the succeeding concrete steps. The representations on the inside of each component and operation are transformed from abstract ones to concrete ones to finally define an implementation as a program code. As a typical example, internal data and their processing are represented with sets and set operators (e.g., union) in abstract steps. The functionality of all the operations is thus rigorously defined. Through refinement, the operations are transformed into implementable arrays and loops on them in concrete steps. The observable functionality is not changed from the first step. In this type of refinement, it is easy to define and reuse patterns of refinement design that frequently occur [6]. In addition, refinement design does not significantly impact the quality of the process or model, e.g., which set operators are made concrete as loops in which order.

With Event-B, by contrast, the refinement mechanism aims at gradually obtaining the whole specification of a complex system. In other words, Event-B can be used to obtain the input to B-Method. A model in Event-B defines states and events in the target system. Events can represent a variety of state changes such as uncontrollable state changes of the environment, and reactive actions by controllable software components. States and events represented in abstract steps can be partial or underconstrained, which are not only conceptual and far from realization.

Consider a system for room security. Abstract steps may only state the foundational aspects of the system, e.g., users move from one room to another physically connected by corridors. Through refinement, the notion of an access list is introduced and the movement is restricted to occur only if a user has permission. Then the notions of card keys, card readers, and a central server are introduced with additional events of the authorization procedure. Constraints introduced in abstract steps are strictly inherited: a movement can occur only if two rooms are physical connected and the user has a permission. However, aspects (dimensions) of the state space, events, and the represented functionality of the system are flexibly extended to obtain the whole specification in a stepwise manner. In this refinement mechanism, refinement design affects the quality of the model as discussed later in this chapter.

This direction for stepwise modeling and verification of an entire system is becoming increasingly significant. This is because current and emerging systems are very complex containing various elements even at the specification level, e.g., Cyber-Physical Systems [19] and Internet-of-Things [5].

2.2 Example of Event-B Models

We now give a concrete example of an Event-B model. We only describe the minimum essence of its refinement mechanisms. Readers who are interested in Event-B can check other resources such as a book [3] or website.[1]

We use the room security system mentioned above (a variation of Chap. 16 in [3]). The (partial) model of this system is shown in Fig. 1, which contains four components. The first one, Context0, is a *context* component that defines carrier sets (sets used as types) and constants as well as axioms regarding them. The other three are *machine* models that define the states and events (state changes) of the target system with different abstraction levels.

The machine M0 is the most abstract step in this model. States of the system are represented as *variables*. In this case, one variable represents areas in which each user is currently located. State changes are defined as *events*, in this case, only one event that represents a movement by a user. Each event has guard conditions, in the WHERE clause, specifying under what conditions the event is triggered. It also has actions, in the THEN clause, specifying what effects the event has on the states (variables). In M0 the event move represents movement of a user to a destination dst area (move_act0) only if there is a physical connection (door) between his/her current location and the destination (move_grd0).

The machine M1 refines M0 and thus defines a more concrete description of the system. Specifically, the notion of an access control list (ACL) is introduced, which is represented as a variable acl as a relation between users and areas. Now the location of each user is restricted within areas permitted in the ACL, defined as the *invariant* inv1 (constraint that the system states should always satisfy). The move event is refined to further constrain it by an additional guard condition on acl (move_grd1). The original guard condition and action in M0 are inherited without any change. Additional events should be introduced to manage the ACL, such as granting permission for a user to enter an area, which is omitted in this figure.

---

[1] http://www.event-b.org/.

```
CONTEXT Context0

SETS AREAS, USERS
AXIOMS
CONSTANTS DOORS ⊆ AREAS × AREAS
```

```
MACHINE M0 SEES Context0

VARIABLES
location ∈ USERS → AREAS

EVENTS
move
    ANY
        user ∈ USERS
        dst ∈ ROOMS
    WHERE
        (location(user), dst) ∈ DOORS    ··· move_grd0
    THEN
        location(user) := dst    ··· move_act0
```

```
MACHINE M1 REFINES M0

VARIABLES
location ∈ USERS → AREAS
acl ⊆ USERS × AREAS // added

INVARIANTS
location ⊆ acl    ··· inv1

EVENTS
move REFINES move
    ANY
        user ∈ USERS
        dst ∈ ROOMS
    WHERE
        (location(user), dst) ∈ DOORS    ··· move_grd0
        dst ∈ acl[{user}]    ··· move_grd1 // added
    THEN
        location(user) := dst    ··· move_act0
```

```
MACHINE M2 REFINES M1

VARIABLES
location ∈ USERS → AREAS
acl ⊆ USERS × AREAS
authorized ⊆ USERS →partial AREAS // added

INVARIANTS
authorized ⊆ acl    ··· inv2

EVENTS
move REFINES move
    ANY
        user ∈ USERS
        dst ∈ ROOMS
    WHERE
        (location(user), dst) ∈ DOORS    ··· move_grd0
        authorized(user) = dst    ··· move_grd1'
                                      // replaced
    THEN
        location(user) := dst    ··· move_act0
        authorized := user ⊲ |authorized    ··· move_act1
                                      // added

authorize
    ANY
        user ∈ USERS
        dst ∈ ROOMS
    WHERE
        (location(user), dst) ∈ DOORS    ··· auth_grd0
        dst ∈ acl[{user}]    ··· auth_grd1
    THEN
        authorized(user) := dst    ··· auth_act0
```

The operator ⊲| denotes domain restriction.

Fig. 1 Example of Event-B model

Correctness or consistency of Event-B models is defined in the form of proof obligations, i.e., properties that should be proved. We introduce a few of essential proof obligations defined in Event-B.

- **Invariant Preservation**. Occurrence of each event must not lead to a state that violates the invariants (assuming that the invariants and guard conditions held before the occurrence). In M0, the move event does not lead to violation of inv1 as it is "guarded" by the guard condition move_grd1.
- **Guard Strengthening**. Guard conditions of an event that refines an event in the preceding abstract step must be equivalent to or stronger than (i.e., must imply) the guard conditions of the refined event. In M1, the guard conditions of the move event are stronger than those in the preceding step M0, i.e., inheriting move_grd0 and adding a new one move_grd1. This proof obligation of guard strengthening is necessary so that the concrete step does not invalidate the proof on invariant preservation in the abstract step. For example, suppose that in the following step after M1, the guard condition move_grd1 is removed. As this condition con-

tributes to invariant preservation of `inv1`, removing it invalidates the proof in
`M1`, thus making it uncertain that the invariant always holds in the concrete step.

- **Action Simulation**. Similarly, actions of an event that refines an event in the
  preceding abstract step must inherit the effects on the states from the actions of
  the refined event.
- **Equality of Preserved Variable**. Event actions in a concrete step cannot change
  values of the variables inherited from the abstract step unless the event refines
  an event that changes the values in the abstract step. In other words, one cannot
  introduce additional events that freely change values of inherited variables not
  constrained by the proof obligations of action simulation. This is necessary for the
  same reason, i.e., not to invalidate proofs on invariant preservation in the preceding
  steps.

A platform for Event-B, called Rodin, provides relevant functionality to modeling
as well as generating and verifying proof obligations.[2]

We now explain the last step `M2`, shown in Fig. 1, which refines `M1`. In `M1`, the
`move` event has a guard condition `move_grd1` as if the user would check the ACL.
This description was abstract and prescriptive, which is far from realization. Step `M2`
is one step forward to a realizable specification and takes into account an authorization
action before movement. A new variable `authorized` is introduced to represent
users who are authorized for access to certain areas. A new event `authorize`
is introduced to represent the authorization action by the system (set the user as
authorized as in action `auth_act0`). Now, the `move` event does not check the ACL
directly in its guard conditions (`acl` in `move_grd1` of `M1`). Instead, it checks the
authorization status (`authorized` in `move_grd1`' in `M2`). The invariant `inv2`
requires that authorization to enter an area be given to a user only if the user has
permission. Notable points regarding the proof obligations are described below.

- Guard Strengthening in `move`. The guard condition `move_grd1` in `M1` is replaced
  with `move_grd1`' in `M2`. The invariant `inv2` ensures that the new guard holds
  only if the original guard holds (the concrete event can occur only when the abstract
  event can occur).
- Invariant Preservation. The newly introduced `inv2` is not violated because of the
  guard condition `auth_grd1` in the event `authorize`, which is equivalent to
  `move_grd1` in the abstract event in `M1`.
- Equality of Preserved Variable. The new event `authorize` does not refine any
  event of the previous step, i.e., `M1`. It must not change the inherited variables
  `location` and `acl` and actually does not. Thus, this event does not break the
  proof of the invariant preservation of `inv1` in `M1`.

Step `M2` is still not a realizable specification. The variable `authorized` is conceptual, and we need to introduce specific realizations such as door locks, a central
software controller, and communication between them. We need further steps after
`M2`. We also have not fully demonstrated how flexible the refinement mechanism is,

---

[2]https://sourceforge.net/projects/rodin-b-sharp/.

either. For example, some variables are not inherited to concrete steps. In our example, `authorized` may be such a one, as the concept is realized with door locks. As another example, it is possible to split or merge events by refinement. However, we now stop talking about concrete models and discuss refinement design.

3   Refinement Design and Its Impact

In this section, we discuss how refinement design affects the quality of an Event-B model, which has been somewhat implicit.

We use the very small example given in Sect. 2.2 to discuss the differences in refinement design. Even in this small example, we can think of different possibilities of refinement design and discuss their impact.

3.1   Comprehensibility and Verifiability

As one example, we can consider another "packed" version of the Event-B model in Fig. 1 where the same system is defined within just one step. This version is shown in Fig. 2. This version is almost the same as the last step M2 in Fig. 1. However, it contains inv1, which was originally separated into another abstract step (M1). It also lacks the guard move_grd1 (commented out), which originally supported preservation of inv1 and was replaced with move_grd1'.

One may still keep move_grd1 in the one-step version. But this representation is prescriptive, not describing what actually occurs, as a user directly sees the ACL. Thus, this condition should be considered as a "theorem" that can be derived from move_grd1', the realizable representation of the guard. Event-B has a notion of theorem to allow for this distinction regarding predicates that can be derived from the other invariants.

The one-step version can be considered as a result of *merging* the three steps in the original model (or in reverse *decomposing* a step). This change in refinement design obviously affects the **local complexity** within each step. We have more elements (variables and predicates) newly introduced in MALL of the one-step version than in M2 of the original version (though the difference is only inv1 in the very small example). Thus, the one-step version has more local complexity, while this is mitigated in multiple steps of the original version. Instead, the original version has more global complexity in managing multiple steps and their consistency.

Why does local complexity matter? Let us focus on the essential invariant of the system: a user resides only in the areas specified in the ACL (inv1). The proof on preservation of this invariant is intrinsically stepwise because the user (or the guard condition of the move event) does not directly check the ACL but requests authorization (from the software controller). In other words, there is a chain of reasoning.

Fig. 2 Packed version of
Event-B model

```
MACHINE MALL

VARIABLES
location ∈ USERS ⇸ AREAS
acl ⊆ USERS × AREAS
authorized ⊆ USERS →_partial AREAS
INVARIANTS
location ⊆ acl      ⋯ inv1
authorized ⊆ acl    ⋯ inv2

EVENTS
move
    ANY
        user ∈ USERS
        dst ∈ ROOMS
    WHERE
        (location(user), dst) ∈ DOORS      ⋯ move_grd0
        authorized(user) = dst    ⋯ move_grd1'
        // dst ∈ acl[{user}]    ⋯ move_grd1
    THEN
        location(user) := dst    ⋯ move_act0
        authorized := user ⩤ authorized    ⋯ move_act1

authorize
    ANY
        user ∈ USERS
        dst ∈ ROOMS
    WHERE
        (location(user), dst) ∈ DOORS    ⋯ auth_grd0
        dst ∈ acl[{user}]    ⋯ auth_grd1
    THEN
        authorized(user) := dst    ⋯ auth_act0
```

- The guard move_grd1 in the move event ensures that the user moves to the area only if the movement is authorized.
- The fact the user is authorized to move to an area means that the user-area pair is contained in the ACL (inv2).
- This fact is ensured by the guard auth_grd1 in the authorize event.

The original version explicitly represents the stepwise reasoning, while the one-step version makes it implicit. Thus, it is somewhat more difficult to understand or replay the proof in the one-step version. First, it is necessary to guess the causal structure, i.e., which predicates (invariants and guards) are relevant to satisfaction of a certain predicate. Second, the above chain is implicit: specifically, the guard move_grd1 is implicit (unless this is added as a theorem). In this way, increasing local complexity matters in terms of comprehensibility by human engineers as well as verifiability by human engineers or automated provers. The impact of the differences is trivial in this small example but can be large in practice (discussed in Sect. 4).

The above discussion does not mean that granular steps are always better in refinement design. It is simply bothersome to manage multiple steps, e.g., switch between steps to understand something. The number of model elements can increase as `move_grd1` in the example. The number of proof obligations also increases as it is necessary to consider proof obligations for each step. This point can be explained as a result of making the chain explicit. Obviously, it does not make sense to decompose a step or make the reasoning chain explicit if the step or chain can be handled reasonably by human engineers or automated provers.

It is necessary to investigate the good balance of local and global complexity aspects between two extremes of one monolithic step that is too complex and too many trivially tiny steps. This point seems analogous to modular design in programming languages. The good balance can be explored only empirically as done for programming languages. Unfortunately, we do not have sufficient data sets to discuss this point empirically for refinement design as we do not currently have large "open-source" model repositories.

3.2 Reusability

Another point is how refinement design affects reusability. As we consider a type of modularization, it is natural to discuss whether a certain module, or description of a certain aspect (concepts and constraints), can be exported to be used in another model.

Each refinement step is therefore not really a reusable module as it depends on the preceding steps. For example, in the original version of the Event-B model in Fig. 1, `inv1` is defined in `M1` and inherited by means of proof obligations of Guard Strengthening and Action Simulation. If we only extract `M2` by removing the REFINES clauses, the invariant on the variable `location` is lost. We can only reuse a sequence of successive steps from the initial most abstract step in a straightforward manner, e.g., "`M0` and `M1`" and "`M0`, `M1`, and `M2`." We demonstrate this point below by focusing on two scenarios for the simple example.

First, suppose a system in which users move around (without any restriction) guided by a software controller to obtain items or reach the destination area. In this case, we may reuse `M0` from the original model and refine it to introduce guidance by the software, and so on. In the one-step version, this reuse is at least not so straightforward because we need to identify relevant elements (variables, events, and predicates) to be extracted.

Second, suppose a system in which users are authorized for areas but they give control commands to the areas instead of moving there. In this case, we may reuse the variable `authorized` and the structure of two events "`authorize` then `act` (originally `move`)." This reuse is not straightforward even in the original version. This is because the authorization aspect is introduced in a later step (`M2`) in the original version while we want to forget what was introduced in the preceding steps such as `location`. We may imagine a different version of the step preceding (refined by) `M2` that only deals with the aspect we want to reuse this time. This version is shown

Fig. 3 Different possibilities
of abstract step of Event-B
model (with renaming for
generalization)

```
MACHINE M1_ANOTHER

VARIABLES
acl ⊆ USERS × AREAS
authorized ⊆ USERS →_partial AREAS
INVARIANTS
authorized ⊆ acl    ···  inv2

EVENTS
act
    ANY
        user ∈ USERS
        target ∈ ROOMS
    WHERE
        authorized(user) = target    ···  act_grd1'
    THEN
        authorized := user ◁| authorized   ···  act_act1

authorize
    ANY
        user ∈ USERS
        target ∈ ROOMS
    WHERE
        target ∈ acl[{user}]    ···  auth_grd1
    THEN
        authorized(user) := target   ···  auth_act0
```

in Fig. 3, in which some names have been changed to match the new context (`move`
into `act`, `dst` into `target`).

In this way, the ordering affects reusability: in what order concepts and constraints
are introduced. Earlier steps should generally be more generic to facilitate reuse,
though it is sometimes difficult to predict beforehand what types of reuse will be
necessary.

4 Our Experiences: Planning and Refactoring of Refinement Design

We have been investigating engineering methods for refinement to investigate quality
aspects such as those discussed in Sect. 3. Specifically, we investigated methods
for explicitly manipulating refinement design. In this section, we summarize our
approach and insights obtained from our experiences. We simplify the discussion
by addressing only an essential part of Event-B. Interested readers may refer to the
original papers [15–17].

4.1 Explicit Manipulation of Refinement Design

To simplify the discussion, let us represent a refinement design as a sequence of steps in which variables and invariants are gradually introduced. In the example in Sects. 2 and 3, the steps of the original version (Fig. 1) can be captured as follows:

$$[\quad < \{location\}\{\} >, \quad < \{acl\}\{inv1\} >, \quad < \{authorized\}\{inv2\} > \quad]$$

The one-step version (Fig. 2) contains all the model elements in one step, including inv1 as a theorem.

$$[\quad < \{location, acl, authorized\}\{inv1, inv2\} > \quad]$$

The other version, which introduces the authorization first (Fig. 3 without renaming), is described as follows:

$$[\quad < \{acl, authorized\}\{inv2\} >, \quad < \{location\}\{inv1\} > \quad]$$

Our approach involves explicitly considering the manipulation of these designs. We enumerate potential designs, such as the above three, which we call **refinement planning**. We also consider changing the design of an existing model into another one, which we call **refinement refactoring**. We discuss our experiences in these two directions below, including the following points:

- What are valid refinement designs among the possible permutations of steps that represent the distribution of model elements?
- How can an automated tool support the task of refinement design?
- What happens in discharged proof obligations when we change the refinement design of an existing model to refactor?
- Is it possible to fully automate the refactoring of refinement design?

4.2 Refinement Planning

4.2.1 Technical Approach

It is easy to imagine all possible permutations of the steps, which represent how the model elements (now simply variables and invariants) are distributed, as shown in Sect. 4.1; however, some are obviously invalid. For example, suppose we swap the order of introducing the invariants, inv1 and inv2, in the original design of the example.

$$[\quad < \{location\}\{\} >, \quad < \{acl\}\{inv2\} >, \quad < \{authorized\}\{inv1\} > \quad]$$

This design is invalid in the sense that it has an error in the second step: `inv2` contains the variable `authorized` but this variable is not yet defined there. This point means that we need to apply a principle: *when a model element is introduced in a step, the elements it depends on must be introduced in the step or preceding steps.*

The notion of dependencies is key. The above example uses a dependency type of "a variable is necessary to describe a predicate." In this chapter, we do not get into more detail of what types of dependencies are considered given the full details of Event-B.

Let us consider another version of the original design by further decomposing it.

$$[\quad < \{location\}\{\} >, \quad < \{acl\}\{\} >, \quad < \{\}\{inv1\} >, \quad < \{authorized\} >, \quad < \{\}\{inv2\} > \quad]$$

The second and third steps of the original versions are now decomposed by delaying the introduction of the invariants into another step. In the new version, the second step does not entail proofs regarding the newly introduced variable `acl` except for type check. Thus, it allows for arbitrary changes in the variable value. Essential proofs appear in the third step; then, the possible changes in the variable value are investigated for the first time. In this sense, we may think that the second step is "nonessential," which suggests the following principle: *When a model element depends only on the elements that are introduced in the current or preceding steps, it should be introduced in the current step.*

This principle is helpful to avoid nonessential steps and consolidate model elements, i.e., putting model elements in the same step when they are relevant in terms of the dependencies. However, this principle is optional as its violation does not cause any error. For example, some of the invariants may be separated to distinguish generic reusable ones from very specific ones.

It should be noted that the design space is not a mere enumeration of all the possible orders. Suppose there are two invariants a and b that depend on other elements as shown in Fig. 4, e.g., a depends on p, which depends on q. The figure shows two plans regarding which invariant to introduce first, a or b. If we take a first as in the left side, we introduce p and q together in the first step. The remainder is handled in the second step. On the other hand, if we take b first, as on the right side, we need to introduce everything due to the dependencies. Then a should be introduced together in the first step by following the second principle. We can clearly see that the second plan fails to mitigate the local complexity.

The two principles shown above are fundamental and do not depend on specific domains. Such principles can come from the rules of Event-B, which have been radically omitted in this chapter, or from empirically well-known guidelines on abstraction or proof structuring. For example, a possible principle is to introduce variables that are relevant to goals, or target of control, such as user location, before those for realization such as card key and door lock. Domain-specific knowledge will help define principles that reflect insights into what are more stable and common (reusable) and what are fragile and specific (not reusable). Design principles or strategies can generally be represented as constraints on grouping and ordering of model elements.

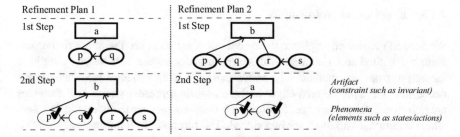

Fig. 4 Example of dependency consideration in refinement plans

- An element e_1 must/should be introduced together with another element e_2.
- An element e_1 must/should be introduced no later than another element e_2.

The user of our planning method can provide constraints to customize the planning. The input of constraints can be intuitively partial, e.g., "let's introduce a before b." Our method can then automatically complement the necessary relevant elements according to the dependencies.

4.2.2 Use Cases

It is necessary to have the list of model elements as the input to define the design space of refinement. We first expect that the whole specification is available as the input for Event-B modeling. This means that we consider that the primary use case of Event-B is to validate a given specification expected to contain the necessary elements. Although there can be defects such as lack of a proper guard condition to support an invariant, the essential concepts and constraints are clear. In other words, we do not consider exploratory use of Event-B to elicit requirements or even develop new ones.

In addition to the model elements obtained from the whole specification, our refinement planning method requires the information of dependencies. This is basically trivial as the default principles only take into account "necessary to define" dependencies, which can be automatically extracted. Other principles may require tagging on model elements such as "goal" and "means."

Given the list of model elements, our refinement planning tool can support the user to explore possible refinement designs. This tool applies default principles such as the two introduced in this chapter. The user may also add grouping or ordering constraints. The tool then outputs all the possible refinement designs under the given constraints. Sometimes the design space is still too large to explore. In this case, the user may give further constraints after looking at the design space derived from the currently given constraints.

4.2.3   Experiences and Insights

We defined generic principles to define default constraints used with our refinement
planning method and tried them with several models. Figure 5 shows an example of
the output from this method. The nodes represent what variables have been introduced thus far. The directed edges and their labels represent which new variables
are introduced. Thus, an edge represents a refinement process to obtain the next step
where additional variables are introduced. The top node in the figure represents a
virtual step where nothing is defined. The next node is the first step of refinement.
This graph suggests that we can only start with introducing D and N first due to
the dependency, i.e., other variables depend on them. Different edges from one node
represent different possibilities of refinement. The graph thus represents different
refinement plans, which finally reaches the same node at the bottom where all the
variables are introduced.

This example was based on a model in the Event-B book [3], and the result made
it explicit that there are possible refinement designs other than the one shown in the
book, which was implicitly selected by the author.

This example in Fig. 5 is based on a relatively small model and comprehensible in
size. There are cases in which the refinement design space is too large and difficult
for human engineers to capture. The size of the design space depends not only on
the size of the target model but also on the "density" of the dependencies. The
more constrained the design is, the smaller space we have. When the model is large,
it is sometimes necessary to give more constraints, especially dividing the model
elements into groups so that planning targets each group. When the density of the
dependencies is low, the design space is too large. However, this means that engineers
have almost free choice of the design.

We also conducted a sensitivity test regarding the quality of the input. The planning
results were mostly stable even when we had defects (e.g., missing guard condition)
in the input list of model elements.

Fig. 5 Example of graph for
refinement plans

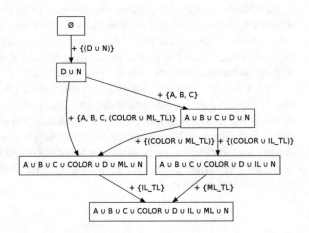

4.3 Refinement Refactoring

4.3.1 Technical Approach

In the general context, refactoring (of program code) refers to the activity that changes the design of code structure without changing its behavior or breaking its correctness to improve quality aspects such as maintainability and reusability. Refactoring is essential because it is very difficult to work on quality aspects while working to have code that runs properly. Another reason is that we cannot perfectly predict what types of evolution or reuse we will have as demands for these emerge through experiences. Refactoring is typically achieved through a combination of small changes, e.g., extracting a method, so that it is easy to ensure that each change will not break the current behavior.

We can imagine the usefulness of refactoring regarding refinement design in a similar manner. We start with a model in which all the proof obligations are discharged and obtain another one with a different refinement design. With our approach, we define two types of small changes that allow for arbitrary change of the refinement structure by combination.

One of the change operations is merging refinement steps. The input is three steps of M_a, M_b, and M_c where M_b refines M_a and M_c refines M_b:

$$[\quad < \{v_{a1}, v_{a2}, \ldots\}, \{p_{a1}, p_{a2}, \ldots\} >, \quad < \{v_{b1}, v_{b2}, \ldots\}, \{p_{b1}, p_{b2}, \ldots\} >,$$
$$< \{v_{c1}, v_{c2}, \ldots\}, \{p_{c1}, p_{c2}, \ldots\} > \quad]$$

Then we merge steps M_b and M_c:

$$[\quad < \{v_{a1}, v_{a2}, \ldots\}, \{p_{a1}, p_{a2}, \ldots\} >, \quad < \{v_{b1}, v_{b2}, v_{c1}, v_{c2}, \ldots\},$$
$$\{p_{b1}, p_{b2}, p_{c1}, p_{c2}, \ldots\} > \quad]$$

This merge operation is straightforward as it does not break existing proofs. We omit a detailed discussion here but as all the necessary predicates are preserved and we can therefore discharge the proof obligations of the new model without modifying the model.

The other change operation is decomposition of refinement steps, the inverse of the above merge operation. Given two steps of M_a and M_c, where M_c refines M_a, we derive M_b that refines M_a and is refined by M_c. This operation is not trivial as discussed below.

The choice of elements for M_b is not identical and the user needs to determine what elements should be extracted according to the objective of refactoring. However, a naive choice easily leads to a step with error, e.g., when a variable necessary for a chosen invariant is not chosen. Such invalid choices can be prevented by checking with the dependencies as in the case of planning (Sect. 4.2.1). It is also necessary to choose common elements in M_a and M_c into M_b as they are inherited through the refinement steps.

Create Sub-Refinement	Merge With Direct Predecessor	Select All	Reset	Auto Select			
Element	Content			Status	Note		Comment
⌄ ▣ Invariants							
☐ inv1	mCard ∈ D ⤖ P				not theorem		
☐ inv2	mAckn ⊆ D				not theorem		
☑ inv3	BLR = dom(mCard) ∪ ran(dap) ∪ red ∪ mAckn			User-Selected	not theorem		
☐ inv4	dom(mCard) ∩ (ran(dap) ∪ red ∪ mAckn) = ∅				not theorem		
☐ inv5	mAckn ∩ (ran(dap) ∪ red) = ∅				not theorem		
⌄ ▣ Variables							
☑ sit				Auto-Selected (Fixed)			
☑ dap				Auto-Selected (Fixed)			
☑ red				Auto-Selected (Fixed)			
☐ BLR				Necessary for [inv3]			
☑ mCard				User-Selected			
☐ mAckn				Necessary for [inv3]			
⌄ ☐ Events							
⌄ ☐ INITIALISATION					not extended, ordinary		
⌄ ☐ Actions							
☐ act1	sit = P×{outside}						
☐ act2	dap = ∅						
☐ act3	red = ∅						
☐ act4	BLR = ∅						
☐ act5	mCard = ∅						
☐ act6	mAckn = ∅						

Fig. 6 Tool interface to support choice for step decomposition

Figure 6 shows the graphical interface of our tool. Some of the elements are automatically chosen as common elements, and one invariant is chosen by the user. There is a warning about elements (variables) that must be accompanied with the current choice, which may be automatically chosen by the tool (according to the interaction preference). There is also a suggestion of elements that should be introduced as these elements depend only on the chosen elements (the second principle in Sect. 4.2.1).

It is difficult to preserve existing proofs with the decomposition operation. Let us go back to our discussion in Sect. 3.1. The original version of the Event-B model has granular steps, and thus requires local proofs using the variables in each step. Specifically, it was necessary to give the proof:

the user only enters an area permitted in the ACL, given that he/she moves only if entrance to the destination area is permitted in the ACL (inv1 and move_grd1).

In the one-step version, this proof becomes implicit or hidden in the reasoning process of human engineers or automated provers because the proof target is:

the user only enters an area permitted in the ACL, given that he/she moves only if entrance to the destination area is authorized (inv1 and move_grd1').

Therefore, decomposing a step may require making a local proof explicit.

In other words, move_grd1 is optional in the one-step version. Thus extracting elements from the one-step version cannot lead to M1 of the original version, unless we somehow add the missing guard condition. We call such predicates complementary predicates, which are necessary to discharge local proofs that emerge after step decomposition.

Fortunately, we have countermeasures to this problem. As a theoretical foundation, we can prove that there always exists a complementary predicate to complete the proofs in the generated step M_b. This also suggests automated generation of complementary predicates. Another practical approach is to reuse logs of existing proofs in the original version before step decomposition because it is likely that the implicit predicates were found and used internally by the automated prover.

With the merge and decomposition operations, arbitrary (valid) changes of the refinement design are supported. To radically change the design, it is possible to merge all steps and then extract steps one by one.

4.3.2 Use Cases

There are several potential use cases of the refinement refactoring method since the objectives of refactoring vary. This method can also be used for maintenance or reverse engineering in a similar manner as program slicing [10]. We present typical use cases below.

- Complexity Mitigation. We can decompose a large step to improve comprehensibility and verifiability, as discussed in Sect. 3.1, or maintainability in general.
- Extraction of Reusable Part of Model. We can extract reusable parts by moving designated elements to earlier steps.
- Extraction of Abstract Explanation. We can extract the essence we want to focus on by filtering out unnecessary details.

4.3.3 Experiences and Insights

First, we refer to an experimental result of verifiability by complexity mitigation through step decomposition in Table 1. In this experiment, a large step was decomposed into four steps. The **Model Elements** column shows the number of variables and invariants introduced in each step (variables may be removed as well). The original step was somewhat large, with 72 invariants, in which it is difficult to see how each invariant is relevant to other predicates such as guard conditions.

The **Complementary Predicates** column shows the number of unique complementary predicates and the total number of complementary predicates. The number of unique complementary predicates affects the effort for discharging proof obligations and is somewhat limited against the number of proof obligations. In almost all cases, it was also possible to reuse logs of existing proofs in the original model.

The **Proof Obligations** column shows the number of manually discharged proof obligations and the total number of proof obligations. The numbers for only proof obligations of the invariant preservation type are in parentheses. Even though the total number increased slightly (from 1127 to 1159) by step decomposition, the number of manual efforts reasonably decreased (from 175 to 147). Specifically, the number of manual efforts decreased for the invariant preservation type (from 163 to 134). Proof obligations of this type are often difficult as they work on new constraints introduced

Table 1 Example of statistics regarding verifiability

	Model elements		Complementary predicates	Proof obligations
	Variables	Invariants	Unique/Total	Manually discharged/Total (those for INV)
Original	$-2+10$	72	–	175/1127 (163/1088)
Decomposed 1	$-1+3$	7	4/17	6/112
Decomposed 2	$-1+3$	17	8/17	30/261
Decomposed 3	$+2$	14	2/3	30/202
Decomposed 4	$+2$	34	0/0	81/584
All decomposed	$-2+10$	72	14/37	147/1159 (134/1088)

Step 1   <u>Persons</u> somehow *move* between <u>locations</u> according to the *authorization of persons to locations*.

Step 2   *Physical connections* between <u>locations</u> are introduced. Persons *move* between physically connected locations.

Step 3   *Doors* with <u>red/green</u> lights are introduced. Doors somehow <u>authenticate</u> persons.

Step 4   <u>ID cards</u> are introduced. *Doors* read cards and <u>communicate</u> with a controller by <u>messages</u> to <u>authenticate</u>.

Step 5   Physical movements of *doors*, *persons*, and <u>lights</u> are considered. <u>Communication</u> is a reaction to a physical event.

(the underlined parts are extracted while the slanted parts are abandoned)

Fig. 7 Example for extraction of reusable part

in each step. Step decomposition contributes to mitigating this difficult part. Thus, controlling the local complexity has a reasonable effect on verifiability.

Second, we show another experiment result for extraction of reusable part in Fig. 7. The underlined elements are extracted for reuse while the italicized elements are abandoned. The problem is a full version of our simple example in Sects. 2 and 3. As in the simple example, the original steps took into account the user movement as the key starting aspect. However, the expected reuse will abandon the user movement, and the elements to be reused are scattered among the steps.

In this case, we can simply merge all the steps automatically, then extract elements to reuse in the unit of steps according to the refinement design of the new model. In this case study, almost everything went automatically except for selection to provide a refinement design for the new model. No complementary predicate was necessary, probably because the proofs were simple or supporting predicates were explicit.

5 Related Work

In this section, we give an overview of the literature relevant to the points discussed thus far.

5.1 Engineering Disciplines in Proof-Based Approaches

It is somewhat obvious that we need engineering disciplines for various types of internal quality, i.e., quality of models that are important for engineers, when we consider the use of formal methods in team-based work in a large project. Since such use is limited compared with the use of programming languages, there have been few studies on explicitly investigating internal quality with formal methods.

The term "proof engineering" was used in a report of industrial application of formal verification for Pentium processors [13]. The authors reported that lack of clear principles resulted in an extensive amount of proof rewriting work, which suggests the significance of proof design and engineering. The discussion focused on structuring and formulation, or modularization, and mentions its use for comprehensibility and reusability.

The significance of proof engineering was again emphasized in another large study on verified microkernel, in which over 400,000 lines of proof scripts in Isabelle/HOL were constructed [14]. The questions raised by the author also include those regarding Integrated Development Environment (IDE) including the refactoring functionality, as well as management of abstraction and modularity.

The issues discussed in this chapter are very relevant to the demands discussed in those studies. Whereas those studies used generic theorem provers, we applied the Event-B formalism. Thus, our work has focused on more specific "refinement engineering," which especially deals with control of abstraction levels.

Experience in industrial application projects of Event-B is reported in [22]. Various approaches to reusability were mentioned explicitly for some of the projects, which suggests the significance of reusability. There was one project in which preliminary design for reusability was difficult. Our approach flexibly allows for later consideration of reusability, in terms of refinement steps, and even trial and error for the refinement design when making a design choice is not so obvious.

5.2 Refinement Support in Event-B

Refinement is the key aspect in Event-B, and there have been many studies on supporting it. A set of case studies by different teams for the same problem resulted in different refinement designs [7]. Trial and error for refinement design was also reported by some teams. Through such experiences, the work in [26] discussed specific guidelines in the domain of control systems.

It is common to use partially or semi-formal models as the input to construct formal models. Naturally, there have been studies considering the refinement design at the level of such input models. Tree-based models, typically models for Goal-Oriented Requirements Analysis, have been considered as the input for Event-B [20, 24]. Typical patterns have been discussed on those tree models.

In those studies, refinement design was discussed in intuitive terms by experiences rather than trials to systematically or formally capture the essence. Our study was unique in that we tried to make various aspects of refinement designs much more explicit: such as principles or constraints to eliminate invalid steps and the impact of differences in refinement designs.

We have an intensive program on education and practical studies for the industry [12]. Our experience showed the demands and effectiveness of explicit practices. We therefore had strong motivations to tackle to make the implicit practices of refinement design explicit.

5.3 Other Approaches for Quality of Formal Models

The quality of formal models includes more aspects than what we discussed in this chapter.

There are other approaches for modularization in Event-B to tackle complexity, or to improve comprehensibility, verifiability, and reusability. Decomposition of systems (not refinement steps) into components is another significant approach [4, 8]. This approach can be said horizontal modularization, whereas refinement is vertical modularization. An effective combination of these two types of decomposition should be investigated.

Refactoring for formal models has been investigated such as a study for UML+OCL [9] and one for Alloy [11]. The study in [25] is more relevant as it considers refactoring of proofs. However, it is based on manipulation of general proof structures and tactics. Our approach is unique in its specific focus on refinement, or control of abstraction levels.

Modularization by refinement steps is not the only approach to increase comprehensibility of formal models. Practices for comprehensibility include naming conventions and separation of elements with different roles, as naturally invented in industrial application [18]. Other possible approaches include view transformation or visualization as well as traceability recovery. We previously investigated the visualization of Event-B in state charts with clear traceability between refinement steps [21]. We also investigated traceability recovery by presuming predicates that are relevant to a given predicate (e.g., which guard conditions in a concrete step are relevant to each guard condition in the abstract step) [23].

The direction of our work should be effectively complemented from these other directions in practice.

6 Prospects

In this chapter, we discussed the problem of refinement design and presented our approach for explicitly exploring and manipulating possible refinement designs.

Specifically, we reported our experiences on refinement planning and refactoring to support engineering activities on refinement.

Refinement planning is an intrinsically difficult problem as the design space can be very large and involves fuzzy human decisions. Nevertheless, we believe that potentials have been demonstrated as a unique approach of systematically and explicitly capturing refinement designs.

The effectiveness of our refinement refactoring method has been demonstrated. It potentially has many use cases not limited to the presented ones, and it is easy to apply with highly automated support.

We focused on the control of abstraction levels as the key to tackle the increasing complexity of emerging systems. We believe that this point is not specific to Event-B and can be investigated in a general context or for other formalisms.

We will continue to investigate what information should be made explicit to support engineering activities on refinement, or on general proof-based development methods.

References

1. J.-R. Abrial, *The B-Book: Assigning Programs to Meanings* (Cambridge University Press, Cambridge, 1996)
2. J.-R. Abrial, Formal methods in industry: achievements, problems, future, in *The 28th International Conference on Software Engineering (ICSE'06)* (2006), pp. 761–768
3. J.-R. Abrial, *Modeling in Event-B: System and Software Engineering* (Cambridge University Press, Cambridge, 2010)
4. J.-R. Abrial, S. Hallerstede, Refinement, decomposition, and instantiation of discrete models: application to Event-B. J. Fundam. Inform. **77**(1–2), 1–28 (2007)
5. L. Atzoria, A. Ierab, G. Morabito, The internet of things: a survey. Comput. Netw. **54**(15), 2787–2805 (2010)
6. F. Badeau, A. Amelot, Using B as a high level programming language in an industrial project: Roissy VAL, in *ZB 2005: Formal Specification and Development in Z and B* (2005), pp. 334–354
7. F. Boniol, V. Wiels, Y. Aït-Ameur, K.-D. Schewe, The landing gear case study: challenges and experiments. Int. J. Softw. Tools Technol. Transf. **19**(2), 133–140 (2017)
8. M. Butler, Decomposition structures for Event-B, in *The 7th International Conference on Integrated Formal Methods (IFM 2009)* (2009), pp. 20–38
9. G. Engels, B. Opdyke, D.C. Schmidt, F. Weil, An empirical study of the impact of OCL smells and refactorings on the understandability of OCL specifications, in *ACM/IEEE 10th International Conference on Model Driven Engineering Languages and Systems (MODELS 2010)* (2007), pp. 76–90
10. K.B. Gallagher, J.R. Lyle, Using program slicing in software maintenance. IEEE Trans. Softw. Eng. **17**(8), 751–761 (1991)
11. R. Gheyi, P. Borba, Refactoring alloy specifications. Electron. Notes Theor. Comput. Sci. **95**, 227–243 (2004)
12. F. Ishikawa, N. Yoshioka, Y. Tanabe, Keys and roles of formal methods education for industry: 10 year experience with top SE program, in *The First Workshop on Formal Methods in Software Engineering Education and Training, FMSEET 2015* (2015), pp. 35–42
13. R. Kaivola, K. Kohatsu, Proof engineering in the large: formal verification of pentium® 4 floating-point divider. Int. J. Softw. Tools Technol. Transf. **4**(3), 323–334 (2004)

14. G. Klein, Proof engineering considered essential, in *The 19th International Symposium on Formal Methods (FM 2014)* (2014), pp. 16–21
15. T. Kobayashi, Supporting planning and refactoring of refinement structure of Event-B models. Ph.D. thesis, The University of Tokyo (2017)
16. T. Kobayashi, F. Ishikawa, S. Honiden, Understanding and planning Event-B refinement through primitive rationales, in *The 4th International ABZ 2014 Conference* (2014), pp. 277–283
17. T. Kobayashi, F. Ishikawa, S. Honiden, Refactoring refinement structures of Event-B machines, in *The 21st International Symposium on Formal Methods (FM 2016)* (2016)
18. T. Kurita, F. Ishikawa, K. Araki, Practices for formal models as documents: evolution of VDM application to "Mobile FeliCa" IC chip firmware, in *20th International Symposium on Formal Methods (FM 2015)* (2015)
19. E.A. Lee, Cyber physical systems: design challenges, in *The 11th IEEE International Symposium on Object Oriented Real-Time Distributed Computing (ISORC 2008)* (IEEE, 2008), pp. 363–369
20. A. Matoussi, F. Gervais, R. Laleau, A goal-based approach to guide the design of an abstract Event-B specification, in *The 16th IEEE International Conference on Engineering of Complex Computer Systems (ICECCS 2011)* (2011), pp. 139–148
21. D. Morita, F. Ishikawa, S. Honiden, Construction of abstract state graphs for understanding Event-B models, in *Symposium on Dependable Software Engineering - Theories, Tools and Applications (SETTA 2017)* (2017)
22. A. Romanovsky, M. Thomas (eds.), *Industrial Deployment of System Engineering Methods* (Springer, Berlin, 2013)
23. S. Saruwatari, F. Ishikawa, T. Kobayashi, S. Honiden, Extracting traceability between predicates in Event-B refinement, in *The 24th Asia-Pacific Software Engineering Conference (APSEC 2017)* (2017)
24. K. Traichaiyaporn, T. Aoki, Refinement tree and its patterns: a graphical approach for Event-B modeling, in *The 2nd International Workshop on Formal Techniques for Safety-Critical Systems (FTSCS 2013)* (2013), pp. 246–261
25. I.J. Whiteside, Refactoring proofs. Ph.D. thesis, The University of Edinburgh (2013)
26. S. Yeganefard, M. Butler, A. Rezazadeh, Evaluation of a guideline by formal modelling of cruise control system in Event-B, in *The 2nd NASA Formal Methods Symposium (NFM 2010)* (2010), pp. 182–191

Constructing Rigorous Sketches for Refinement-Based Formal Development: An Application to Android

Shin Nakajima

Abstract Event-B allows us to develop descriptions incrementally with its refine-ment mechanism. Correctness of the resultant artifact is ensured by construction. This refinement task can be transparent if we have a refinement plan that mentions how an initial specification is elaborated into a target artifact. Unfortunately, we do not have such a target artifact before starting the refinement-based development. We have informal or narrative documents on the target only. In order to resolve this chicken-and-egg situation, we firstly introduce an iterative process to use Alloy for studying the documents. Its outcome is possibly under-constrained, but unambiguous Alloy descriptions, which acts as a rigorous sketch of the target for us to make a refinement plan. The proposed modeling method was assembled as educational materials for Event-B.

1 Introduction

Constructing a whole model is not easy if the software artifact is large and complex. A common practice is to build a detailed model gradually, starting from a simple one to a model close to the target. This notion of stepwise development is also useful to ensure correctness of programs. It is referred to as *correct by construction*, which allows to let the correctness proof and model grow hand in hand [5]. A whole development process is decomposed into a series of small refinement steps. Each step connects an abstract model with a refined one, and the correctness of the refinement is ensured. Because the refinement-based approaches are promising to achieve expected reliability levels of software-intensive systems, formal methods including the B-method [2] or Event-B [3] provide the notion of refinement in the first place.

Although a goal of refinement steps is obtaining a model faithfully incorporating requirements, these requirements in practice are mostly informal, or semi-formal at best. Requirements are desirable to be well-organized so that traceability between

S. Nakajima (✉)
National Institute of Informatics, Chiyoda-Ku, Tokyo, Japan
e-mail: nkjm@nii.ac.jp

© Springer Nature Singapore Pte Ltd. 2021
Y. Ait-Ameur et al. (eds.), *Implicit and Explicit Semantics Integration
in Proof-Based Developments of Discrete Systems*,
https://doi.org/10.1007/978-981-15-5054-6_15

331

requirements and models is transparent. A standard approach, in Event-B [3], is to adopt requirements document (RD) consisting of a set of simple statements; each refers to either functional or other types of requirements. Existing works propose a way of rewriting industrial documents to be well-formed RD [18], a domain-specific methodology for structuring RD [19], or a tool support for tracing requirements into Event-B models [7]. In addition, a *pre-formal* notation is desirable for mapping requirements to Event-B [6]. However, requirement documents in practice take divergent forms, including UML-like graphical notations, or using styles in favor of operational interpretations. These are quite different from the original recommended style of requirements appropriate for RD.

In theory, as mentioned before, refinement is a process to build a model gradually, from an abstract one to a concrete model by adding details. In practice, however, a refinement process itself involves *trial-and-error*, and refinement steps proceed in an adaptive way as we incorporate a fragment of requirements into the model. Resultant models may be formal, but awkward, for which later refactoring is desirable. In some cases, formal models become unnecessarily complex, and thus correctness proofs at a certain step require a huge manual guidance, or even the steps are stuck because of a failed proof. We may abandon such a process and proceed again, from the start, to go through alternative steps.

Refinement-based development can be made predictive and avoid trial-and-error processes if we follow a refinement plan as a guide. A refinement plan, not necessarily formal, mentions how an initial model is elaborated into a final one through a series of refinement steps. Figuring out a refinement plan requires some pieces of knowledge about the final artifact, although it contradicts with the fact that no final model exists before refinement-based development starts. We need to resolve this *chicken-and-egg* situation.

This paper illustrates a two-staged method for refinement-based modeling with Event-B. Firstly, we use Alloy [10] to study given technical documents iteratively involving trial-and-error. Alloy is a lightweight formal method [9] and supports these iterative formalization activities. We obtain, as a result, possibly under-constrained, but unambiguous Alloy models. Then, using the Alloy models as a rigorous sketch of the final artifact, we figure out a refinement plan with an appropriate initial model. By following this refinement plan, we construct Event-B models verified using the RODIN tool [17]. Thus, we can avoid cumbersome trial-and-error activities in the refinement-based development with Event-B/RODIN.

2   Refinement-Based Modeling with Event-B

Event-B [3] is a formal method to support the refinement-based development practice. A central language element is *event*; it is a basic unit to be refined. An Event-B event is conceptually a tuple of enabling conditions (guard conditions) and actions consisting of generalized substitutions. Both the enabling conditions and generalized substitutions refer to a set of variables. Variables may be accompanied with invariants declared explicitly.

Refinement in Event-B is defined basically in terms of forward simulation relation between two events, abstract and concrete. The relation provides a sufficient condition for refinement (Chap. 14 of [3]). Moreover, refinement checking follows the *posit-and-prove* paradigm; we introduce abstraction relations as well as both abstract and concrete events, and then ensure correctness of the refinement by discharging proof obligations referring to these pieces of information (Chap. 5 of [3]). The proof obligations include GRD (guard strengthening) and SIM (simulation). Below, instead of presenting the mathematical formulation, we will see roles of refinements from specifiers' viewpoints. Event-B, indeed, allows two distinct uses of refinements, vertical refinement and horizontal refinement.

Vertical refinement is a classical notion of data refinement [8] and is a rigorous basis for stepwise developments of programs. Abstract variables in an abstract event are replaced by concrete variables. Through vertical refinement steps, we elaborate an initial abstract specification to obtain concrete descriptions at a program level. An abstract event is a specification, against which a refined concrete event is checked. When such a check is successful, the correctness of the implementation with respect to the specification is ensured. In a special case where an abstract event is taken as a property to be satisfied, the refinement can be a way to check whether a refined concrete event satisfies the property.

In horizontal refinement, abstract variables are retained and possibly new variables are added. This style is sometimes called superposition refinement [4] because the steps add new functional features incrementally. In a standard practice with Event-B, each statement in a given requirements document (RD) is incorporated into a new refined event. In addition, the horizontal refinement allows introducing a new event to incorporate a new piece of RD. Horizontal refinement steps stop when a whole RD is taken into account in a final set of refined events. All the events are at the same abstraction level. This is in contrast with the vertical refinement in which the refinement goes down to an executable program level, from an initial abstract specification.

An important observation is that a refinement process consists of successive steps, and that their order is often significant. Furthermore, refinement steps are indeed a mixture of both vertical and horizontal refinements. Because the two refinement methods play different roles, planning in advance is needed to make it explicit how each method is used in the whole refinement steps. For vertical refinement, the order is mostly clear because the steps go toward a program implementation level. In horizontal refinement, however, the order of adding new features is flexible if adding a particular feature does not interfere with existing ones. Alternatively, we may decide the order so that each refinement checking, namely, correctness proof, is simple enough, ideally amenable to automatic proving. These pieces of information on the refinement order are desirable to be made explicit.

We introduce a notion of refinement plan, which involves an initial model, a target artifact, and pre-planned refinement steps linking the initial one with the target description. Because the refinement concerns with ensuring the correctness of the target artifact, the plan itself is a piece of reason for the correctness. It is valuable in view of documenting the development process and is as important as a product aspect, which refers to model descriptions of software artifact and properties to be checked.

Please note that plans need not be formal, but can be informal, because human specifiers follow the plan. We do not expect that refinement plans are a kind of meta-level directives to control refinement steps automatically.

3   Two-Staged Method

As discussed in Sect. 2, a refinement plan is valuable because the information therein can be a guideline which we follow so that we can conduct refinements in a predictive style, but not in an ad hoc manner.

A refinement plan describes how an initial model is elaborated into a final one through successive refinement steps. A plan need not be detailed enough to record all the refinement steps precisely,[1] but is an informal document to illustrate a refinement process as a whole. A rough sketch of a final model, although not definitive, is needed to figure out such a plan. This contradicts with the fact that no final model exists before refinement starts. We present in this section a new method to facilitate figuring out refinement plans by resolving this *chicken-and-egg* situation.

Requirements' documents, in practice, take divergent forms. Documents are informal using diagrams, such as UML, or are operational describing a set of use cases. These pieces of information together describe a target artifact from various perspectives. A consistent description of a final artifact is not there but must be reconstructed by collecting fragmented pieces of the information. It requires us to study the requirements documents iteratively and to draw a sketch in a trial-and-error manner. We use a tool-supported lightweight formal method [9] to construct such a sketch, a model rigorous enough for automatic analysis. Then, we obtain an unambiguous sketch whose functional behavior is validated with the tool. The sketch can be a basis for us to figure out a refinement plan with an appropriate initial specification description.

Figure 1 illustrates our proposed two-staged method. In the first stage, we study given documents with a help of the Alloy tool. The outcome is possibly under-constrained, but unambiguous descriptions, which acts as a *rigorous* sketch of the target. Then, given an appropriate initial model, we figure out a refinement plan, a series of refinement steps.

At this point, we figure out how the two notions of refinements, vertical and horizontal, are used. We may take into account the role of initial or intermediate models; a model may refer to a property to be checked, or to a tiny system model to be elaborated (see Sect. 2).

In the second stage, we use the RODIN tool to construct an Event-B model. We follow a refinement-based development process. Because we use the refinement plan

---

[1] Our early work [13] introduces a refinement planning sheet, referring to detailed steps.

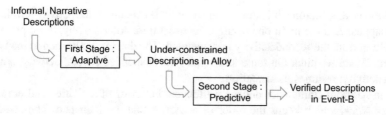

Fig. 1 The two stages

as a guide, the process can be predictive. We may expect that the final Event-B model is acceptable to faithfully represent what we expect, because its functional behavior is validated in the first stage with Alloy and correctness of refinements is ensured in the second stage with RODIN.

We, here, present a rationale to employ Alloy in the first stage. Alloy [10] is a model-based formal specification language with a scope-bounded automatic analysis method. The mathematical model is based on relational logic, with which we can represent formal descriptions in a set-theoretic style. Thus, there is not a large syntactical gap in formulas between Alloy and Event-B. Although the scope-bounded analysis is not complete (Chap. 5 of [10]), Alloy descriptions are analyzable, finding *models* of logical formulas that the descriptions are defining. Because this analysis method searches for possible *models* to satisfy the formulas, Alloy can return appropriate answers even when the formulas are under-constrained. In such cases, the answers contain redundant instances, but we can ensure, by inspection, that the obtained results do not conflict with our expectations. Alternatively, we can add further constraints to the Alloy descriptions so that the tool returns exactly what we expect. In this way, we can validate functional behavior with the automated analysis method. Because of these, Alloy is considered an appropriate formal method for *drawing* a sketch of a target artifact.

4 A Case Study

We illustrate the proposed two-staged method with a case study of an application to Android.

4.1 Android Application

Imagine that we see the home screen of an Android smartphone, a specific application icon appearing on the display. Clicking the icon invokes the main activity of the application. Then pushing a GUI button, provided by this main activity, starts another activity to occupy the screen. The main activity is suspended at this time. When we press the BACK button while this second activity is executing, the activity terminates to disappear and the main one resumes its execution. Alternatively, if we press the HOME button, the second activity is suspended. The display is returned to the home screen. When we click the same icon again, the suspended second activity, not the main activity, resumes its execution.

The Android framework defines a lifecycle behavior of activities and employs a *back stack*, which keeps the order of activities that are launched. The standard Developers Document [1] explains the lifecycle behavior in a narrative way, and the descriptions are scattered in many pages. Definitions of basic terminologies are not clearly given, and they need further interpretations to remove ambiguities. After an iterative process of studying the document, we found that the lifecycle of activities was essentially a state transition machine (Fig. 2). Invoking a callback method is considered as a transition event and is attached to a transition arc. In the state transition machine description, on completing a callback method, an activity is in a stable lifecycle state.

The Developers Document introduces an alternative view of activities. An activity is accompanied with a specific attribute regarding its visibility on the screen. Firstly, an activity is *visible* when a human user can see its GUI image on the screen but is *hidden* otherwise. Secondly, a visible activity is *foreground* when it is running and can accept GUI events from users. These events are generated by users' actions, for example, a finger touch on the screen. An activity is *background* otherwise.

Those attributes and states in the lifecycle behavior are related, but their relations are not clearly mentioned in the Developers Document. According to Fig. 2, an activity is running only when it is in the *resumed* state. It, further, implies that a foreground activity is definitely on top of the back stack. However, in the *started* state, an activity is visible, but background, while it is on top of the back stack as well. Therefore, an activity location in the back stack and activity attributes of foreground/background is not one-to-one.

Both the lifecycle behavior and attributes are two different views on the states of an activity. Attributes are, indeed, more or less related to intuitive views of user interactions. The lifecycle behavior is best encoded as a state transition system, an operational design description. Because of these, we may regard the attribute changes as properties to be checked against the system description of the lifecycle behavior.

4.2 Alloy Descriptions

Although the state transition machine is a concise description of the target, the standard document [1] does not include such as the one in Fig. 2. We extracted, from the scattered descriptions in many pages, what we thought essential. Its correctness may be checked whether it can reconstruct usage scenarios documented in the other pages. Furthermore, the diagram is still informal and may be ambiguous. Through an iterative process of constructing Alloy models and checking them, we elaborate details of the diagram to be unambiguous.

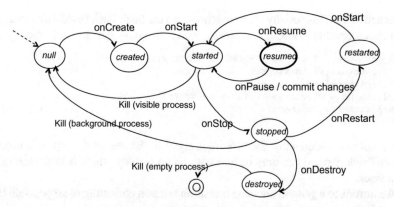

Fig. 2 State transition machine of lifecycle

4.2.1 Lifecycle Behavior

We construct Alloy descriptions of the state transition machine in Fig. 2. Firstly, we define `State`, a set of constants, each of which denotes a particular state. We followed a standard method of introducing a set of constants in Alloy.

```
abstract sig State {}
one sig null extends State {}
one sig created extends State {}
. . .
```

We may regard that `null` and `created` are declared to be two constants of type `State`.

World consists of a `BackStack` element and a set of `Activity`. A map `status` maintains states of each activity referred to by `activities`.

```
sig World {
  back : BackStack,
  activities : set Activity,
  status : activities ->some State
}
```

A state transition caused by a callback method is encoded as a *state transformer* of two `World` instances. Below is `onCreate` which is also checking the back stack and manipulating it as needed.

```
pred onCreate ( w1, w2 : World, x : Activity ) {
  (w1.status)[x] in null
  (some w1.back.indices) implies not((w1.status)[top[w1.back]] = resumed)
  w2.status = w1.status ++ (x -> created)
  push[w1.back, w2.back, x]
}
```

The `++` symbol stands for an overriding operator, and thus the state of the specified activity is assigned to `created` in the post-state without changing the states of the other activities. This activity is also pushed onto the back stack in the post-state. The next shows another example, onDestroy, which also manages the back stack (pop).

```
pred onDestroy ( w1, w2 : World, x : Activity ) {
   (w1.status)[x] in stopped
   top[w1.back] = x
   w2.status = w1.status ++ (x -> destroyed)
   pop[w1.back,  w2.back]
}
```

The onDestroy ensures that the activity is equal to the one on the top. It is because this callback method can only be executed on an activity which is at the top of the back stack.

We introduce a general one-step transition relation consisting of all possible transitions between a pair of pre-state and post-state.

```
pred trans ( w1, w2 : World, x : Activity ) {
     onCreate[w1, w2, x] or onStart[w1, w2, x]  or onResume[w1, w2, x]
  or onStop[w1, w2, x]   or onPause[w1, w2, x]  or onRestart[w1, w2, x]
  or onDestroy[w1, w2, x]
}
```

4.2.2   Scope-Bounded Analysis Scenarios

We use a technique of scope-bounded checking to see how the Android application activities behave dynamically. It is to explore all possible transition sequences from an initial state within a given scope. We here consider a case for the scope in which the number of World instances is five.

```
one sig ScenarioOne {
  w1, w2, w3, w4, w5 : World
}{
  all w : World | w in w1+w2+w3+w4+w5
}

pred ichi ( s : ScenarioOne, x : Activity ) {
     init[s.w1.back] and s.w1.status = x -> null
  and trans[s.w1, s.w2, x] and trans[s.w2, s.w3, x]
  and trans[s.w3, s.w4, x] and trans[s.w4, s.w5, x]
}

run ichi for 2 but 5 World, 5 BackStack
```

In the above, we assume that the initial state is empty in which no activity is in World. The command run instructs the Alloy analyzer to search for a sequence of four transitions. We use a solution enumeration technique to obtain multiple answers. The first may be the one in Fig. 3a, and the next one is shown in Fig. 3b. They are two possible instances of transitions from the initial state.

Occasionally, we want to choose a particular state to be a start, from which the bounded analysis is conducted. Since a global state, in our Alloy model, is determined completely by World, we set up such an initial world.

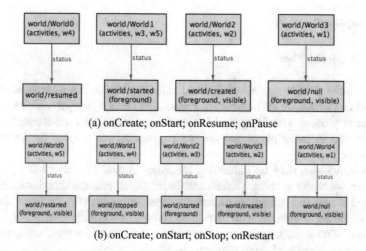

(a) onCreate; onStart; onResume; onPause

(b) onCreate; onStart; onStop; onRestart

Fig. 3 Alloy GUI snapshot

```
pred toS ( w : World, x : Activity, s : State ) {
  some b : back | {
    init[b] and push[b, w.back, x]
    w.status = x -> s
  }
}
```

Next, we introduce a property P to be checked. The example below is to check whether a specified activity is in the *restarted* state.

```
pred P ( w : World, x : Activity ) {
  w.status[x] = restarted
}
```

We can check whether P is satisfied or not at each transition state.

```
pred ichi ( s : ScenarioOne, x : Activity ) {
      toS[w1, x, resumed]
  and trans[s.w1, s.w2, x] and trans[s.w2, s.w3, x]
  and trans[s.w3, s.w4, x] and trans[s.w4, s.w5, x]
  and (   P[s.w1, x] or P[s.w2, x] or P[s.w3, x]
      or P[s.w4, x] or P[s.w5, x])
}
```

This example is satisfied for transition sequences such as in Fig. 3b. The method is, indeed, a special case of bounded model checking (BMC) of linear temporal logic (LTL) formulas, $\Diamond$P where $\Diamond$ is an eventually operator. Thus, $\Diamond$P states that the property P is eventually satisfied.

We check scenarios so as to ensure that the description is just as intended. If the check fails, we may fix faults in Alloy models. At the end, we have unambiguous Alloy descriptions of the target artifact. We refer to a companion paper [15] for details about the descriptions and further analyses using Alloy.

4.3 Refinement Plan

4.3.1 A Global Plan

We now have an Alloy model of Android activities, which is a rigorous sketch for Event-B model that we shall construct. Please note that the Alloy model is a result of an iterative process. The Alloy model incorporates, in itself, all requirements, which are scattered in many pages and are not expressed succinctly in the Developers Document.

One naive idea may be translating the Alloy model into Event-B equivalent. However, we prefer to using a refinement-based development method, starting with an initial model and following refinement steps so that the correctness of the final Event-B model is ensured with respect to some given criteria. We now have two problems, what an appropriate initial model is, and how refinement steps, possibly mixing vertical and horizontal refinements, are conducted.

As explained in Sect. 4.1, the Developers Documents present two alternative views of activity behavior, lifecycle behavior and attribute changes. We regard changing attributes as properties to be checked against the system description of the lifecycle behavior. Unfortunately, the Developers Documents do not present concisely when attributes are changed, and thus the property to be checked is not well defined. We introduce a series of refinement steps starting with an almost trivial property so that a refined one is detailed enough to be checked against the lifecycle behavior.

We divide a whole refinement step into two. The early part is focusing on refining *properties*, for which we use the vertical refinement because properties are elaborated and detailed. Independent of these property descriptions, we construct lifecycle behavior that is comparable to the rigorous sketch written in Alloy. Starting with a bare state transition machine, we may use the horizontal refinement method to add back stack features. Then, we insert a refinement step to connect the property and state transition machine. The step is intended to ensure that the state transition machine satisfies the required properties of chaining attributes, which is depicted as a view shift in Fig. 4.

Fig. 4 A global plan

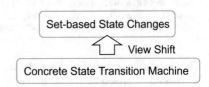

4.3.2 A Detailed Plan

As in Fig. 4, we divide the whole refinement steps into two parts. This strategy is a kind of top-down decomposition of a whole refinement step. Now, we consider refinement plans for each.

Firstly, we adopt set-based abstract representations to encode changes in attribute values. Each activity belongs to a set, and belonging to a particular set indicates that an activity instance has a particular attribute value. Then, changing values is encoded as moving the activity instance from a set to another. This view is useful when we gradually elaborate the sets. In fact, we start from a *big* set and then gradually divide it into detailed subsets. These elaboration steps are data refinement. Figure 5a illustrates that they consist of two refinements involving three machines (M0, M1, and M2).

Secondly, we construct a concrete state transition machine. Each state in the machine is uniquely identified with a named label. Changing states is encoded as rewriting the label from an old to a new. This view is useful when we enumerate all the states beforehand. As shown in Fig. 5b, we start with a simple transition system (L0) and then add features referring to the back stack (L1). The horizontal refinement is suitable because of the addition.

Last, the view shift (Fig. 4) connects the two models, M2 and L0, by the vertical refinement. Because we regard M2 to represent abstract behavior viewed from outside, this refinement step is inserted to ensure that the concrete state transition system satisfies the required property of M2.

4.4 Event-B Descriptions

We now start constructing Event-B models following the refinement plan. Firstly, we define Context C0 that essentially introduces a set Activity, its elements denoting an activity instance.

SETS
 Activity

Machine M0, seeing Context C0, introduces two of set-valued variables, *new* and *current*. They refer to subsets of Activity and are disjoint with each other (Fig. 5a).

VARIABLES
 new
 current
INVARIANTS
 inv1 : current $\subseteq$ Activity
 inv2 : new $\subseteq$ Activity
 inv3 : current $\cap$ new = $\varnothing$

Create event in Machine M0 is creating a new activity instance that becomes *current*. The event is defined so that an element (x) is moved from *new* to *current*. The activity referenced by *x* changes its status.

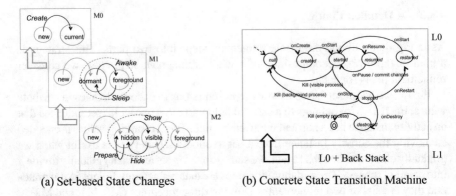

(a) Set-based State Changes   (b) Concrete State Transition Machine

Fig. 5 Refinement Steps

Create $\triangleq$
any
  x
where
  grd1 : x ∈ new
then
  act1 : new := new \ { x }
  act2 : current := current ∪ { x }
end

Next, we refine M0 to have M1. We elaborate the set *current* divided into two *dormant* and *foreground*. As the variable names suggest, instances in *foreground* are running, but those in *dormant* are suspended. Two events Awake and Sleep change their execution status and are defined similar to Create above. Below, inv1 is a gluing invariant to connect variables in M0 and M1, *current* in M0, and *dormant* and *foreground* in M1. We introduce an additional constraint to state that the size of the *foreground* set is 0 or 1, which indicates that at most one activity instance can be running.

INVARIANTS
  inv1 : partition(current, dormant, foreground)
  inv2 : finite(foreground) ∧ card(foreground) ≤ 1

The refinement M2 is defined similarly. The set *dormant* is further decomposed into *hidden* and *visible*. Three events, Show, Hide, and Prepare, change status of activity instances as shown in Fig. 5a.

We now construct a concrete state transition machine, which is essentially representing the diagram in Fig. 2. We introduce a set Status in Context C10, each of whose element refers to a particular state of the machine. These elements are constants.

SETS
  Status
CONSTANTS
  created, started, . . .
AXIOMS
  axm1 : partition(Status, {created}, {started}, . . .)

Behavioral aspects of the state transition machine are encoded as events in Machine L0, seeing both C0 and C10. A new variable manages state changes of activity instances.

VARIABLES
   status
INVARIANTS
   inv1 : status $\in$ Activity $\nrightarrow$ Status

State transition is encoded similar to Alloy descriptions in Sect. 4.2. As an example, an event onCreate defines a state transition from the null to created states. Below is an Event-B version. Note that the Event-B snippet is almost a direct translation of the Alloy counterpart. Section 5 will compare these descriptions.

onCreate $\overset{\triangle}{=}$ **refines** Create
any
   x
where
   grd1 : $(x \mapsto \text{null}) \in$ status
then
   act1 : status := status $\Leftarrow$ { x $\mapsto$ created }
end

We show here a final version of onCreate event defined in Machine L1. It adds features relating to the back stack and retains grd1 and act1, namely, L0 and L1 are related by horizontal refinement. This event is comparable to the Alloy version of onCreate.

onCreate $\overset{\triangle}{=}$ **refines** onCreate
any
   x
where
   grd1 : $(x \mapsto \text{null}) \in$ status
   grd2 : $(ms > 1) \Rightarrow ((y \mapsto \text{resumed}) \notin \text{status})$
then
   act1 : status := status $\Leftarrow$ { x $\mapsto$ created }
   act2 : body := body $\Leftarrow$ { ms $\mapsto$ x }
   act3 : ms := ms + 1
end

We now look at the refinement relation between Machines M2 and L0. Because there is a view shift, we introduce a gluing invariant to connect elements in the two machines. As it is a large formula, we show here its part only.

INVARIANTS
   inv2 : $\forall x \cdot (x \in \text{Activity}) \wedge ((x \in \text{new} \Leftrightarrow (x \mapsto \text{null}) \in \text{status})$
       $\wedge (x \in \text{hidden} \Leftrightarrow ((x \mapsto \text{stopped} \in \text{status}) \vee (x \mapsto \text{restarted} \in \text{status}))$ ...

All the proof obligations (POs) in those refinement steps were discharged. POs in machos M0, M1, and M2 are mostly trivial. For machine L1, 14 out of 17 POs were automatically discharged. In machine L0, however, there are 24 POs, half of which required manual proof. It is because the gluing invariants responsible for the view shift from Machine M2 to L0 are more complicated than the others. Thus, the ratio of manual proof was large.

5   Discussions

We recall our motivation for using Alloy. In B-method, refinement is starting from an abstract model to reach a concrete model comparable to executable programs [2]. Because we are familiar with the computational models of executable programs, figuring out a final model before starting refinements is not difficult. In Event-B, however, a final refined model is not necessarily an executable program, but usually represents a set of requirements. Their abstraction levels are arbitrary, and thus we need to have a kind of *mental* computational model before starting refinements. We used Alloy to study given requirements for obtaining a rigorous sketch of the final model.

Alloy is a lightweight formal method. We construct a formal model and check it, which gives us a quick feedback whether the model faithfully represents what we need to have. The automated analysis is based on model finding, with which we are able to use Alloy for bounded model checking. The formal analysis is helpful to validate the constructed model. After checking a set of scenarios, the level of our confidence in the behavioral model can be acceptable, and thus we adopt it as a rigorous sketch for a refinement-based development with Event-B.

In using refinement-based development, the importance is widely recognized as the step from informal requirements to a formal model. In Event-B community, a standard approach is assuming Requirements Document (RD), which is a collection of one-sentence requirement statement in English. Su et al. [18] proposed a way of rewriting industrial documents to be well-formed RD. ProR [7] is a tool for tracing requirements, as a form of RD, into Event-B models. Yeganefard et al. [19] and Gimehlich et al. [6] adopt Problem Frames as a *pre-formal* notation before RD. Our approach is to use Alloy as such a pre-formal tool,[2] although it is a lightweighted formal method itself.

We now compare models in Alloy and Event-B using the lifecycle behavior. In fact, state transitions are encoded similarly. For Alloy, we introduce the following snippets to maintain that a certain activity is in a particular state.

```
activities : set Activity,
status : activities ->some State

pred onCreate ( w1, w2 : World, x : Activity ) {
  (w1.status)[x] in null
  w2.status = w1.status ++ (x -> created)
}
```

Event-B counterpart takes essentially a similar approach.

```
VARIABLES
    status
INVARIANTS
    inv1 :  status ∈ Activity ⇸ Status
```

---

[2]Event-B is formal, but the others are all pre-formal here.

```
onCreate  ≜ refines Create
any
    x
where
    grd1 : (x ↦ null) ∈ status
then
    act1 : status := status ⩤ { x ↦ created }
end
```

Note slight differences in Alloy and Event-B. The Event-B uses a partial function (⇸), while the Alloy version uses a total function (->) and its domain is `activities`, a subset of Activity. Updating the name of the state is similar to use overriding operators, `++` in Alloy and ⩤ in Event-B. These differences come from differences in language specifications in these two formal methods.

In regard to combining Event-B with Alloy, Matos and Silva [12] proposed to encode Event-B descriptions into Alloy so that they make use of scope-bounded analysis of Alloy for bounded model checking Event-B descriptions that use set-theoretic constructs. We do not consider such syntactical translations. As seen in the above simple snippets, we may need a lot of rules to bridge detailed differences between Alloy and Event-B. We adopt the Alloy model for a rigorous sketch of the final refined model of Event-B.

Last, because we employed the scope-bounded method to analyze the Alloy descriptions, we had enough confidence in the correctness of the Event-B descriptions. When RODIN automatic prover fails for a certain PO, we need not worry about faults hidden in the descriptions. Those failures are due either to probable weakness of the automatic prover or to the descriptions being under-constrained. It is not easy for software engineers, who are not familiar with interactive prover, to consider two different issues at the same time, one for a possible fault in model descriptions and another for a probable weakness of the automatic prover. Our conjecture is that the proposed two-staged method may lower a potential barrier for software engineers to use refinement-based development methods such as Event-B.

6 Conclusion

We used the presented material[3] in an introductory course on formal methods, a revised course of what is reported in [14]. In addition to teaching the basics of Event-B and RODIN tool usages, the course includes material for obtaining skills in modeling. Since Event-B is refinement-based method, we view that making a refinement plan is one of the major concerns. Thus, we introduced the proposed method as a *Learning by Doing* (e.g., [11]) style material. Last, although the Android example was small and concise, choosing it was successful because it drew much students' attention. Actually, analyzing the lifecycle behavior is technically interesting as it is related to energy consumption behavior of the Android applications (e.g., [16]).

---

[3] It includes the Alloy descriptions and the RODIN archive.

Acknowledgements The idea of the two-staged method grew out of a position statement of this author at an ICFEM 2014 panel discussion session, *Are Formal Engineering Methods and Agile Methods Friend or Enemy?*, organized by Professor Shaoying Liu of Hosei University, to whom we express our sincere thanks for motivating us to consider the subject matter.

References

1. Android, http://developer.android.com
2. J.R. Abrial, *The B-Book - Assgining Programs to Meanings* (Cambridge University Press, Cambridge, 1996)
3. J.R. Abrial, *Modeling in Event-B - System and Software Engineering* (Cambridge University Press, Cambridge, 2010)
4. R.-J. Back, K. Sere, Superposition refinement of reactive systems. Form. Asp. Comput. **8**(3), 324–346 (1996)
5. E.W. Dijkstra, The Humble Programmer – ACM Turing Award Lecture (1972)
6. R. Gmehlich, C. Jones, Experience of deployment in the automobive industry, in [17] (2013), pp. 13–26
7. S. Hallerstede, M. Jastram, L. Ladenberger, A method and tool for tracing requirements into specifications. J. Sci. Comput. Programm. **82**, 2–21 (2014)
8. J. He, C.A.R. Hoare, J.W. Sanders, Data refinement refined - resume, in *Proceedings of the ESOP'86* (1986), pp. 187–196
9. D. Jackson, J. Wing, Lightweight formal methods. IEEE Comput **29**(4), 21–22 (1996)
10. D. Jackson, *Software Abstractions – Logic, Language, and Analysis*, Rev. edn. (The MIT Press, Cambridge, 2012)
11. P.G. Larsen, J.S. Fitzgerald, S. Riddle, Learning by doing: practical courses in lightweight formal methods using VDM++, CS-TR-992, University Newcastle upon Tyne (2006)
12. P.J. Matos, J.M. Silva, Model checking Event-B by encoding into alloy (an extended abstract), in *Proceedings of the ABZ 2008* (2008), p. 346
13. S. Nakajima, A refinement planning sheet, in *Rodin User and Developer Workshop 2010*, Dusseldorf (2010)
14. S. Nakajima, Using alloy in introductory courses of formal methods, in *Proceedings of the 4th Workshop on SOFL+MSVL* (2015), pp. 97–110
15. S. Nakajima, Analyzing lifecycle behavior of android application components, in *Proceedings of the COMPSAC Workshop 2015* (2015), pp. 586–591
16. S. Nakajima, Model-based analysis of energy consumption behavior, in *Trustworthy Cyber-Physical Systems Engineering* (CRC Press, Boca Raton, 2016), pp. 271–305
17. A. Romanovsky, M. Thomas (eds.), *Industrial Deployment of System Engineering Methods* (Springer, Berlin, 2013)
18. W. Su, J.-R. Abrial, R. Huang, H. Zhu, From requirements to development: methodology and example, in *Proceedings of the ICFEM 2011* (2011), pp. 437–455
19. S. Yeganefard, M. Butler, Structuring functional requirements of control systems to facilitate refinement-based formalization. ECEASST **46** (2011)

Printed in the United States
by Baker & Taylor Publisher Services